Jonathan R. Bryan

Thomas M. Scott

Guy H. Means

2008

Mountain Press Publishing Company

Missoula, Montana

First Printing, February 2008

Cover image: *Geologic map of Florida.* —From Scott et al. (2001) and courtesy of the Florida Geological Survey

Back cover images: left, *Gainer Springs Group, Econfina River, Bay County;* top right, *the Anastasia Formation, Martin County;* bottom right, *sand dunes at Deer Lake State Park.*

Library of Congress Cataloging-in-Publication Data
Bryan, Jonathan R.
Roadside geology of Florida / Jonathan R. Bryan, Thomas M. Scott, Guy H. Means. — 1st ed.
p. cm.
Includes bibliographical references and index.
ISBN 978-0-87842-542-6 (pbk. : alk. paper)
1. Geology—Florida—Guidebooks. 2. Florida—Guidebooks. I. Scott, Thomas M. II. Means, Guy H. III. Title.
QE99.B79 2008
557.59—dc22
2007049853

PRINTED IN THE UNITED STATES

Mountain Press Publishing Company
Missoula, Montana
2008

A Florida Geological Survey field camp at Aspalaga Bluff on the Apalachicola River, 1909. Left to right: *Elias Sellards (Florida's first state geologist), Roland Harper, Herman Gunter (Florida's second state geologist).* —Courtesy of the Florida Geological Survey Photo Archives

The authors dedicate Roadside Geology of Florida *to our parents and families for their influence and support. Our families have put up with years of stopping at outcrops, collecting more fossils than one could possibly need, carrying rock hammers in luggage, and bringing tons of rock home from vacations! We also dedicate this volume to the great geologists who came before us in the southeast, setting the stage for our "southern" geologic education.*

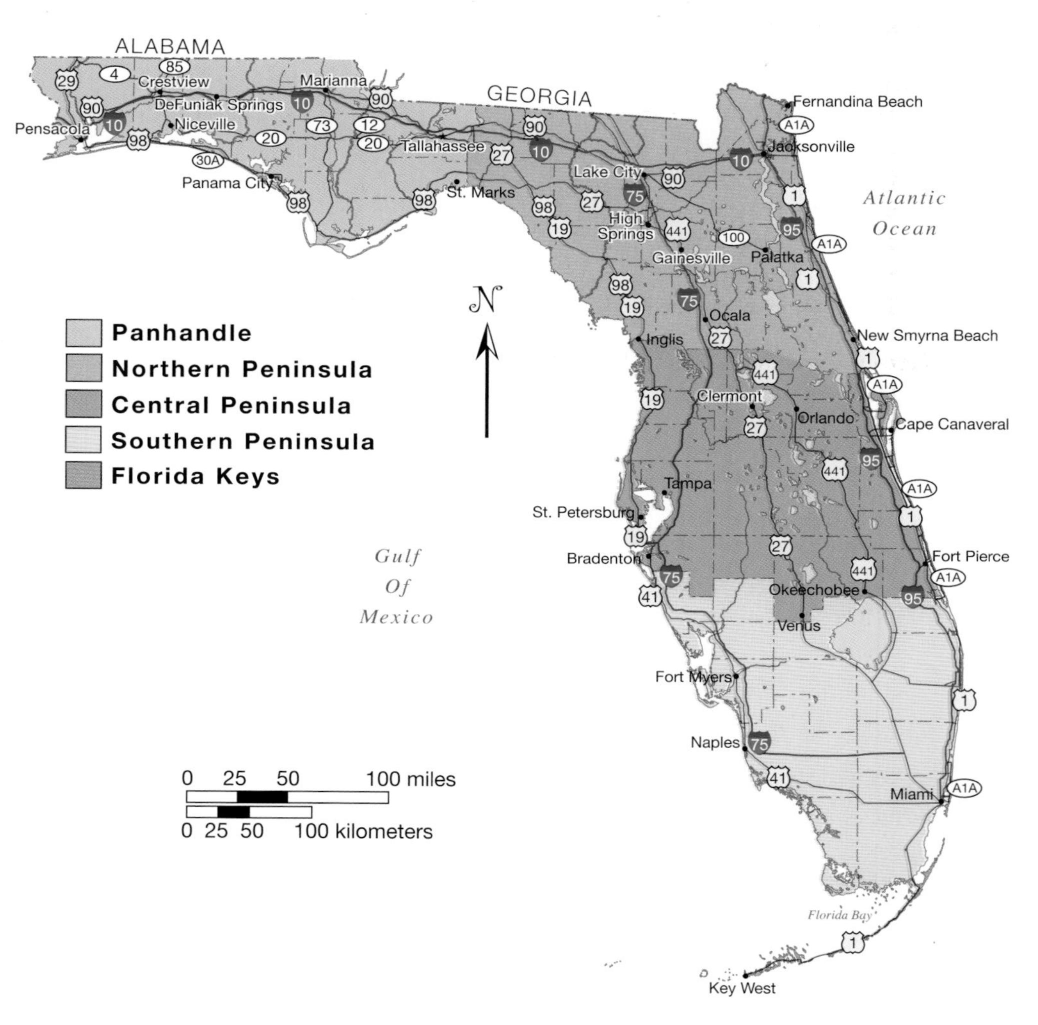

Roads and sections of this book. —Map by Mountain Press

Contents

PREFACE

If you were asked to reflect on images and ideas associated with the state of Florida, the subject of geology probably would not be one of the first things to come to mind. Walt Disney World, orange juice, alligators, manatees, the Everglades, and miles of beautiful beaches, maybe—but not rocks. While it is true that the Sunshine State offers no mountains, Grand Canyon, volcanoes, or dinosaurs, only a narrow-minded geologist with very dirty field glasses could fail to appreciate the geologic wonders of this land.

Florida has many distinct geological features. Of all the fifty states, with the exception of volcanic Hawaii, Florida is the land that most recently emerged from the sea, and it is covered with geological evidence of its submarine past. Because of its relatively recent exposure and the near absence of significant tectonic deformation, Florida has little topographic relief. In fact, Florida has the lowest highest point of all fifty states, topping out at 345 feet above sea level in the western panhandle! The state has some 1,350 miles of coastline, more than the rest of the entire eastern shore from Georgia to Maine. These Florida coasts include the pure quartz sands of the barrier islands and dune fields in northwest Florida, and the tropical coral reefs and islands of the Florida Keys. These beaches are, with no exaggeration, among the most beautiful in the world. But the lure of Florida's shores creates special environmental challenges for a rapidly growing population seeking proximity to this attractive coast.

Florida also has more than seven hundred springs, most of which are boiling with relatively cool (68 to 75 degrees Fahrenheit), azure waters from one of the largest and most productive limestone aquifers in the world—the Floridan aquifer system. Some of these springs even emerge offshore in the Gulf of Mexico through porous limestones of the continental shelf. Florida has considerable cave and cavern development, as well, including one of the largest underwater caves in the world—Wakulla Springs. And for the paleontologist, Florida is a gold mine. Its record of invertebrates of Cenozoic time (especially the last 40 million years) is nearly unmatched anywhere in the world, and it has one of the best mammal records of the last 25 million years in North America.

The purpose of this volume is to familiarize the Florida highway traveler with some of the most interesting and important geological features of the state. One can observe this geology from the highway, in easily accessible parks or roadside areas, at springs, along rivers or canals, and along coastal or even offshore areas, such as the Florida Reef Tract. Of course, we couldn't cover all of the best geologic exposures in this book. Some of the most scientifically valuable outcrops are remote and not easily accessible, and others are at extremely dangerous locations. Some require special permission for entry, and others are simply off-limits. Perhaps the best and safest way to see more geologic exposures is to join one of the regular field trips led by various geological and paleontological organizations in the state (see **Appendices**), or tag along with a more experienced field geologist or fossil collector. Florida is well known for its avocational fossil collectors, many of whom are very knowledgeable and generous with their time and expertise.

Florida is a deceptively long state. A drive from one end to the other—from Pensacola to Key West—is more than 800 miles, with most of the Florida Panhandle located in the central time zone. For the purposes of this road guide, we divided the state into the Panhandle, Northern Peninsula, Central Peninsula, Southern Peninsula, and the Florida Keys. Road logs, which are geological narratives, follow major highways through these regions, usually between larger cities. There are additional logs for side roads that lead to accessible areas of interest. The text emphasizes geological history, the paleontological record, and the processes that have shaped the modern landscape. And because of Florida's recent emergence from the sea and the continued importance of marine and coastal processes in the ongoing formation of the state's geology, we give oceanographic environments and processes special consideration. Springs and canoe-accessible streams expose some of the best natural rock outcrops in the state, so some of these areas receive as much attention as the more easily accessible highway exposures. Throughout the book, we include other noteworthy information, including historical, archaeological, and ecological points of interest.

The user of *Roadside Geology of Florida* should be aware of some discrepancies concerning county and state roads. Commercial maps and those of the Florida Department of Transportation occasionally disagree about whether a road is a state road or county road. Even within the Florida Department of Transportation maps there are scattered inconsistencies. We have driven the routes and attempted to give the correct designation.

Acknowledgments

A project such as this cannot be accomplished without the help of many friends. Amanda Kosche of Crestview produced most of the outstanding computer-generated graphics and assisted with many other digital challenges. Pamela Hynes, librarian at Okaloosa-Walton College, tracked down many references, some old and obscure, during the research phase. Sylvia Bryan reviewed the text and made numerous corrections, grammatical and otherwise. Roger W. Portell of the Florida Museum of Natural History provided digital images of fossil and living invertebrates and reviewed an early draft of the text. Richard C. Hulbert, also of the Florida Museum of Natural History, provided digital images of some vertebrate fossils. Rick Green of the Florida Geological Survey provided several digital images of field localities. Lauren Davis assisted with photos from the Florida State Archives in Tallahassee.

Special thanks to Jennifer Carey of Mountain Press for getting this project started, and for her encouragement during the early stages of work. Special thanks also go to James Lainsbury of Mountain Press for his patient and thorough editing of the text. James reined us in and saved us from many grammatical, stylistic, and other literary errors.

The completion of *Roadside Geology of Florida* coincides with the one-hundredth anniversary of the permanent establishment of the Florida Geological Survey in Tallahassee. (Official geological survey work began as early as 1852, but it was never continuously funded by the state until 1907.) Congratulations to the Florida Geological Survey as we commemorate and celebrate one hundred years of geologic research in Florida, and as we enter into a second century of geologic work on one of the most unique geologic terrains in North America!

Era	Period	Epoch	Began Millions of Years Ago	Geological and Paleontological Events in Florida
CENOZOIC	QUATERNARY	*Holocene*	0.01	Sea level rises (Holocene Transgression) about 390 feet and Florida assumes present configuration. Many modern environments (sinkholes, springs, rivers, estuaries, dunes, beaches, barrier islands, and shallow marine settings), which resemble those of the past, form. Coral reefs and limestone still forming in southern Florida along Florida Reef Tract. Excellent archaeological records of Archaic, Woodland, and Mississippian periods.
CENOZOIC	QUATERNARY	*Pleistocene*	1.8	The Great Ice Age. Florida largely exposed as land, with a cool, semiarid climate. Sea level fluctuates dramatically as glaciers advance and retreat, resulting in the formation of many terraces. Coquina, ooid bars, coral reefs, and rich shell beds form. Diverse land mammal fauna includes horses, mastodons, mammoths, bison, bears, and saber-tooth cats. Paleoindians arrive. Many large mammals become extinct at end of Pleistocene time.
CENOZOIC	NEOGENE	*Pliocene*	5	Much like Late Miocene time. Deltas, beaches, and barrier islands common. Mollusk-rich beds deposited, and some reef beds form in southern Florida. Opening of Panamanian Land Bridge results in the Great American Biotic Interchange and greatly alters ocean circulation in Gulf of Mexico.
CENOZOIC	NEOGENE	*Miocene*	23	Florida repeatedly submerged and exposed as sea level fluctuates. Patch reefs and tropical mollusks thrive early on, but cooler water species indicate conditions changed in Late Miocene time. Ocean circulation intensifies in gulf as Gulf Trough fills with sediment, diverting water flow. Diverse terrestrial mammal fauna and other vertebrates present, including horses, camels, rhinos, elephants, and carnivores. The great white shark *Charcharodon megalodon* roamed the sea.
CENOZOIC	PALEOGENE	*Oligocene*	34	High sea level. Much of Florida a carbonate platform. Abundant larger foraminifera and other invertebrates present. Patch reefs widespread. In Late Oligocene time Gulf Trough drains, becoming an estuary, and much of Florida exposed as dry land as global sea level falls more than 300 feet. First land mammals (horses, saber-tooth cats, and others) appear.
CENOZOIC	PALEOGENE	*Eocene*	56	High sea level. Most of Florida a carbonate platform with larger foraminifera. Mollusks, bryozoans, echinoids, archaeocete whales, sea turtles, sea snakes, sea grasses, and early dugongs present. Suwannee Channel narrows and forms Gulf Trough.
CENOZOIC	PALEOGENE	*Paleocene*	65	High sea level. Most of Florida a carbonate platform with a central evaporative lagoon. Abundant larger foraminifera and some reefs present. Quartz-rich sediments and clays deposited in panhandle.
MESOZOIC	*Cretaceous*		145	Very high sea level. Most of Florida an extensive carbonate platform, with abundant larger foraminifera, bryozoans, brachiopods, oysters, ammonites, small crinoids, and rudist bivalves, which form reefs. Suwannee Channel forms. Oil deposits forming in southern peninsula.
MESOZOIC	*Jurassic*		200	Gulf of Mexico opens. Extensive limestone, sandstone, salt, oil, and gas deposits. Fossil reefs—mounds of cyanobacteria, algae, and sponges—deposited in panhandle. Ammonite cephalopods present.
MESOZOIC	*Triassic*		251	Pangaea begins to break apart. Rift basins (grabens) form along eastern seaboard of North America and extend into Florida. African bedrock, originally part of Gondwana, remains sutured to North America during rifting, becoming Florida bedrock.
PALEOZOIC	*Permian*		299	Pangaea forms. No known rock record in Florida.
PALEOZOIC	*Carboniferous*		359	Megacontinents Gondwana and Laurasia begin full collision and form Pangaea. Florida bedrock is sutured to North America.
PALEOZOIC	*Devonian*		416	Shales and red sandstones with brachiopods, cephalopods, crinoids.
PALEOZOIC	*Silurian*		444	Eurypterids (sea scorpions) and other marine fossils preserved in black shales in northern Florida.
PALEOZOIC	*Ordovician*		488	Sandstones, siltstones, and black shales deposited with many marine invertebrate fossils, including brachiopods, conularids, and graptolites.
PALEOZOIC	*Cambrian*		542	Formation of Osceola Granite and St. Lucie Metamorphic suite (possibly Ordovician in age).
LATE PRECAMBRIAN			1,000	No known rock record in Florida.

Geologic timescale.

Introduction

The Foundations of Eden

They came in the past, they come today, and they will surely come in the future. And those who come are all looking for something—escape, solitude, sunshine, relaxation, romance, refuge, retirement, or maybe just birds, flowers, manatees, and other wildlife. Perhaps all of these are only modern variations of Juan Ponce de Leon's legendary fountain of youth, or Hernando de Soto's gold, or William Bartram's "sylvan pilgrimage." For most of its cultural history, La Florida (the "land of flowers"), so named by Ponce de Leon in 1513, has certainly been romanticized—mosquitoes, cockroaches, snakes, alligators, hurricanes, and sinkholes notwithstanding—as an Edenic paradise, an idyllic, subtropical land of warm breezes, palm trees, orange groves, and healing springs. In fact, the first sight of a remote, crystal blue, boiling springhead in a northern Florida forest, with thick beams of summer sun piercing the canopy, is truly stunning. It's a sight that suggests an earthly paradise. In fact, there were once towns in Alachua and Levy counties named Paradise, Adam, and Eve. And, evidently, the founders of Eden Gardens State Park at Point Washington on Choctawhatchee Bay couldn't resist the temptation.

Some Floridians, in fact, have taken this Edenic imagery literally. According to the late E. E. Callaway of Blountstown, the true Garden of Eden was located in northern Florida, between the towns of Bristol and Chattahoochee, near what is now the Apalachicola Bluffs and Ravines Preserve, also known as Alum Bluff. Callaway claimed that this beautiful landscape was the only place on Earth that fit the biblical description of Eden in Genesis 2:10–14, where one river "divided and became four rivers." According to Callaway, the river that flows out of Eden is the Apalachicola, which is joined at the Georgia state line by four other rivers: the Chattahoochee River, Fish Pond Creek, Spring Creek, and Flint River. These four rivers correspond to the Pishon, Hiddekel, Euphrates, and Gihon rivers in the Genesis story. And according to Genesis, the Pishon River, the equivalent of the Chattahoochee, skirts the land of Havilah, where the gold was good. Callaway was certain that Havilah referred to the gold-bearing district in northern Georgia, known to geologists as the Dahlonega gold belt, near the headwaters of the Chattahoochee. Thus Eden was in southern Georgia and the Garden of Eden in northern Florida.

Alum Bluff on the Apalachicola River, E. E. Callaway's Garden of Eden.

Callaway further proposed that Noah built his ark in the Florida Garden of Eden, using the rare and native Torreya (*Torreya taxifolia*) tree—the "gopherwood" of Genesis 6:14—which is found in cool steephead ravines near the Apalachicola River. According to the Bible, though, the garden was located near the mouths of the Tigris and Euphrates rivers. Callaway easily explained this confusion: Noah's ark drifted from Florida to the Middle East, where, after he disembarked, Noah named the rivers after the original Georgian and Floridan rivers!

To a geologist, there is a wonderful irony in this curious piece of Florida lore. The fact is that the deep bedrock of Florida, unlike most of the rest of North America, was once part of a much larger landmass that included the area of the biblical Eden. Geologists call this landmass Gondwana. This ancient southern megacontinent comprised present-day Africa, South America, Antarctica, Australia, India, and the Middle East.

Florida is part of both the Gulf and Atlantic coastal plains, physiographic provinces that mostly consist of sedimentary strata of Mesozoic and Cenozoic age (251 million years ago to the present) that are gently inclined into the Gulf and Atlantic. The strata were deposited during a long history of sea level fluctuations. The origin of the 400-mile-long Florida Peninsula—probably the most familiar physiographic feature of the North American continent and one of the most distinctive peninsulas on Earth—was for many years a source of much geologic speculation. Some geologists hypothesized that the Appalachian Mountain trend continued in the subsurface as the bedrock of Florida. Others proposed that a continual southerly accretion of coral reefs created much of the peninsula (through the "unceasing labors of the coral animal," according to an opponent of this theory, Angelo Heilprin), with the Florida Keys and modern Florida Reef Tract being the latest growth phases. The famous earth scientist Alfred Wegener, the primary architect of the theory of continental drift in the early twentieth century, thought the southeasterly orientation of the Florida Peninsula relative to the rest of the continent indicated a sort of lagging behind of the peninsula as the greater part of North America drifted to the west. The continental drift theory posited that continents moved independently of each other across the Earth's surface, pushing their way through a static oceanic crust. None of these hypotheses are correct, although each contains an inkling of truth.

Florida does have deep bedrock that is similar to Appalachian rocks, but it originated in Africa. The Appalachian trend continues westward under the Mississippi Embayment, the basin of the Mississippi River, and into the Ouachita Mountains of Arkansas and Oklahoma. And while partly originating through reef construction, much of the subsurface of Florida consists of thick layers of limestone that compose a relatively flat-topped structure called the Florida Platform. This structure and the islands of the Bahamas formed in similar ways. The Florida Platform was constructed under the sea over many millions of years as calcium carbonate—the remains of various sea organisms—was deposited over the Gondwanan bedrock. Later, shallow ocean currents and rivers transported quartz-rich sediments eroded from the Appalachians southward, covering the limestone. Together with the limestone, the sediments formed a sedimentary connection between the North American continent and the platform. And Florida is indeed slowly drifting west, along with the rest of North America, as part of the passive margin of the eastern seaboard (see **Florida's Plate Tectonic Context** in **Geological and Paleontological History**). But it is in no way lagging behind.

The Florida Peninsula—or the "great pointed paw" and "southern cul-de-sac"—forms the eastern rim of the Gulf of Mexico Basin, reaching from temperate to tropical latitudes. It is only the exposed portion of the Florida Platform, which extends to the 600-foot-deep contour of the continental shelf edges in the Gulf of Mexico and Atlantic Ocean. More than half of the Florida Platform remains submerged beneath the waters of the Gulf of Mexico, and a minor area is submerged beneath the Atlantic. As sea level fluctuated in the past, more and less of the platform was exposed. During the height of the Great Ice Age, the exposed peninsula was about twice as wide as it is presently. The central axis of the platform extends northwest to southeast, approximately paralleling the present-day west coast of the peninsula.

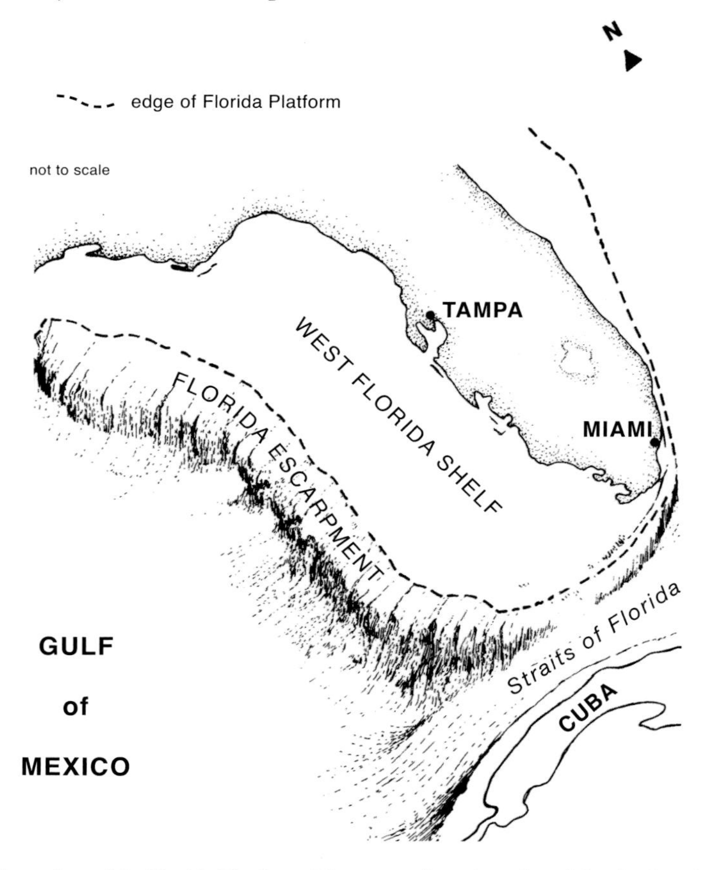

Oblique view of the Florida Platform. The steep submarine edge of the platform is the Florida Escarpment. —Modified from Lane (1986) and courtesy of the Florida Geological Survey

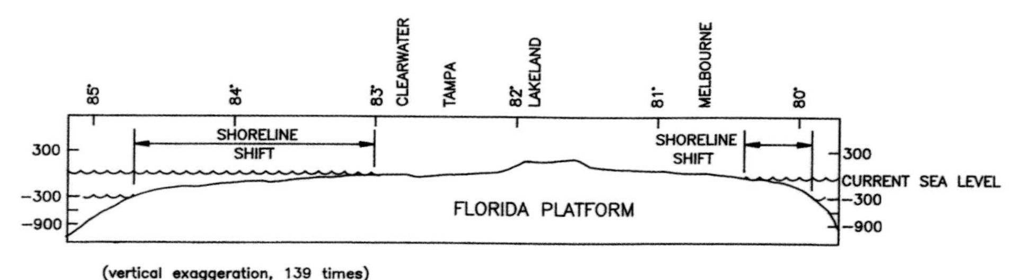

Profile of the Florida Platform at 28 degrees north latitude (middle of the peninsula, from Clearwater to Melbourne), showing how much sea level has risen since the end of Pleistocene time. The central high ground includes the Lake Wales Ridge. —From Cooke (1939) and courtesy of the Florida Geological Survey

All surface exposures in Florida are sedimentary rocks of Cenozoic age. The oldest outcrops are part of the Avon Park Formation, which was deposited during Middle Eocene time, over 40 million years ago. Rocks of Late Eocene and Early Oligocene age, deposited between 40 and 30 million years ago, are well exposed in the north-central portion of the state and the eastern panhandle. Rocks and sediments of Miocene and Pliocene time, 23 to 1.8 million years ago, are even more widespread. Deposits laid down in Pleistocene time, 1.8 million to 10,000 years ago, blanket most of the state, particularly near coastal areas. Because most of Florida's surficial geologic record has essentially remained undisturbed by tectonic forces, the stratigraphy is relatively simple. It is composed of what geologists call "layer-cake stacks" of relatively horizontal sedimentary strata. The complexity that is visible in outcrops is largely a function of weathering, erosion, subsidence, karst processes, and variability in the amount and composition of sediment that was deposited.

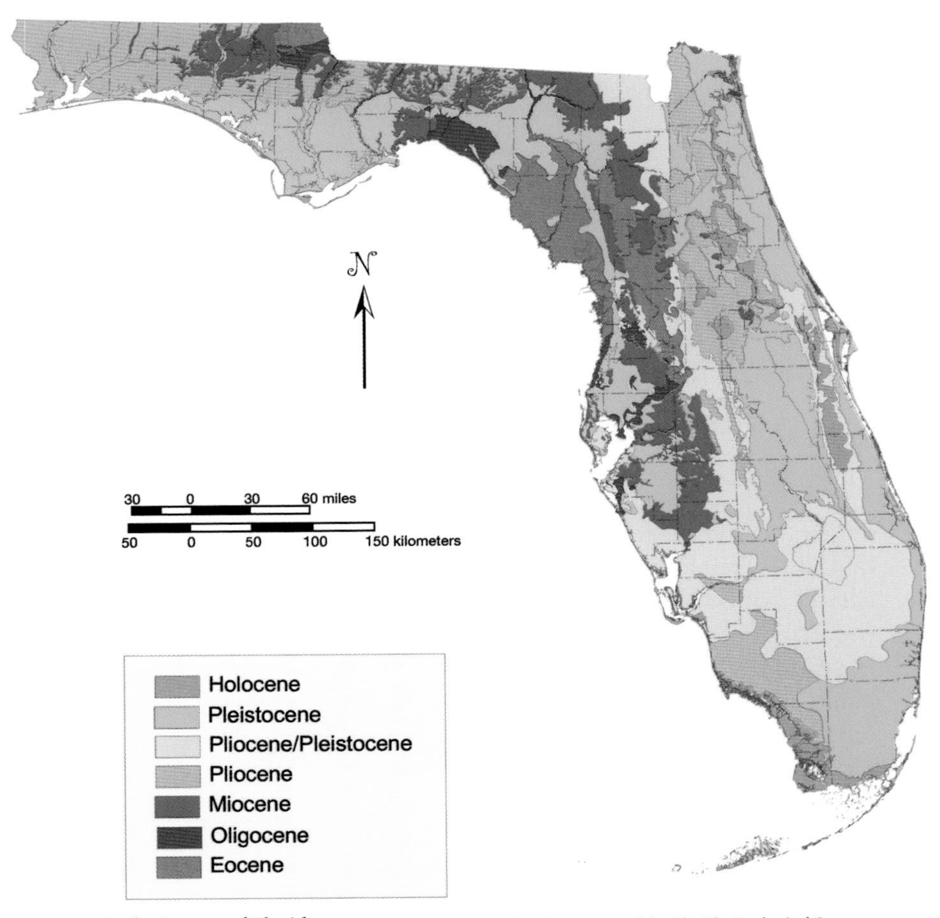

Geologic map of Florida. —From Scott et al. (2001) and courtesy of the Florida Geological Survey

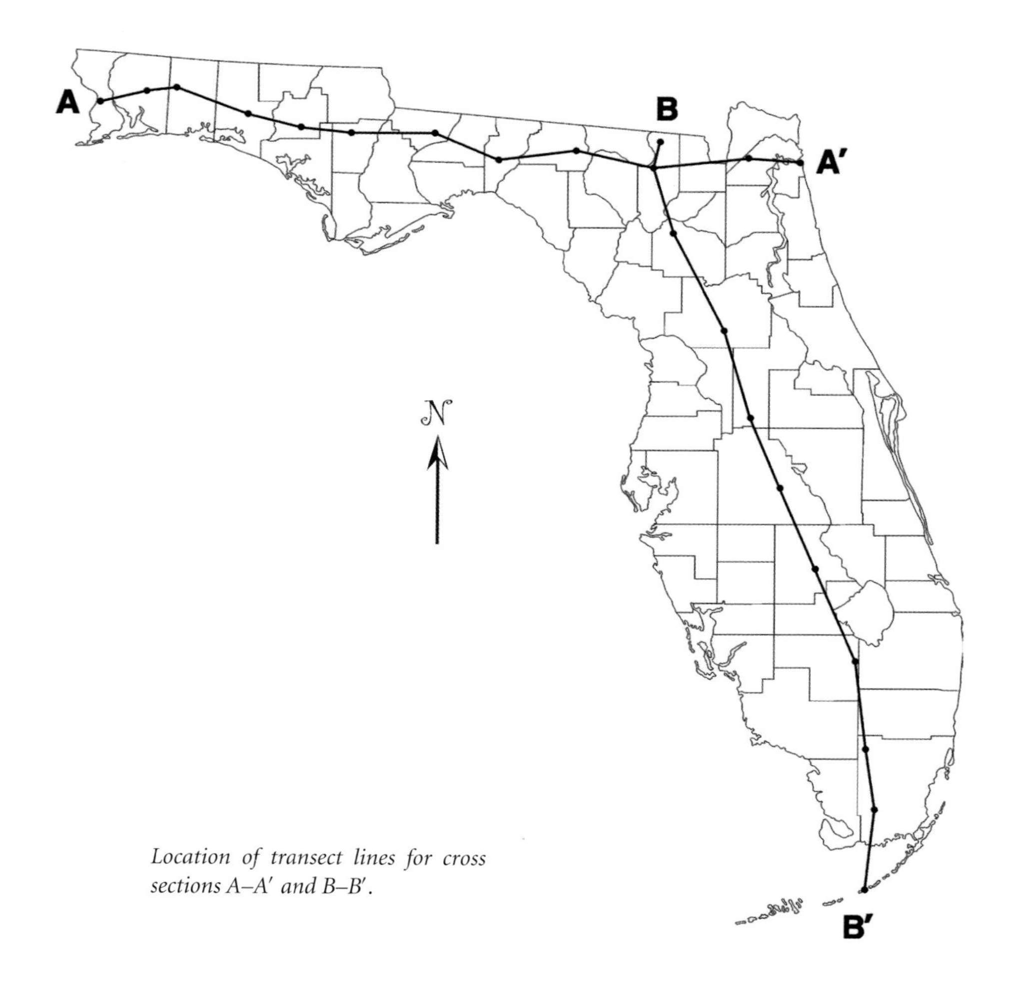

Location of transect lines for cross sections A–A′ and B–B′.

Geologic Fieldwork

To experience the geology of Florida, you must know exactly where to look and when to go. Because of its low topographic relief and restricted surface exposures, the traditional haunts of Florida field geologists have been river valleys and bluffs. Rivers naturally cut deep into rock during their rush to the sea, exposing and washing rock outcrops along the banks. Riverbanks are some of the best natural areas to observe rock outcrops, and a canoe, kayak, or motorboat is one of the most useful field vehicles for Florida geologists. Sinkholes, limestone quarries, open-pit surface mines, canals, and some highway construction sites also have exposed impressive sections of strata. Many geologic features are exposed underwater in springs and spring-fed rivers and can be easily observed because of excellent water clarity. For years paleontologists

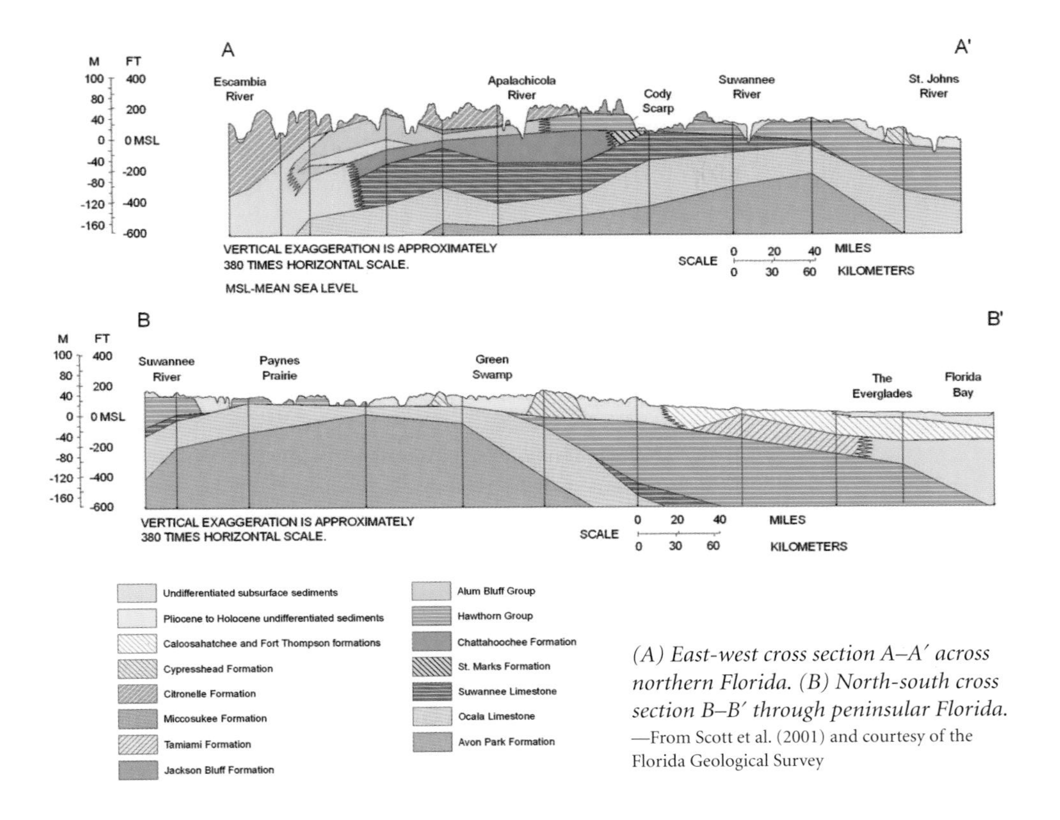

(A) East-west cross section A–A′ across northern Florida. (B) North-south cross section B–B′ through peninsular Florida. —From Scott et al. (2001) and courtesy of the Florida Geological Survey

have utilized snorkeling and scuba gear to collect a wealth of fossil material from Florida's rivers and springs. Also, most Florida geologists take to the field only in late fall, winter, or early spring, when the temperature is pleasant, humidity is low, bugs are few, rivers are low (fall and spring), and vegetation is not so dense.

As with many geologic exposures, but especially those in the humid southeastern United States, Florida outcrops are frequently modified by weathering, erosion, river sedimentation, and ever-expanding vegetation. Also, many quarries and natural exposures regularly change ownership and may lie tantalizingly just out of reach, securely guarded with padlocked gates and posted with warm greetings, such as "NO TRESPASSING. VIOLATORS WILL BE PROSECUTED TO THE FULL EXTENT OF THE LAW." Needless to say, such obstacles may effectively discourage the geology enthusiast. But with adequate permission and preparation, one can eventually reach most field areas. Fortunately, many of Florida's most notable geologic features are preserved in easily accessible state parks.

Florida has a wonderful history of early scientific exploration by many naturalists with botanical, zoological, archaeological, and geological interests. The explorations of John James Audubon and New England botanists John and William Bartram, for example, are well documented. Early geologists working in the state include Pennsylvanian Revolutionary War veteran and surveyor Andrew Ellicott, who mapped the 31st parallel and the northern boundary of the state between 1796 and 1800 and commented on the limestone construction of the peninsula. In the mid- to late 1800s, many renowned geologists and paleontologists worked in Florida, including Timothy A. Conrad, Michael Tuomey, Nathaniel S. Shaler, and Alabama State Geologist Eugene A. Smith. In 1857, Louis Agassiz and Joseph LeConte developed a theory on the origin of the peninsula by coral reef accretion. In 1886, another Pennsylvania scientist, Angelo Heilprin, undertook an extensive geological survey of the west coast of Florida and the Okeechobee wilderness.

In the late nineteenth and early twentieth century, such notable geologists as William Dall, Gilbert Harris, Alcide d'Orbigny, Raphael Pumpelly, C. W. Cooke, Thomas Wayland Vaughan, W. Storrs Cole, George Mansfield, Julia Gardner, and Katherine Palmer carried out geologic and/or paleontologic work in Florida. Vertebrate paleontologists Joseph Leidy, George Gaylord Simpson, and Alfred S. Romer were the first to examine Florida's rich record of fossil mammals. And with the establishment of the Florida Geological Survey, Florida geology became a profession. Today, numerous geologists from state, federal, and academic institutions are actively researching the mysteries of Florida's geology.

Common Minerals, Rocks, and Fossils

There are many interesting, and some unusual, minerals and rocks in the state, some of which are of great economic value, such as phosphorite, which has an enormous variety of uses (especially in detergents and fertilizers). Ilmenite, rutile, zircon, staurolite, garnet, tourmaline, corundum, spinel, topaz, and other minerals occur as sand-sized grains and have been concentrated in heavy-mineral sands, which are found throughout much of the state and mined in the northeastern peninsula. But these mineral sands are generally not the kinds of specimens sought by rock hounds. Perhaps the most beautiful and abundant mineral that collectors can obtain in Florida is calcite, which occurs in many caves, limestone cavities, and some fossil deposits. There are a number of rarer minerals, usually found in phosphate deposits, such as vivianite and wavellite, but only the most discerning collectors find these. Common rock varieties in Florida, such as sandstone, limestone, coquina, dolostone, and chert, can be collected in abundance.

Curiously, Florida's state gemstone is moonstone—a transparent or translucent feldspar of pearly or opaline luster. It may be colorless or various pale shades of yellow, gray, or blue. Although not found in Florida (some of the best moonstone comes from Sri Lanka and Burma), the state legislature adopted moonstone as the state gemstone in 1970 to commemorate the Apollo 11 mission, which landed on the moon on July 20, 1969. This and subsequent lunar expeditions, as well as the space shuttle voyages, were, of course, launched from the Kennedy Space Center at Cape Canaveral.

In 1979, the state legislature made a more appropriate geological choice for Florida's state stone—agatized coral, a beautiful variety of quartz called chalcedony. Agate is a banded variety of chalcedony. In agatized coral, the layers of fine crystalline quartz have replaced, or remineralized, coral, which originally was composed of the calcium carbonate mineral aragonite. This process is called *remineralization* or silicification. Agatized coral is commonly found in several areas of the state and is prized by rock hounds and gem enthusiasts. When cut and polished, it makes a beautiful decorative stone.

Although agatized coral is as much a fossil as it is a mineral, Florida does have a state fossil, though currently "unofficial"—not yet approved by the state legislature but recognized by the Florida Geological Survey. It is an Eocene-age sea biscuit called *Eupatagus antillarum.* This echinoderm, which burrowed through sediments on much of the Florida Platform about 40 million years ago, can be found in the lower part of the Ocala Limestone. The Florida Geological Survey also unofficially recognizes calcite as the state mineral and Ocala Limestone as the state rock.

Florida has long been a favorite destination for fossil collectors, who come seeking perfectly preserved Miocene-age mollusks from the Chipola Formation; Pliocene- or Pleistocene-age mollusks from the Jackson Bluff, Tamiami, and Caloosahatchee formations; or Miocene-age shark teeth and other vertebrate fossils from Venice Beach or in river gravels in hundreds of locations around the state. The collecting of vertebrate fossils (excluding shark teeth) on state lands requires a collecting permit, which is issued by the Program of Vertebrate Paleontology of the Florida Museum of Natural History. Beachcombing for modern mollusks and other marine life also attracts many. Almost any beach can be the source of a respectable shell collection, but it is common knowledge that some of the best shelling is along Sanibel and Captiva islands in Lee County. The state shell, adopted in 1969, is the horse conch (*Pleuroploca gigantea*). This giant gastropod, which can reach lengths of 24 inches, is especially common in Florida's shallow coastal waters.

Geological Keepsakes of Florida

Moonstone, the state gemstone.

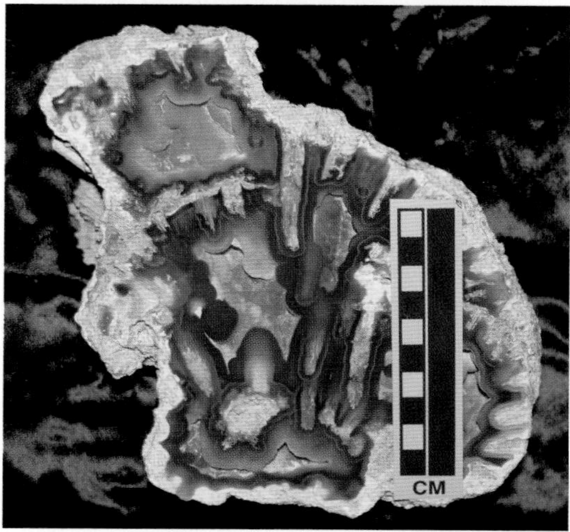

Agatized coral, the state stone. This specimen is from the Tampa Member of the Arcadia Formation of Late Oligocene age.

The sea biscuit Eupatagus antillarum, *the unofficial state fossil. This specimen is from the lower Ocala Limestone of Late Eocene age.*

Heavy-mineral sands (magnified about 25 times), Florida Panhandle.

Calcite, the unofficial state mineral. This specimen is 12 centimeters across and from the Ocala Limestone of Eocene age.

Fossiliferous Ocala Limestone of Eocene age, the unofficial state rock.

*The horse conch (*Pleuroploca gigantea*), the state shell.*

Sedimentary Rocks

All rocks exposed at the surface in Florida are sedimentary—rocks that form as sediments are buried, compacted, and cemented into a solid material. The cements that "glue" the sediments together are typically calcium carbonate, silica, or iron oxide minerals, all of which may precipitate from water that percolates through the sediments. Sedimentary rocks may also form by the direct precipitation of minerals from seawater, for example, the salts halite and gypsum and the mineral dolomite. Many of Florida's sedimentary "rocks" are, in fact, still sediments because they have not been fully cemented. Even a single formation can be solid in one place but loosely cemented in another. In order to identify most sedimentary rocks, one must usually determine the mineral composition and grain size of the sediments, as well as various other characteristics. Almost any sedimentary rock that contains abundant fossils can be prefaced by the term *fossiliferous.* A listing of sedimentary rock types found in Florida can be found in the **Appendices.**

In addition to their general description and classification, sedimentary rocks are assigned to larger bodies of rock called *formations.* Many nongeologists use the term "rock formation" to refer to any large exposure of rock, an unusual body of rock, an isolated hill, or cave structures such as stalactites. But when geologists speak of formations, they are normally referring to large, thick layers of rock that extend continuously for a considerable distance, usually many miles, and sometimes hundreds of miles. Rock formations consist of the same or similar rock types. There may be some degree of variation, but one rock type is usually dominant. Formations may be subdivided into *members* (for example, the Tampa Member of the Arcadia Formation), or formations may be classified with related formations and called a *group* (for example, the Alum Bluff Group or the Hawthorn Group). Exactly how formations are subdivided into members or associated into groups depends on the nature of the formations in question—their extent, degree of similarity, degree of variability, and other properties. Opinions will differ, but once the rocks are thoroughly described geologists usually agree on an acceptable classification.

A formation is first described from rocks in an area where it is well exposed and accessible, and it is often named for a town or natural feature near the area where it occurs. For example, the Marianna Limestone was first described and named by geologist George C. Matson in 1909 for exposures in the town of Marianna. The Marianna Limestone is a soft, fine-grained, fossiliferous, white to gray limestone. It is well exposed around Marianna and has been quarried there for many years, but it is recognized across southern Alabama and Mississippi, all the way to the Mississippi River at Vicksburg—a distance of more than 350 miles. Isolated, eroded remains of the Marianna Limestone also occur northeast of Marianna as far as Hawkinsville in south-central Georgia. The Ocala Limestone, named for exposures near Ocala, is found across most of the state of Florida as well as much of southern Alabama, Georgia, and South Carolina. Other formations are not as extensive as the Marianna or Ocala limestones, but these two are typical formations.

Unlike igneous and metamorphic rocks, which primarily record Earth's magmatic and tectonic history, sedimentary rocks record Earth's environmental and biotic history. Rocks of all three types reflect a particular tectonic setting, but sedimentary rocks especially are archival in nature.

Sediments are fragments of minerals, rocks, fossils, or even organic particles, ranging in size from microscopic clay particles to giant boulders. Sediments may be transported by wind, water, ice, or organisms, but eventually they are deposited. Large areas where sediments are deposited are called *sedimentary basins* (for example, the Gulf of Mexico Basin), whereas smaller areas of accumulation are called *sedimentary environments* and include beaches, rivers, and lakes. Geologists can usually determine the original sedimentary environment of a rock from various characteristics, such as its mineral composition, grain size, grain-size distribution (called *sorting*), organic carbon content, fossil content, or sedimentary structures that form in the sediment before it hardens, such as cross-stratification, mud cracks, or ripple marks. Sedimentary environments are generally divided into three major settings: continental, marine, and transitional (coastal).

In Florida, all three environments are represented in the sedimentary record. Continental environments include rivers, caves and sinkholes (karst features), lakes, freshwater wetlands, and interior dunes. Transitional or coastal sedimentary environments include deltas, beaches, barrier islands, estuaries, lagoons, and longshore sandbars. Fully marine sedimentary environments are the most common depositional environments represented in the global geological record, and they are especially common in Floridan strata. Sediments that were deposited on the inner to middle continental shelf are volumetrically the most abundant sediments in Florida's sedimentary archive.

The Process of Fossilization

Fossils are the remains or traces of past life. They are usually mineralized skeletal structures, such as bones, teeth, and shells, but soft tissues can be preserved, and tracks and trails are also considered fossils. Some fossils may remain unaltered over time, but most have been mineralized

Common Sedimentary Rocks of Florida

Quartz sandstone from the Citronelle Formation of Pliocene age. Ruler is in centimeters.

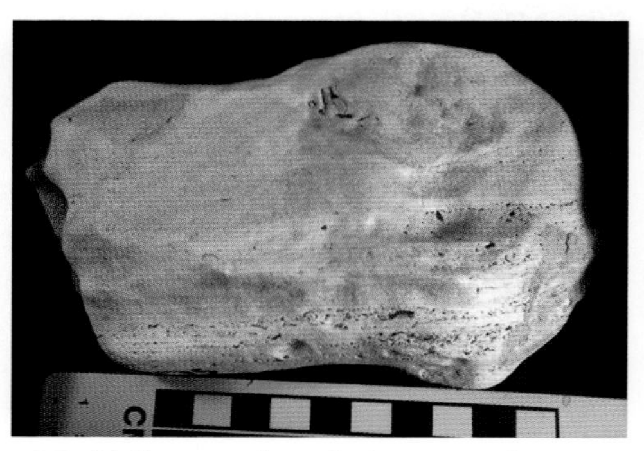

Micritic limestone from the Suwannee Limestone of Oligocene age. Ruler is in centimeters.

Grainstone from the Suwannee Limestone of Oligocene age. Ruler is in centimeters.

Oolitic limestone (magnified 25 times) from the Miami Limestone of Pleistocene age.

Coralline limestone from the Key Largo Limestone of Pleistocene age. Ruler is in centimeters.

Coquina from the Anastasia Formation of Pleistocene age. Ruler is in centimeters.

Phosphorite from the Bone Valley Member of the Peace River Formation of Miocene to Pliocene age. Ruler is in centimeters.

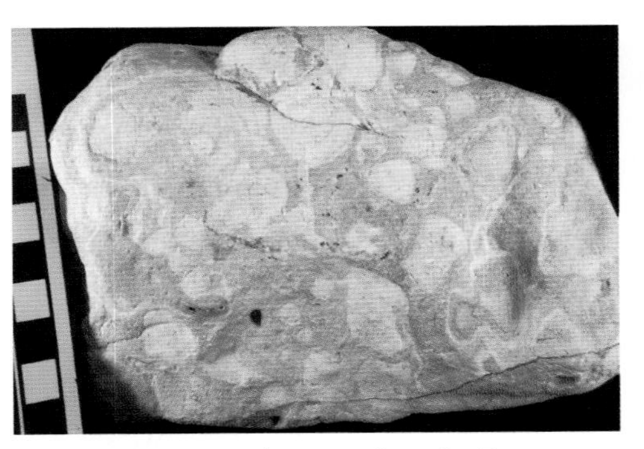

Limestone conglomerate from the Tampa Member of the Arcadia Formation of Oligocene age. Ruler is in centimeters.

Peat of Holocene age from the banks of the Choctawhatchee River. Ruler is in centimeters.

Chalcedony: left, *chert from the Marianna Limestone;* middle, *flint from the Tampa Member of the Arcadia Formation;* right, *and jasper that replaced fossilized coral, found in Suwannee Limestone. Ruler is in centimeters.*

Dolostone from the Marianna Limestone of Oligocene age; note the straight, flat molds where crystals of the mineral gypsum (calcium sulfate) grew in the dolostone but were later dissolved by groundwater. Ruler is in centimeters.

or modified to some degree. Rapid burial favors fossilization, but it is not necessary.

Petrification refers to the complete replacement of organic material by mineral material. In petrified wood, quartz minerals perfectly replicate the original wood—in some cases even preserving growth rings—but not a single cell of wood remains. *Permineralization* refers to the addition of mineral material to a fossil. Many permineralized fossil bones in Florida are black. The original bone is present, but other minerals have been added to the bone, changing its color. *Carbonization* refers to a carbon residue, or outline, left by an organism, usually plants or soft-bodied animals. Many shelled mollusks, especially those in limestone, are preserved as molds or casts. *External molds* preserve the external surface of a shell, as if a shell were pressed into clay and then removed. *Internal molds* preserve the internal contours of a shell, as if a shell were filled with plaster, then the shell was removed from the hardened plaster. *Casts* are the sediment fillings of external molds. *Trace fossils*, also called *ichnofossils*, are burrows, borings, crawling trails, footprints, or other structures that indicate the former presence of an animal.

Fossils are preserved almost anywhere sediments accumulate. The most common sedimentary environment for fossil preservation is the shallow marine realm over the continental shelf. This area receives a high volume of sediment from rivers, deltas, and beaches, and from the production of limestone in the ocean itself. Sedimentary rocks that formed on the continental shelf are the most abundant and best exposed in the entire geological record. So, naturally, most fossils in the paleontological record are animals that lived in this environment, such as sponges, corals, mollusks, bryozoans, brachiopods, echinoderms, and other shelled marine invertebrates.

Most of the exposed rock formations in Florida that are dominantly composed of marine invertebrates formed in middle to inner continental-shelf settings. Examples include the Avon Park Formation; the Ocala, Suwannee, and Marianna limestones; the Alum Bluff and Hawthorn groups; the Tamiami, Caloosahatchee, and Fort Thompson formations; and the Key Largo and Miami limestones. In some cases, the fossils accumulated in an estuary or within the intertidal zone and preserve a mixture of marine and terrestrial species.

If sediment deposition rates are slow, fossils may accumulate and become concentrated before being fully buried. In such cases, some bones or shells may be smooth and rounded from having been rolled in a current, or they may be encrusted by other animals (such as oysters) if they remained on the seafloor for some time. Clam valves are always separate, and shark teeth may be common because sharks constantly shed teeth. Other fossil assemblages form as a result of rapid burial, as in a storm deposit or other natural catastrophe. The Pliocene-age sun stars from the Tamiami Formation are a good example (see **Fossil Sea Stars of Florida—The Panamanian Connection** in the **Southern Peninsula** chapter).

In many limestones, the sediment particles are actually fossils, such as those of larger foraminifera, corals, or mollusks. So the rock itself is 100 percent fossil material. The Anastasia Formation, Key Largo Limestone, Bridgeboro Limestone, and much of the Ocala Limestone are good examples. Some fossil shell deposits in Florida probably formed as tides, currents, storm waves, and even groundwater washed through the shell material and removed sand and clay. The remaining fossils become concentrated to form extremely dense shell beds.

Marine vertebrate fossils are, of course, found with marine invertebrate fossils, but even terrestrial animals are associated with marine sediments. Carcasses may wash into the sea or estuaries through rivers or be washed in during tides or storms. But terrestrial vertebrate fossils in Florida are much more frequently found in lake deposits (Moss Acres), ancient river channels (Love Bone Bed), springs (Wakulla Springs), and sinkholes and cave systems (Thomas Farm). Many vertebrate remains that fossil collectors routinely find are secondary accumulations of material in stream or river channels that eroded out of sediments along the banks. Such fossil concentrations, called *lag deposits*, may result in the mixing of fossils of different ages, depending on what rocks the river is eroding. It is also common for bones from Miocene, Pliocene, or Pleistocene time to erode from offshore rock exposures and wash onto beaches in the surf, such as at Venice Beach.

Structural and Geomorphic Regions

Structural features are aspects of Earth's crust that indicate movement or deformation, usually by tectonic forces. The Florida Platform lies between the rapidly subsiding Gulf of Mexico Basin and the more-slowly subsiding trailing edge of the North American plate under the Atlantic Ocean. Only a few small earthquakes have ever been reported in Florida, and most of them originated elsewhere. But deep within Florida's subsurface, geologists have identified faults with as much as 1,500 feet of displacement along the fault. These faults indicate that Florida was tectonically active in the past. Structural features closer to the surface are much more subdued, with faults having less than 100 feet of displacement. The structural framework of the Florida Platform consists of highs (positive features) and lows (negative features), and these seem to have influenced deposition on the platform either before or after Middle Eocene time. The slope angle, or dip, of the rocks in

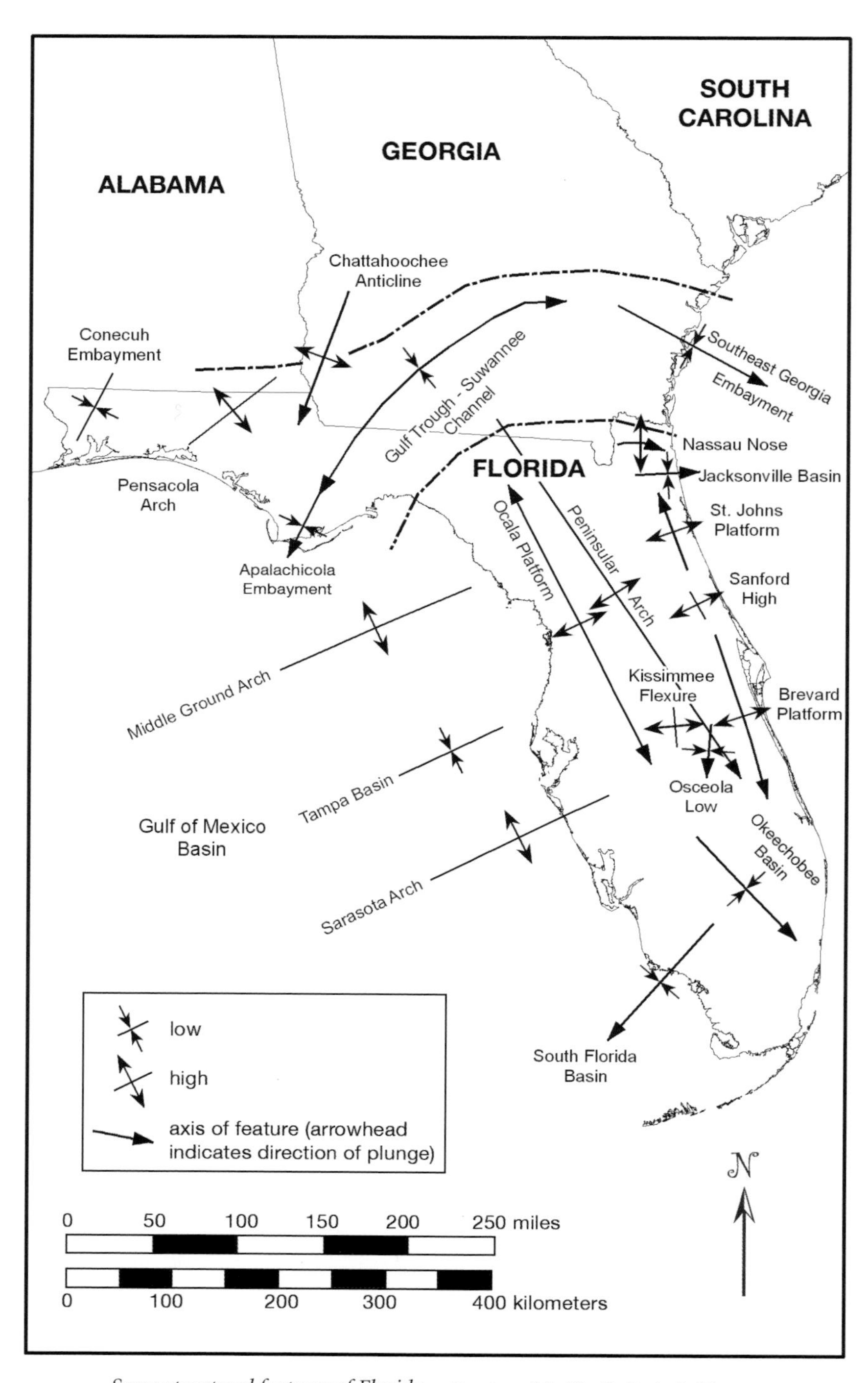

Some structural features of Florida. —Courtesy of the Florida Geological Survey

Florida rarely exceeds 10 feet per mile (about 0.1 degree) and is virtually imperceptible to the human eye.

Florida's underlying structural features dramatically affected the deposition of sediments on the platform beginning around 40 million years ago, after Middle Eocene time. Thinner sediment packages were deposited on the highs, while thicker sediment sequences accumulated in the lows.

The large-scale surface features (geomorphic regions) of Florida consist of east-west-trending highlands in the northern and northwestern portions of the state, north-south-trending highlands extending from Georgia down approximately two-thirds the length of the peninsula, and coastal lowlands. Some areas, such as the Everglades, are remarkably flat, with very little variation in elevation over vast areas. But in the interior, on the Lake Wales Ridge, the Brooksville Ridge, and the Tallahassee Hills, there are areas of rolling hills and valleys with localized relief of nearly 200 feet. These vistas stand in stark contrast to the flat profile often associated with Florida. Several hilltops in the Central Highlands exceed 300 feet above mean sea level, and the highest point in the state, Britton Hill, is 345 feet above sea level. It's in the Southern Pine Hills District near the Alabama-Florida state line.

Geologists recognize ten geomorphic districts in Florida, each of which has distinctive topography and can be readily seen on topographic maps and viewed as one travels through the state.

- Apalachicola Delta District—a Quaternary-age delta plain
- Barrier Island Sequence District—Pliocene- and Quaternary-age barrier islands
- Central Lakes District—mostly karst seepage lakes that developed beneath a cover of sand
- Dougherty Karst Plain District—developed over Eocene- and Oligocene-age limestones
- Everglades District—a freshwater peat marsh that developed over Pliocene- and Pleistocene-age limestones
- Ocala Karst District—developed over Eocene- and Oligocene-age limestones
- Okefenokee Basin District—a peat-forming swamp that formed over Pliocene- and Pleistocene-age sediments
- Sarasota River District—developed over a plain of Miocene- to Pleistocene-age sediments
- Southern Pine Hills District—a Miocene- and Pliocene-age delta plain
- Tifton Uplands District—a Miocene- and Pliocene-age delta plain

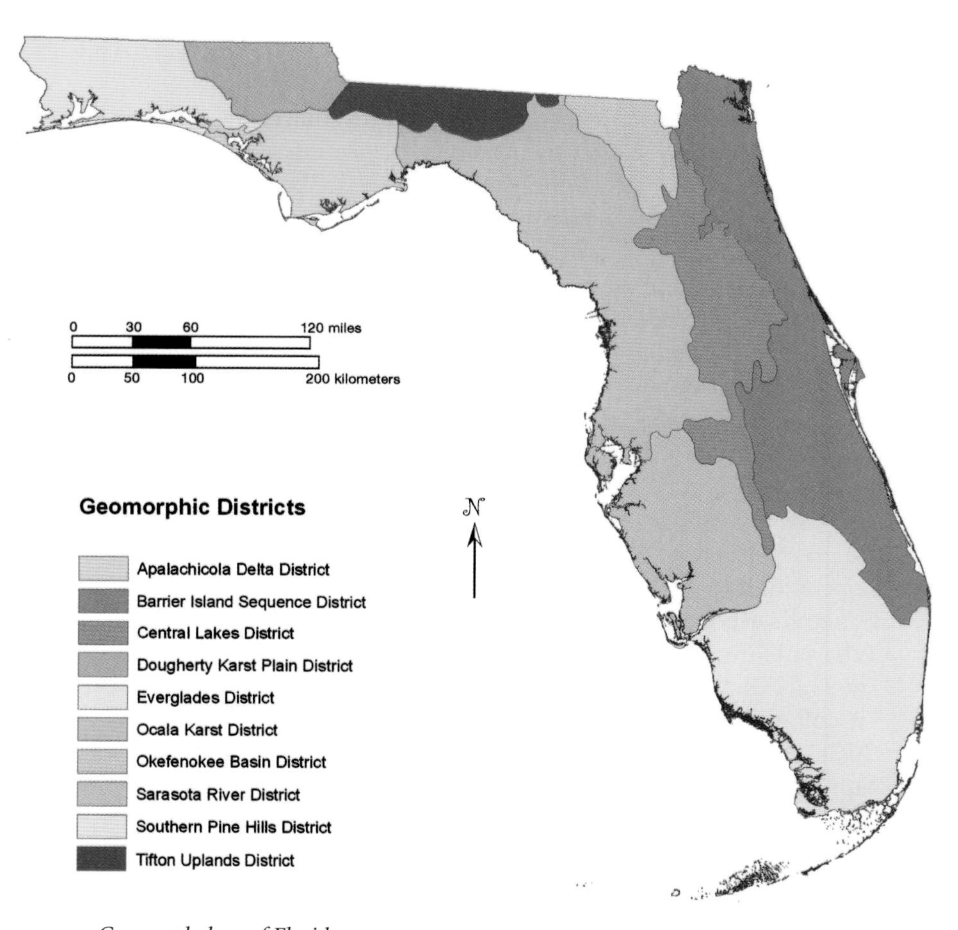

Geomorphology of Florida. —From Scott (2006) and courtesy of the Florida Geological Survey

The Suwannee Channel

The Gulf Stream after the creation of the Central American Barrier, found its way back to the Atlantic sweeping over southern Georgia and northern Florida, and supplying the food needed to build up the great organic beds of the Chattahoochee and Chipola.

—Raphael Pumpelly, 1893

Long ago, geologists working in Florida recognized that a great current must have once flowed in a northeastern direction from the Gulf of Mexico across northern Florida and southern Georgia to the Atlantic. Originally called the Suwannee Strait, this current-swept channel dramatically influenced marine life and sedimentation patterns in northern

Florida from Late Cretaceous to Middle Miocene time (99 to 16 million years ago). The Suwannee Channel proper existed from Late Cretaceous to Middle Eocene time, up to about 49 million years ago. From then to Middle Miocene time, the channel was considerably narrower and is referred to as the Gulf Trough. This persistent oceanic strait opened widely into the Gulf of Mexico near Apalachicola, at what is called the Apalachicola Embayment.

The origin of the Suwannee Channel may be related to deep structural features. Its location nearly coincides with a suture zone—called the Suwannee Suture—where Florida bedrock (originally African) was welded to the North American continent, and it parallels the trend of rift basins of Triassic age, one of which is called the Tallahassee Graben. These features could be related. The suture dates back to the formation of the supercontinent Pangaea. The down-warped suture was rifted apart during the breakup of Pangaea in Triassic time, and it formed a natural low area for a sea channel. In the high seas that came later, clockwise-flowing currents in the ancient Gulf of Mexico, much like the

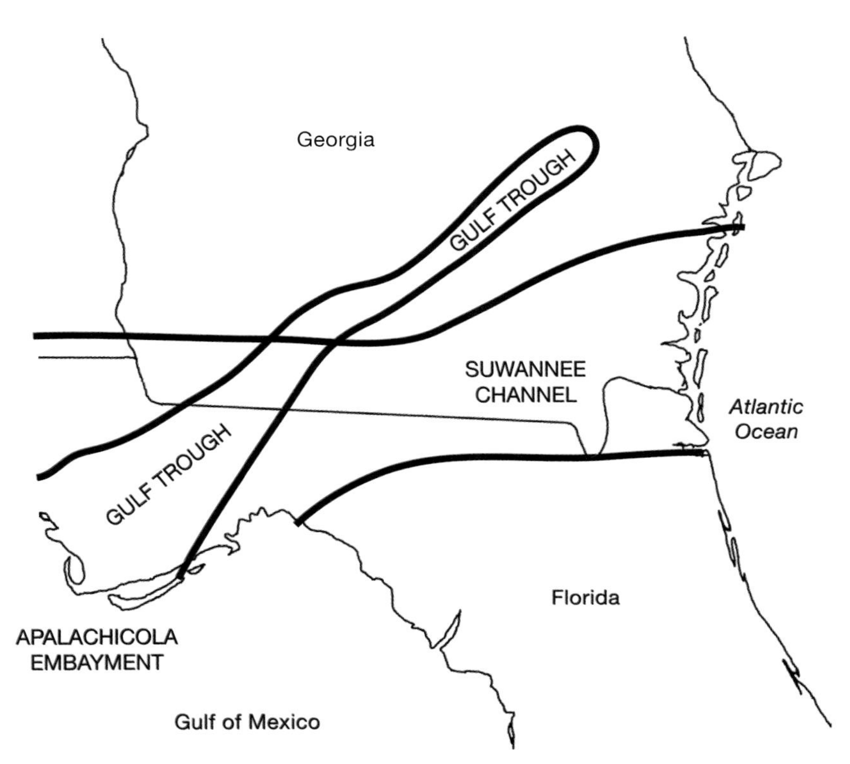

The Suwannee Channel (Cretaceous to Middle Eocene time) and Gulf Trough (Middle Eocene to Middle Miocene time), the former paths of the Suwannee Current.
—Modified from Huddlestun (1993)

modern Loop Current, would have naturally flowed through this area. Sea channels are common across modern carbonate platforms, such as the Bahamas Platform.

During Eocene time, the axis of the channel shifted progressively to the northwest in response to the rising sea. By Late Eocene time, the trough appears to have been its deepest—as much as 600 feet in southwestern Georgia. A thick buildup of sediment was deposited along the flanks of the trough there, forming the Ocala Limestone. During Early Oligocene time, corals and coralline red algae also constructed a prominent rim along the flanks of the trough, forming the Bridgeboro Limestone on top of the Ocala. By Late Oligocene and Early Miocene time, the Gulf Trough was probably no more than a long estuary, and it was completely filled with sediment by Middle Miocene time. The Gulf Trough is now detectable only in the subsurface. However, there may be a modern offshore remnant of this ancient seaway. A region in the northeastern Gulf of Mexico known as DeSoto Canyon lies along the same trend as the Gulf Trough and Apalachicola Embayment, and it may represent the last vestiges of this ancient channel system.

GEOLOGICAL AND PALEONTOLOGICAL HISTORY

Florida's Plate Tectonic Context

The Floridian peninsula is a region where geological action has been gentle, slow, and very uniform.

—W. H. Dall, 1892

The theory of plate tectonics is the central, organizing theory in the geological sciences. Without an understanding of the dynamics of crustal movement, it is impossible to understand seismic and volcanic activity or reconstruct past environmental and climatic conditions on Earth. Plate tectonics is sometimes confused with the older theory of continental drift. Continents are indeed "drifting," but not independently of the ocean floor as the older theory maintained, and the two theories are quite different in regard to causes of movement and the nature of crustal deformation.

A full review of the theory of plate tectonics is beyond the scope of this book. But in short, the theory states that the surface of Earth moves because its interior is very hot. The movement of heat energy upward from the interior results in the fragmentation of the hard exterior shell of rock at Earth's surface, which is called the *lithosphere.* The lithosphere includes the crust, or surface, and the uppermost part of a deeper layer called the *mantle*. Over time, the lithosphere breaks apart into large segments called *plates*, which are in constant motion, averaging 0.5 inch to 4 inches of movement per year. Plates are created continually in areas where magma is rising, such as in volcanic mountain chains like the Andes or along the volcanic mid-ocean ridge, where oceanic plates are spreading apart as extruded magma is added to them. But plates are also being destroyed where they sink back into the Earth, such as what occurs along the extensive ocean trench system around the margins of the Pacific Ocean. At these margins one plate is forced under another, and its rock is heated and melted, becoming magma. Most seismic and volcanic activity occurs where moving plates make contact.

Florida is part of the trailing edge of the westward-drifting North American plate, far removed from most tectonic action. The trailing edge of a plate is also called a *passive margin* because it is not directly colliding with another plate, so there is minimal seismic and volcanic activity. One of the nearest major plate boundaries to Florida is the Mid-Atlantic Ridge, in the middle of the Atlantic Ocean. This is the giant volcanic scar along which the supercontinent Pangaea started breaking apart 230 million years ago. Florida has been leisurely enjoying its passive margin status since at least Jurassic time, about 200 million years ago. But the Sunshine State was not always in such a comfortable tectonic setting.

Going Back in Time . . . Deep Time

Scientists often deal with phenomena that are beyond our normal range of perception. Chemists speak of atoms, which we cannot see. Physicists talk of forces that are invisible to us. Biologists tell us of complex DNA molecules that control the development of life, but we can only imagine what these look like through models or pictures. Astronomers speak of billions of galaxies in the universe, each with billions of stars, but we cannot possibly imagine these numbers or the size of the universe.

Similarly, geologists speak of geologic time, millions and even billions of years of Earth history, which John McPhee once referred to as "deep time." This amount of time is equally beyond our ability to grasp since our personal experience of time usually extends only two or three generations. Family history or history books may extend this range just a bit, but when considering geologic timescales we quickly find ourselves beyond the reach of our imaginations. Geologists have long known that the Earth is old, with a complex history of changing environments and changing life. This we can easily deduce by the thick sequences of fossil-rich strata in the Earth's crust. The order in which these strata occur, and the succession of epochs (time periods) in which the various extinct animals lived, was broadly understood nearly two hundred years ago.

But scientists could only crudely estimate how old these strata were. Estimates ranged from thousands to millions of years.

To determine the ages of strata, we needed some form of natural clock—something that keeps a record of time and makes an archive of that record in Earth's rocks or fossils. Natural timekeepers include annual tree rings, annual growth bands in some reef corals, annual deposits of glacial ice in continental glaciers, and annual sediment layers in some deep lakes. Some tree ring records extend back 12,000 years, and some glacial ice and lake sediments reach back 40,000 years. But as old as these records are, they are merely a drop in the bucket of geologic time. To tell time in the millions or billions of years, geologists primarily use a natural clock called *radioisotopes.*

Radioisotopes are radioactive varieties of certain elements, such as potassium, rubidium, and lead. These occur naturally in many minerals, especially those of igneous rocks. Once an igneous rock solidifies, the radioactive material begins to decompose, giving off subatomic particles (radioactivity). As a radioactive element (called the *parent*) loses subatomic particles, it decomposes into another element (called the *daughter*). We can easily determine how fast this process (called the *rate of decay*) occurs. When these radioactive materials decompose inside a crystal, the daughter element is usually trapped inside the crystal. By measuring the amount of daughter elements locked in the crystal and knowing the rate at which the parent decays into daughters, we can estimate the age of the mineral.

To illustrate the use of radioactivity to determine the age of rocks, imagine a large hourglass with sand falling through it. Could you determine how long ago the hourglass had been tipped? You could if you knew the rate at which the sand was falling through the glass and the amount of sand that had already fallen to the bottom. The hourglass is like an igneous rock. When the hourglass was tipped over is the equivalent of when the igneous rock solidified from molten magma. The rate at which the "parent" sand is falling through the glass is the rate of radioactive decay. The amount of sand on the bottom is the amount of daughter elements that have been produced since the igneous rock solidified.

Radioisotopes are used to determine the ages of key strata throughout the geologic record. By documenting the succession of strata and fossils through time, and by determining the age of strata using radioisotopes, the modern geological timescale has emerged. Geologists estimate the Earth is 4.6 billion years old, having used a combination of radioisotope dates from the oldest rocks on Earth (3.8 to 4 billion years old), the Moon (up to 4.2 billion years old), and meteorites. Meteorites—rocks that fall to Earth from our solar system—are consistently dated at 4.5 to 4.7 billion years old. Meteorites are considered the oldest materials in our solar system, and they approximate the age of all of the planets, which formed from the same materials. Earth's oldest rocks are not quite 4.6 billion years old because the Earth continues to recycle and melt its crustal materials through erosion and plate tectonic processes. Nevertheless, several large fragments of Earth's continental crust are at least 3.5 billion years old.

But how do we determine the age of strata and fossils in Florida, given that there are no radioactive igneous minerals at the surface? We know that the Ocala Limestone, for example, formed at the end of Late Eocene time, between 40 and 34 million years ago. How did we arrive at this age? The Ocala Limestone was deposited in the middle of a continental shelf environment over ancient Florida. This we know by the ecological requirements of the marine invertebrate fossils in the Ocala Limestone. Many of these same fossils occur in the Pachuta Formation in Alabama, which lies in the same stratigraphic position as the Ocala Limestone. But sediments of the Pachuta Formation were deposited in slightly deeper water than the Ocala, over the middle to outer continental shelf. Because of the deeper marine environment, the Pachuta Formation has many tiny deepwater microfossils that occur around the world in strata that also lie in the same position in the sequence of Earth's strata. By these fossils, we see that the Pachuta Formation, and its equivalent strata in Florida—the Ocala Limestone—are the same age as strata in northern Italy, which have been dated using radioisotopes.

Exotic Terrane—Late Precambrian and Paleozoic Time (1,000 to 251 Million Years Ago)

Florida's geological history begins as early as late Precambrian or earliest Cambrian time with the formation of the Osceola Granite complex and the St. Lucie metamorphic complex of rocks. These rocks lie deep below the Florida Platform. Sandstones, siltstones, and black shales of Ordovician, Silurian, or Devonian age were deposited in what is called the Suwannee Basin, which lies under the northern peninsula and panhandle. These sediments were probably derived from the Osceola and St. Lucie rocks as they eroded. What geologists know of this bedrock comes from rock cuttings and cores obtained from deep petroleum exploration wells. In the northern peninsula, these cores have been taken at depths ranging from 2,625 feet to 7,546 feet, and in the southern peninsula rocks have been sampled from as deep as 2.4 miles below the surface.

Through numerous lines of evidence, geologists have determined that this deep bedrock of Florida originally formed as part of the continental margin of western Africa, near modern Senegal and Guinea. This

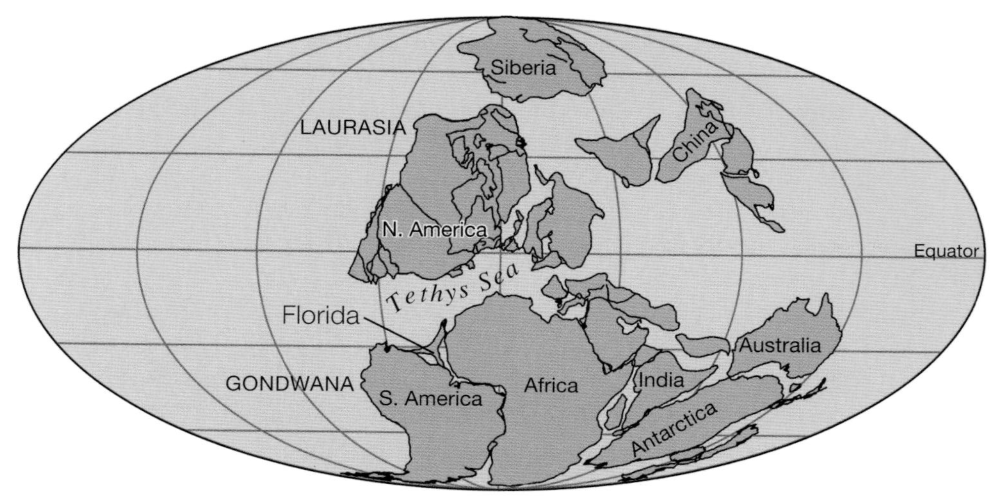

The megacontinent Laurasia represents primarily the northern hemisphere continents of today, and the megacontinent Gondwana the southern hemisphere continents. Laurasia and Gondwana were separated by a sea called Tethys, but they merged in late Paleozoic time to form the supercontinent Pangaea. Florida is shown here at its original home—between northeastern South America and northwestern Africa. When Pangaea broke apart, the bedrock of Florida was left attached to the North American plate. —Illustration by Amanda Kosche

Panhandle	Northern Florida	Southern Florida	Time	Age (millions of years)
			MESOZOIC	
Selma Group	Lawson Formation	Lawson Fm.; Rebecca Shoals Dolomite	Cretaceous	65
Eutaw Formation	Pine Key Formation	Pine Key Formation; Card Sound Dolomite	Cretaceous	
Tuscaloosa Group	Atkinson Formation	Atkinson Formation	Cretaceous	
unnamed sediments	unnamed sediments	Marquesas Supergroup: Naples Bay Group	Cretaceous	
		Marquesas Supergroup: Big Cypress Group	Cretaceous	
		Marquesas Supergroup: Ocean Reef Group (Includes Sunniland Fm.)	Cretaceous	
		Marquesas Supergroup: Glades Group	Cretaceous	
unnamed sediments; Sligo-Hosston Formation		Marquesas Supergroup: Fort Pierce Fm.: Pumpkin Bay Fm.	Cretaceous	
		Marquesas Supergroup: Fort Pierce Fm.: Bone Island Fm.	Cretaceous	145
Cotton Valley Group	Cotton Valley Formation	Marquesas Supergroup: Fort Pierce Fm.: Wood River Fm.	Jurassic	
Haynesville Formation	Cotton Valley Formation	unnamed volcanic complex	Jurassic	
Smackover Formation			Jurassic	
Norphlet Sandstone			Jurassic	
Louann Salt/Werner Anhydrite	diabase		Jurassic	
diabase			Triassic	200
Eagle Mills Formation	Newark Group		Triassic	

Mesozoic timescale and rock formations of Florida.

was the northwestern margin of the megacontinent Gondwana during Paleozoic time. During late Paleozoic time, Gondwana and Laurasia (another megacontinent) came together and formed the supercontinent Pangaea. When this supercontinent broke apart, a patchwork of fault-bounded slices of rock from Gondwana were left attached to the North American continental block. These granitic, metamorphic, and sedimentary rocks formed the foundation upon which the sedimentary rocks of the Florida Platform were deposited. Such displaced, or exotic, terranes are common in the western United States and Canadian Rockies. The Piedmont Mountains of Georgia and the Carolinas are also

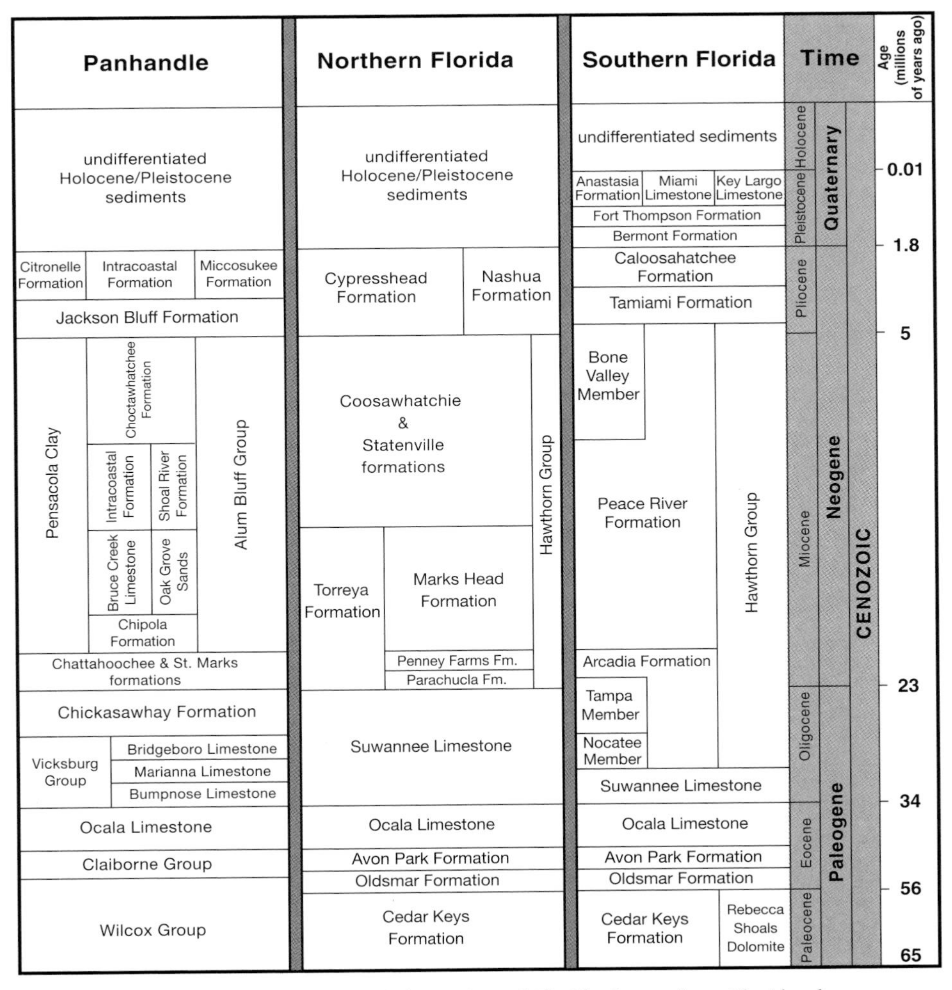

Cenozoic timescale and rock formations of Florida. In northern Florida, the Torreya Formation also extends into the eastern panhandle.

exotic. They were part of an oceanic volcanic complex that was attached to North America during the formation of Pangaea.

A very unusual assemblage of marine invertebrate fossils has been recovered from the Ordovician-age rocks of Florida, including brachiopods (lampshells), conularids (conical fossils of uncertain biological affinities), graptolites (extinct planktonic colonial animals), and trilobites. Silurian- and Devonian-age black shales contain chitinozoans, microfossils of uncertain biological affinities with walls of tough organic material; phyllocarids, shrimplike crustaceans with a bivalve shell; ostracodes, small bivalve crustaceans; and two species of eurypterids, or sea scorpions. Many of these closely resemble fossils that are typically found in Gondwanan bedrock.

When the Gulf Was Young—Triassic and Jurassic Time (251 to 145 Million Years Ago)

It was during Triassic time that Pangaea began to break apart, east from west, creating the great rift basins of the eastern seaboard of the United States. These rifts are crustal tears that were filled with coarse sands, gravels, volcanic rocks, and lake sediments. Similar rifts are developing today in the East African Rift Valley. There is little sedimentary record of this period in Florida, except for a southern extension of a rift system from Georgia. Part of this rift, the Tallahassee Graben, is filled with thick, red, feldspar-rich sandstones and igneous intrusions of basaltic magma. (A graben is a block of rock that has dropped downward by faulting to create a low area that may be filled with sediments.)

As Pangaea continued to separate during Jurassic time, the Gulf of Mexico opened, probably from the Pacific side, as a tropical sea in what was evidently a very arid climate. Extensive salt deposits, or evaporites, of halite and anhydrite—called the Louann Salt—precipitated in the western Gulf of Mexico and part of the what is now the Florida Panhandle. When deeply buried, salt becomes soft and fluid and protrudes upward into overlying strata, forming salt domes. These domes create pockets that can trap oil and natural gas. The elongate Destin Dome, over 40 miles long, is such a feature. Located offshore of the panhandle, it is one of the easternmost expressions of salt dome formation involving the Louann Salt. It contains untapped reserves of natural gas.

In the westernmost panhandle, sand dunes, alluvial fans, and river sands of the Norphlet Sandstone were deposited along the arid, tropical Gulf Coast, but these were soon flooded by a rising sea and covered by the fossiliferous limestones of the Smackover Formation. The Smackover Formation was part of a gigantic carbonate platform and reef complex of Jurassic and Early Cretaceous time that extended from the Bahamas

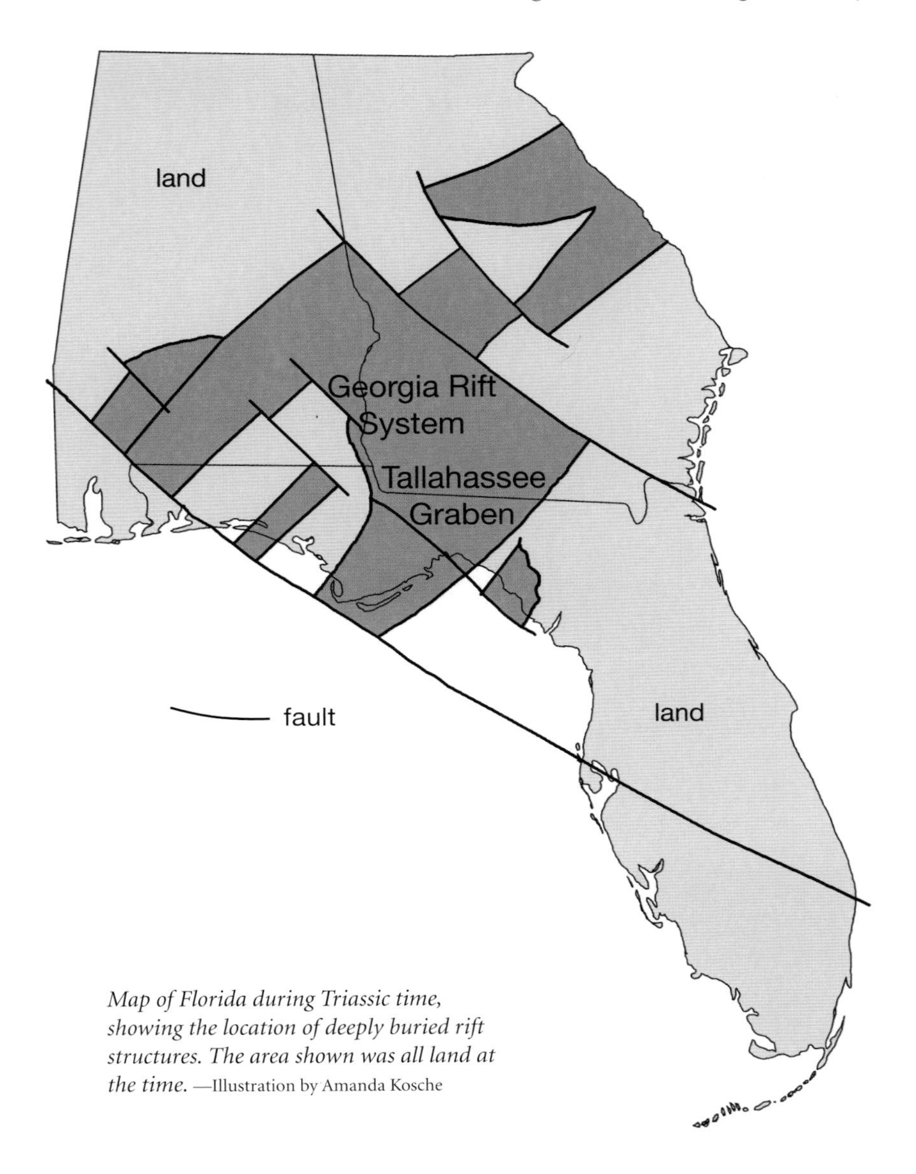

Map of Florida during Triassic time, showing the location of deeply buried rift structures. The area shown was all land at the time. —Illustration by Amanda Kosche

up most of the continental shelf edge of the eastern United States. This limestone factory was truly gigantic—more than four times as long as the modern Great Barrier Reef of Australia.

The Smackover contains reeflike mounds called *stromatolites,* which were constructed by cyanobacteria, sometimes called *blue-green algae.* Sponges were also present, as were tiny, rounded sediment grains called *ooids,* which look like miniature white jelly beans. Ooids form exclusively in shallow, wave-swept tropical waters. Oil and gas have been recovered from the Smackover Formation and the Norphlet Sandstone near Jay.

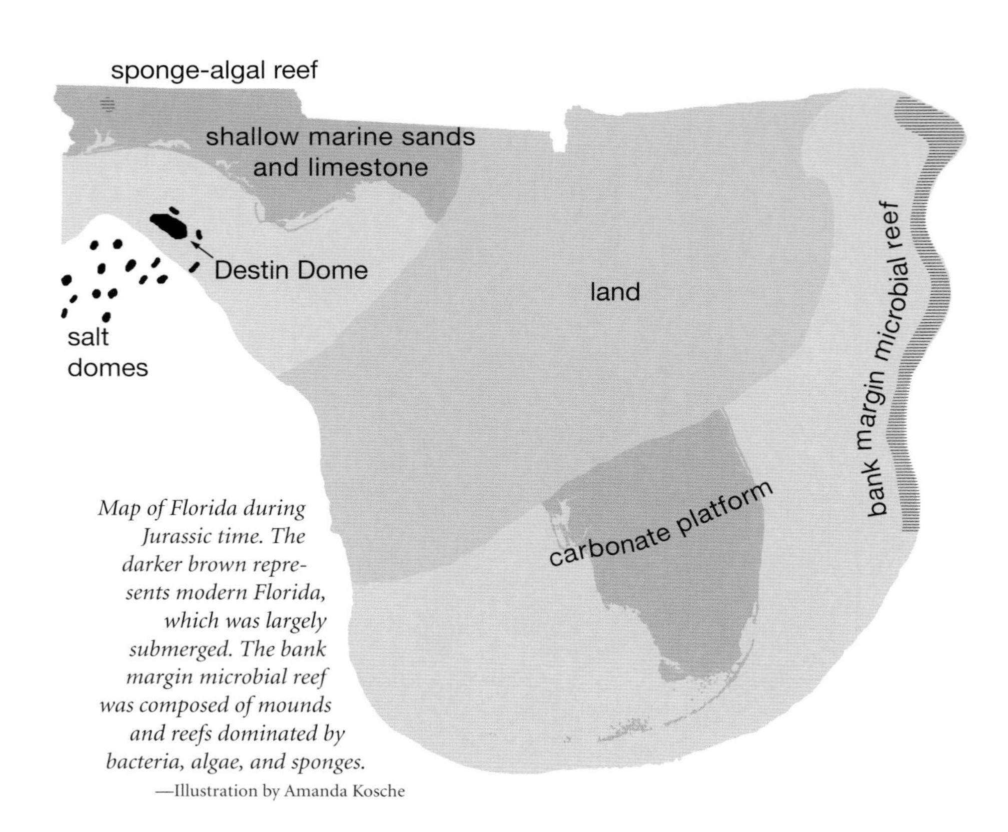

Map of Florida during Jurassic time. The darker brown represents modern Florida, which was largely submerged. The bank margin microbial reef was composed of mounds and reefs dominated by bacteria, algae, and sponges.
—Illustration by Amanda Kosche

Jurassic-age laminated anhydrite (dark layers), dolostone (light layers), and organic-rich layers from a deep oil well near Jay. This specimen was recovered from 15,000 to 16,000 feet below the surface. Specimen is about 7 centimeters tall.

When Rudists Ruled the Earth—Cretaceous Time (145 to 65 Million Years Ago)

In Cretaceous time, the sea reached some of its highest levels in all of Earth's history. Florida was nearly, if not fully, submerged, and it developed an extensive record of marine strata that consists of conglomerates, sands, shales, limestones, and evaporites. Marine invertebrate fossils are common in the many wells that penetrate these deep strata. These fossils include larger foraminifera, such as species of the genus *Orbitolina*; bryozoans; brachiopods; bivalve mollusks, such as species of the genera *Texigryphaea*, *Inoceramus*, and *Chondrodonta*, all of which are unique to Cretaceous-age strata; ammonites (extinct mollusks that were similar to nautiluses) of the genus *Dufrenoya*; and small crinoids, such as species of the genus *Applinocrinus* that lived in the open ocean. Fish bones and even sea turtle bones have also been recovered from deep oil wells.

Throughout tropical latitudes, peculiar oysterlike clams called *rudists* were especially abundant. Along with corals, they were instrumental in

Map of Florida during Cretaceous time. Most of the Florida Platform was submerged, with some land exposed in the northern peninsula. Rudist-dominated reefs were common. Note that rudist reefs extend west of the western edge of the Florida Platform. This reflects that much of the western edge of the platform has eroded since Cretaceous time.
—Illustration by Amanda Kosche

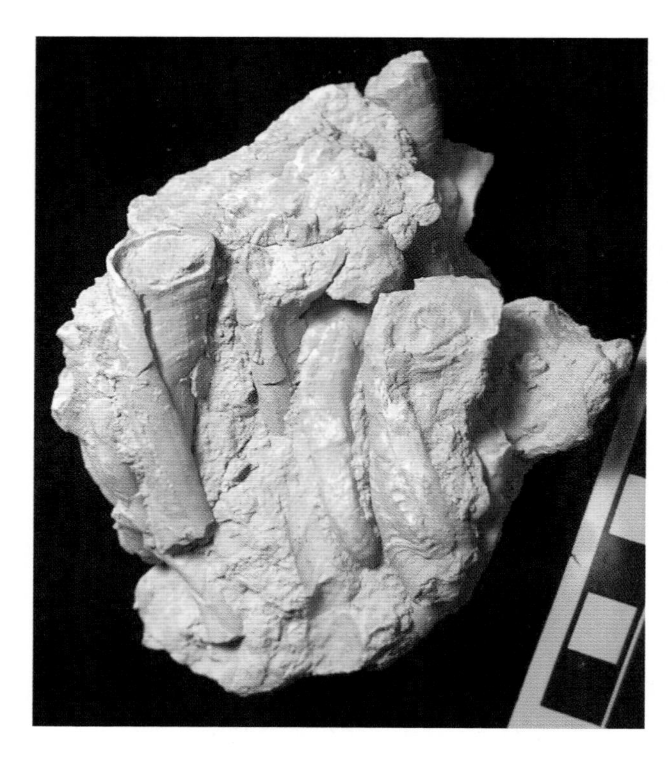

A cluster of rudist bivalves. This specimen is from Texas strata of Cretaceous age, but similar rudists are found in abundance in the subsurface of Florida. Ruler is in centimeters.

the construction of many Cretaceous-age reefs. Some reefs, in fact, were composed almost entirely of rudist clams. Some rudists were gigantic, reaching 4 or 5 feet in length, although most were more modestly sized—only several inches long. One of the rudist's shells was usually conical in shape, and the second acted as a lid for the conical shell. Rudists are found almost exclusively in Cretaceous-age rocks. They formed inner-shelf (shallow) skeletal shoals now found deep in the subsurface of southern Florida, and they also made reefs at the edge of the continental shelf. In 1943, oil was discovered in mounds of rudist shell debris within the Cretaceous Sunniland Formation, deep in the subsurface of Collier County. (Oil is commonly found in subsurface reef deposits because the porous reef limestone is a good trap for migrating oil.)

Sea World—Paleocene and Eocene Time (65 to 34 Million Years Ago)

During Paleocene and Eocene time, Earth was very much in a warm, ice-free, "greenhouse" state, and Florida remained submerged under a shallow sea. The Florida Platform continued to grow as limestone of the Cedar Keys, Oldsmar, and Avon Park formations was deposited. Evaporite minerals were common, indicating that water circulation over the platform was very restricted, which allowed salts to precipitate out of briny waters. It is possible that some of this water restriction was related to an extensive reef that surrounded much of the peninsula. It began forming in Cretaceous time and continued growing in Paleocene time, forming what is now the Rebecca Shoals Dolomite. Paleontologists are uncertain about the identity of the organisms that formed this reef because the original limestone recrystallized into dolostone, and these deeply buried rocks have not been adequately sampled. In the panhandle area, continental shelf sands and shales of the Wilcox Group were deposited during Paleocene and Early Eocene time. But during the high sea levels of Late Eocene time, the Ocala Limestone was deposited across much of the Florida Panhandle and portions of Alabama, Georgia, and South Carolina.

Carbonate rocks (limestone and dolostone) of the Avon Park Formation are the oldest rocks exposed at the surface in Florida. They are 40 to 45 million years old. In outcrops, the Avon Park is a dolostone with

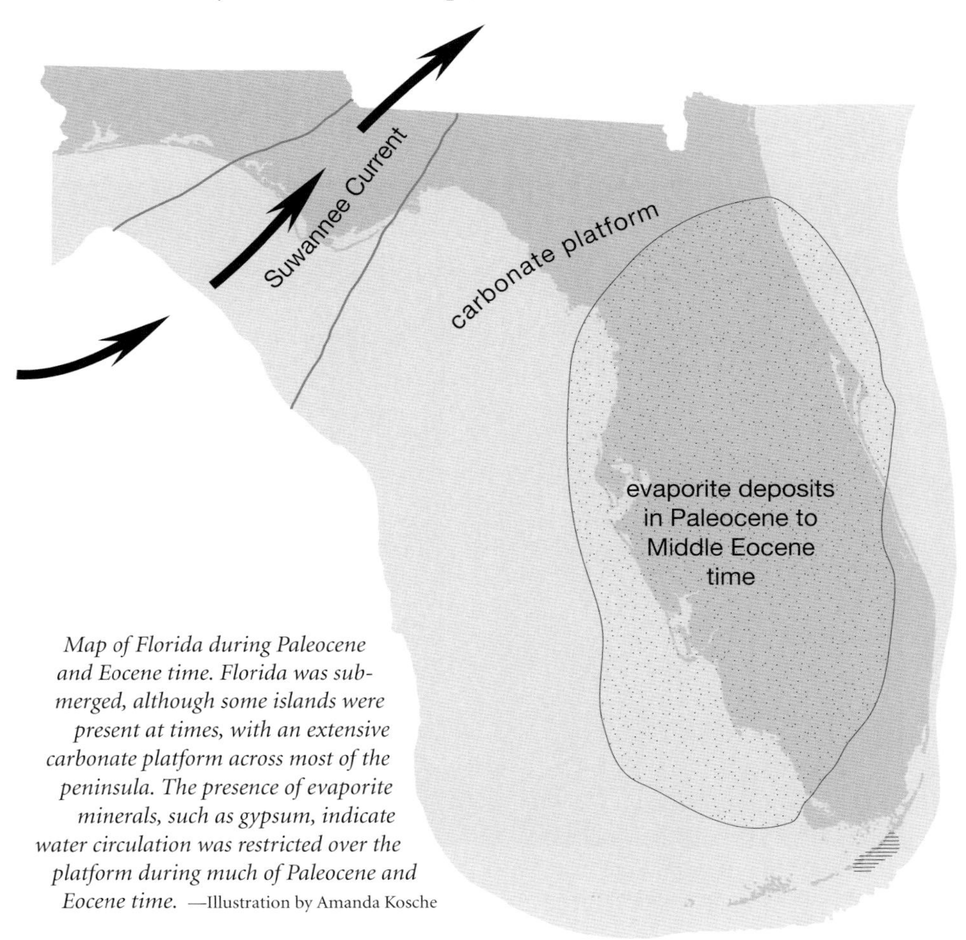

Map of Florida during Paleocene and Eocene time. Florida was submerged, although some islands were present at times, with an extensive carbonate platform across most of the peninsula. The presence of evaporite minerals, such as gypsum, indicate water circulation was restricted over the platform during much of Paleocene and Eocene time. —Illustration by Amanda Kosche

well-preserved sea grass fossils, indicating that these sedimentary rocks were deposited in very shallow waters, probably less than 30 feet deep. It is likely that islands emerged at this time across some of the Florida Platform. During the latter part of Eocene time, extensive deposits of the shells of very large foraminifera formed gravelly shoals across much of the Florida Platform. Most foraminifera are rather small, but some Eocene-age species, such as those of the genera *Nummulites* and *Lepidocyclina*, grew quite large—up to the size of a silver dollar. These shoals are found in the Ocala Limestone and are very similar to the famous *Nummulites* deposits in Egypt, which formed rocks that were used to construct the great pyramids. The Ocala has a rich fauna of invertebrate fossils, including many species of echinoids and mollusks, some of which were closely related to Eocene-age species of the Tethyan (Mediterranean) region. Descendants of these species live in the Pacific and Indian oceans today. Small patch reefs of coral are also found in the Ocala Limestone, and a platform margin reef likely developed along part of the Atlantic side of the Florida Platform.

Some highly unusual fossils have been found in the Eocene-age strata of Florida. A fossilized nut of a tropical palm (*Attalea gunteri*) was found in the Ocala Limestone near the town of Williston. Whether it came

Nummulitic limestone from the Ocala Limestone, collected near Steinhatchee. Note the abundance of the larger foraminiferan Nummulites willcoxi *(arrows). Ruler is in centimeters.*

from a small Florida island, floated from the Caribbean, or drifted from the Gulf Coastal Plain of the southeastern United States (where palm remains dating back to Eocene time are common) before being fossilized will probably never be known. The Avon Park Formation has produced bones from some of the earliest dugongs, those of the genus *Protosiren*, and early whales of the genera *Basilosaurus*, *Zygorhiza*, and *Pontogeneus* have been recovered from the Ocala Limestone. Rare sea turtle and sea snake remains have also been found in strata of this age.

*Fossilized palm nut (*Attalea gunteri*) from the Ocala Limestone. Specimen is 3.2 centimeters long.* —From Berry (1931) and courtesy of the Florida Geological Survey

The Ocala Limestone is the most extensively exposed Eocene-age rock in Florida. This rock is extremely pure, up to 99 percent calcium carbonate, and it is mined extensively in the state for road base and other applications. The Ocala is also one of the primary rock formations that composes the Floridan aquifer system, one of the most productive aquifers in the world and the source of most of Florida's drinking water and springwater.

Tropical Paradise—Oligocene Time (34 to 23 Million Years Ago)

In Oligocene time, limestone continued forming on the Florida Platform, primarily as the Suwannee Limestone, but the limestones formed in much shallower water than did those of Late Eocene time. Much of the Florida Peninsula was probably an island, or islands. In 1910, T. W. Vaughan coined the name Orange Island for this landform. The evidence for this island is the extensive absence of Oligocene-age strata in the northeastern part of the state, and the fact that Oligocene-age limestones of the peninsula were deposited in extremely shallow water and coastal settings. But some geologists think the Suwannee Limestone was originally deposited over much, if not all, of Florida because erosional remnants of it, called *outliers*, occur well away from the areas where the formation is found intact. It is clear that large volumes of Oligocene-age limestone of the Florida Platform were dissolved and eroded.

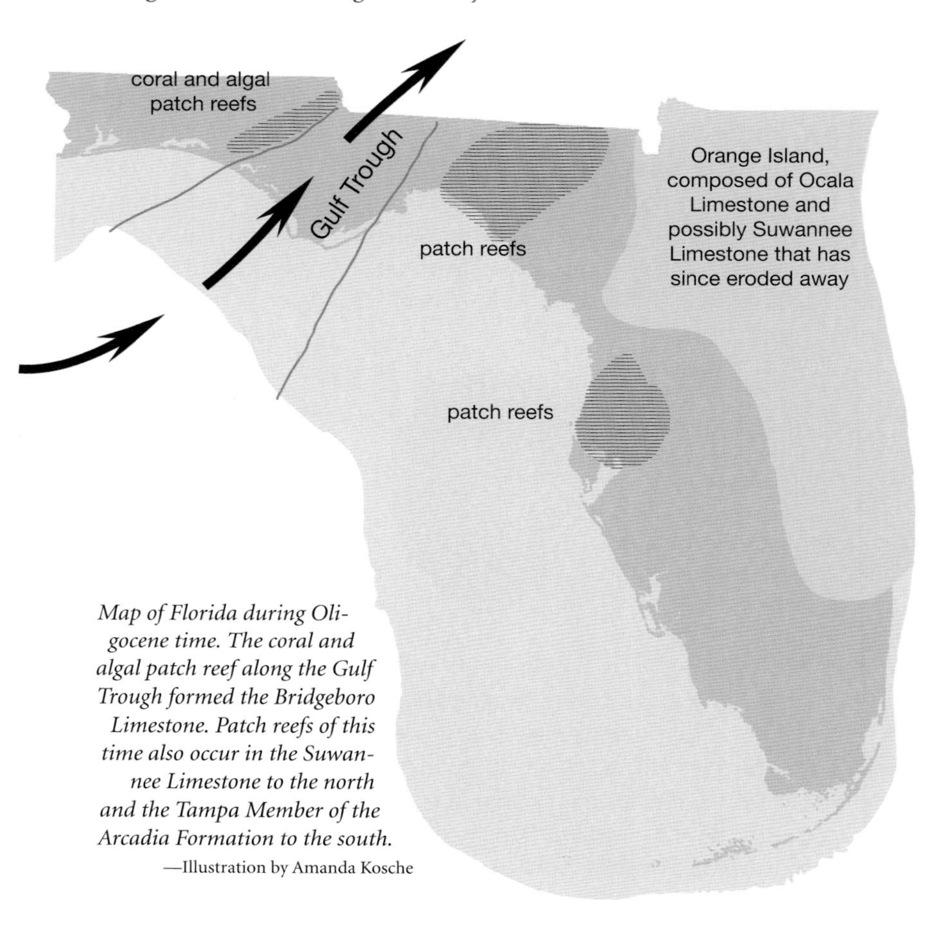

Map of Florida during Oligocene time. The coral and algal patch reef along the Gulf Trough formed the Bridgeboro Limestone. Patch reefs of this time also occur in the Suwannee Limestone to the north and the Tampa Member of the Arcadia Formation to the south.
—Illustration by Amanda Kosche

Throughout Oligocene time, and in the early part of Miocene time, much of the platform was littered with small coral patch reefs. Most of the agatized coral of Florida, so prized by gem enthusiasts, is collected from these patch reefs. Especially productive collecting areas include Tampa Bay (Tampa Member of the Arcadia Formation), the Withlacoochee River in Madison and Hamilton counties (Suwannee Limestone), and the Econfina River area in Taylor County (Suwannee Limestone). Along the northern edge of the Gulf Trough, coral patch reefs and rhodoliths—round nodules composed of concentric layers of coralline red algae—grew in abundance, forming the Bridgeboro Limestone, and muddy calcium carbonate covered a broad shallow seafloor over much of what is now the panhandle, forming the Marianna Limestone.

Marine vertebrates of this time, including sharks, bony fish, dugongs, and a rare sea turtle, have been found in the Marianna Limestone. And in the middle and latter part of Oligocene time, the first land vertebrates appear in small sinkhole deposits in central Florida. Fragmentary remains of a saber-toothed cat, the early horse *Mesohippus*, oreodonts, several other large hoofed-mammals, and many small mammals clearly indicate that a significant landmass had emerged by this time. The appearance of these land mammals in Florida coincides with the well-documented global fall in sea level during mid-Oligocene time—a fall of over 300 feet. This event certainly created some land corridors that connected portions of the Florida Platform with mainland North America, perhaps for the first time. The Gulf Trough became a narrow inlet or estuary, but it still extended far into Georgia. It was during Oligocene time that Earth began to enter a global "icehouse" state, with continental glaciers growing over the Antarctic continent. Sea level has risen and fallen many times since, but it has never returned to levels as high as those of Early Oligocene time. During Late Oligocene and Early Miocene time, uplift in the Appalachian Mountains resulted in a resurgence of erosion that delivered sediments southward, filling the Gulf Trough.

Bush Gardens—Miocene and Pliocene Time (23 to 1.8 Million Years Ago)

During Miocene time sea level regularly fluctuated. At various times during Early and Middle Miocene time, the entire Florida Platform was covered by the sea. During these high sea levels, both quartz-rich and carbonate sediments were deposited, forming parts of the Hawthorn Group. Important economic mineral deposits formed, including phosphorite and fuller's earth clay, in Hawthorn Group formations.

In Late Miocene time, sea level dropped significantly and much of the platform was exposed. A further drop in sea level at the very end of Miocene time exposed even more of the platform. During Pliocene time (5 million to 1.8 million years ago), much of the Florida Platform remained exposed. Only the southern peninsula and the fringes remained submerged. The deposition of both quartz-rich and carbonate sediments continued in these areas until latest Pliocene time, when quartz-rich sediments became the dominant sediment type. During this time, the delta sediments of the Citronelle and Miccosukee formations were deposited in the panhandle, while marine sands of the Cypresshead Formation were deposited across the northern two-thirds of the peninsula.

The Miocene and Pliocene strata of Florida preserve an exceptional record of mollusk fossils and other invertebrates, especially in the Alum Bluff Group in the panhandle, and the famous southern Florida shell beds of the Tamiami, Caloosahatchee, and related formations. These are

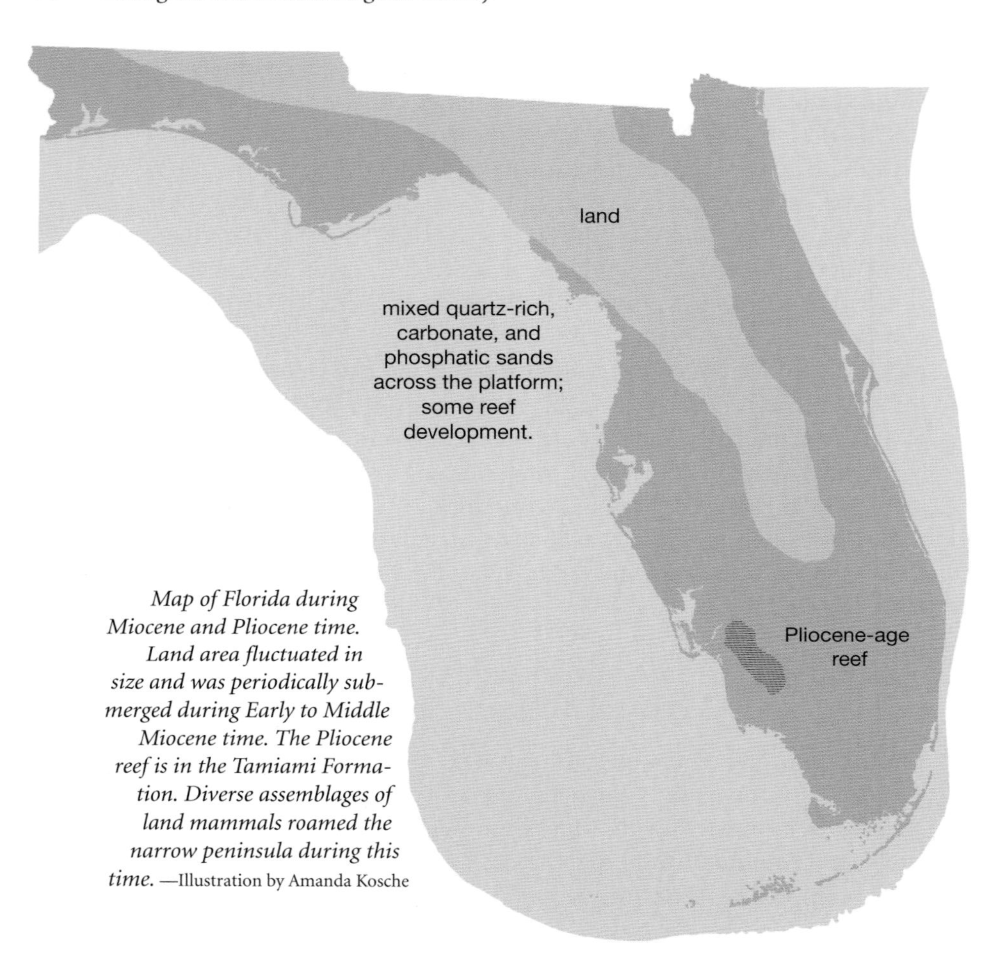

Map of Florida during Miocene and Pliocene time. Land area fluctuated in size and was periodically submerged during Early to Middle Miocene time. The Pliocene reef is in the Tamiami Formation. Diverse assemblages of land mammals roamed the narrow peninsula during this time. —Illustration by Amanda Kosche

some of the most species-rich shell beds in the world. The invertebrate fossils of this time indicate that the marine environment progressively cooled. There were tropical faunas and small coral reefs in Early Miocene time (Chipola Formation in the panhandle and Parachucla Formation of the Hawthorn Group in northern Florida), cooler-water faunas in Middle Miocene time (Shoal River Formation and Oak Grove Sand in the panhandle), and a mix of tropical and cooler-water creatures in Pliocene time (Tamiami Formation in southern Florida).

Although shallow marine sediments from this time are widespread in the state, the record of land vertebrates, as well as some fossilized plant remains, indicates much of the state was periodically above sea level. Fossilized palm fronds are found in the eastern panhandle, and petrified wood, including entire tree trunks, is not uncommon in many areas. Florida's fantastic record of unusual Miocene and Pliocene land mammals includes saber-toothed cats, camels, three-toed horses, shovel-tusk elephants, pronghorn antelopes, hyena-like bear-dogs, rhinos, and many others. Giant flightless birds were also present, and sharks, whales, and dugongs are commonly found. These strange vertebrate fossils are found in a variety of geologic settings, including shallow marine and estuarine deposits, river sediments, and sinkholes, and they indicate that rich coastal ecosystems, grassland savannahs, and other environments existed throughout the peninsula.

It was during Pliocene time that an enormously important geologic and oceanographic event occurred—the tectonic uplift of a narrow strip of land called the Central American Isthmus, or Panamanian Land Bridge. This narrow land connection between North and South America had surprising consequences for Florida. It restricted ocean circulation between the Atlantic (the Caribbean Sea) and the Pacific. This isolated marine faunas on each side and also diverted tropical Atlantic circulation northward, intensifying Gulf Stream flow. The land bridge between the two American continents also allowed a host of land mammals to migrate both north and south.

The north-south exchange of mammals, called the Great American Biotic Interchange, is nowhere more easily detectable than in Florida's vertebrate fossil record of Pliocene and Pleistocene time. Highly unusual South American mammals migrated to North America, and vice versa. Some of the South American immigrants became extinct in North America at the end of Pleistocene time, such as ground sloths and glyptodonts, while others survive to this day, including the opossum, armadillo, and porcupine. Likewise, some North American emigrants survive today in South America. Llamas, a zoological icon for South America, are, in fact, native to North America, and their fossils are commonly found in Miocene, Pliocene, and Pleistocene deposits of Florida. They became extinct in North America at the end of Pleistocene time, along with many other large terrestrial mammals, such as mammoths and mastodons. When calculated at the genus level, about 10 percent of modern North American mammals are of South American origin, and more than 50 percent of modern South American mammals are descendants of North American immigrants.

No Frozen Mammoths—Pleistocene Time (1.8 Million to 10,000 Years Ago)

During Pleistocene time, sometimes called the Great Ice Age, a gigantic continental glacier covered most of Canada and at times reached as far south as Ohio and Kentucky. Florida was never glaciated (and there are no deeply frozen mammoth remains as there are in Siberia), but it did experience cooler, drier climates and fluctuating sea levels as the continental glacier repeatedly grew in size (advanced) and partly melted back (retreated). At times, sea level was nearly 100 feet higher than at present. But near the end of the Ice Age, about 18,000 years ago, sea level was at a record low, perhaps as much as 390 feet below the present level, exposing nearly twice as much of the Florida Platform as is exposed today.

In spite of cooler terrestrial climates, tropical conditions prevailed in the marine realm of southernmost Florida. About 125,000 years ago, the Florida Keys—a fossilized barrier reef—formed during an interglacial period, when the continental glacier was relatively small. Sea level may have been up to 30 feet higher than at present. North of this reef tract, an oolite shoal and a bay behind the shoal formed what is now

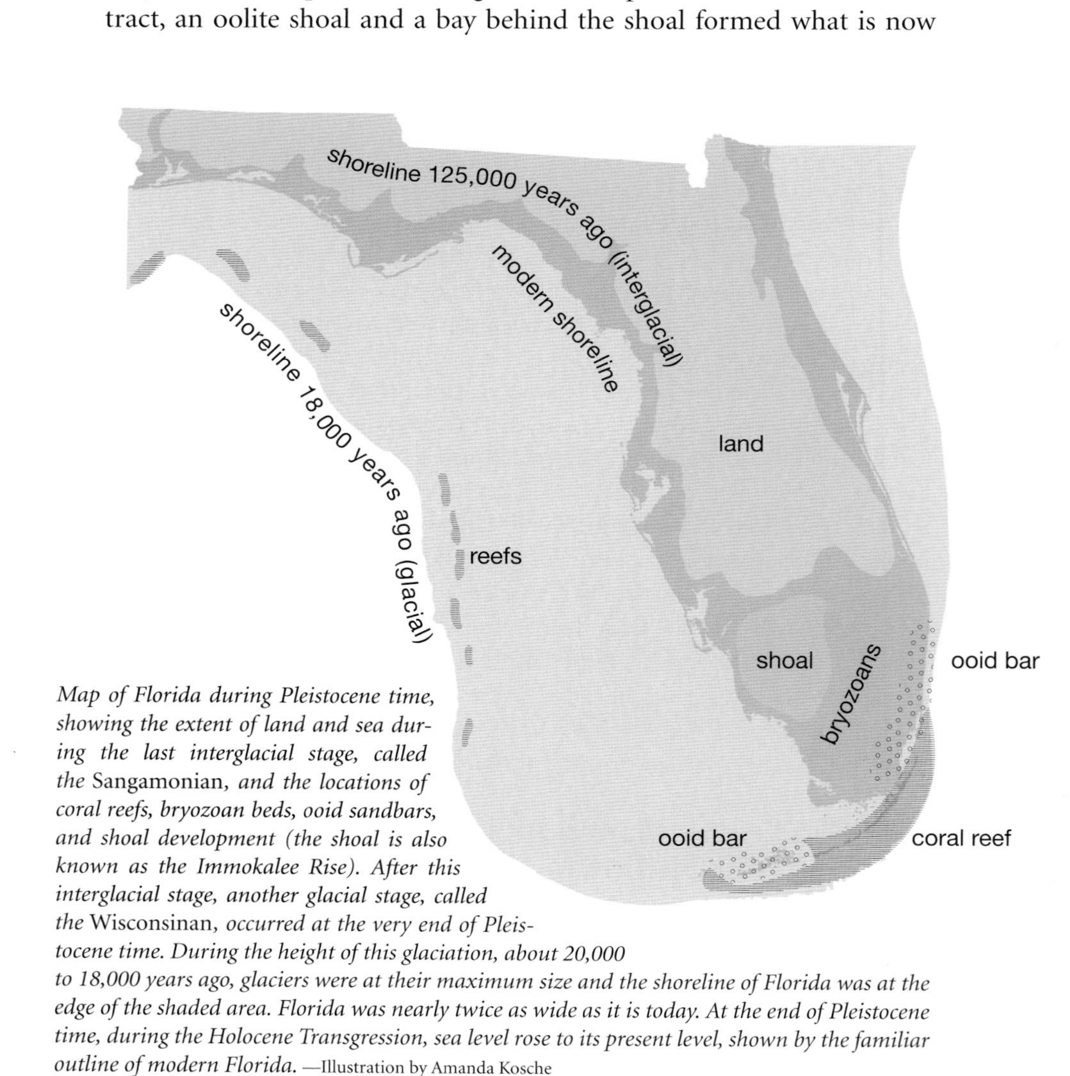

Map of Florida during Pleistocene time, showing the extent of land and sea during the last interglacial stage, called the Sangamonian, *and the locations of coral reefs, bryozoan beds, ooid sandbars, and shoal development (the shoal is also known as the Immokalee Rise). After this interglacial stage, another glacial stage, called the* Wisconsinan, *occurred at the very end of Pleistocene time. During the height of this glaciation, about 20,000 to 18,000 years ago, glaciers were at their maximum size and the shoreline of Florida was at the edge of the shaded area. Florida was nearly twice as wide as it is today. At the end of Pleistocene time, during the Holocene Transgression, sea level rose to its present level, shown by the familiar outline of modern Florida.* —Illustration by Amanda Kosche

the Miami Limestone. The Miami Limestone grades northward into the Anastasia Formation, a mixture of coquina and sand that was deposited along what were then Florida's east coast beaches. The Anastasia Formation composes the backbone of the Atlantic Coastal Ridge, a major linear and topographical rise that parallels Florida's east coast. A number of shell beds also formed around the margins of the southern peninsula during periods of high sea level, and deposits of quartz sand—representing river channels, floodplains, marshes, beaches, and dune fields—can be found across much of the state.

As with Miocene and Pliocene deposits, a great many fossil mammals have been recovered from Pleistocene-age deposits of Florida, including mammoths, mastodons, horses, saber-toothed cats, bear, bison, wolves, giant beavers, giant armadillos, armored glyptodonts, ground sloths, and many other exotic forms. Fragmentary remains of these creatures commonly wash ashore on some beaches, and one can collect them among the rocks and debris that concentrate in the middle of river channels. With the end of Pleistocene time came the extinction of many of the larger Ice Age mammals in North America. The causes for this extinction are still debated, but they undoubtedly include climate change and perhaps overhunting by early Native Americans—the Paleoindians. The modern North American mammalian fauna seems impoverished compared to that of the Great Ice Age.

Coastal Property—Holocene Time (10,000 Years Ago to the Present)

After the global rise of sea level at the end of Pleistocene time, Florida began to assume its modern configuration. Over the past 10,000 years, a relatively thin blanket of sediment has accumulated in many marine and terrestrial environments across the state, including nearshore marine areas, coral reefs, estuaries, river basins, lakes, swamps, marshes, beaches, and dunes. Soils, windblown silt, and other debris have been added to these strata like a layer of dust over the geological furniture. All of these sediment layers are Holocene in age, but geologists rarely give them formational names because they are still accumulating. In short, the sediments have not yet fully entered the geological archives because they are still being written, erased, and rewritten.

The climate of Florida is humid subtropical. In northern Florida, average maximum temperatures range between 79 and 80 degrees Fahrenheit, and minimum temperatures fall between 56 and 59 degrees Fahrenheit. In central Florida, average maximum temperatures range between 80 and 83 degrees Fahrenheit, and minimum temperatures fall between 59 and 63 degrees Fahrenheit. In southern Florida, average maximum temperatures range between 82 and 85 degrees Fahrenheit, and minimum temperatures fall only to 64 to 73 degrees Fahrenheit. Were it not for the maritime influence of the Gulf of Mexico and Caribbean Sea, desert conditions would be expected for Florida's midlatitude location on the eastern margin of a large continent. But in fact Florida is one of the wettest areas in the United States. Average annual precipitation reaches 60 inches or more in the panhandle and roughly 55 inches down most of the peninsula, with a low of 45 inches in Key West—the driest part of the state. Thunderstorms are especially common across the peninsula due to the dual influence of Gulf and Atlantic sea breezes, and there are more lightning strikes in Florida than anywhere else in the United States.

Major Holocene-age terrestrial ecosystems in the state include pine flatwoods (level grasslands with scattered trees), which cover about 50 percent of the land area of the state; dry prairies; scrubby flatwoods; southern hardwoods, or hammocks; forested wetlands, or swamps, which cover about 10 percent of the state; freshwater marshes; and various coastal systems, such as salt marshes and mangrove forests. There are hundreds of endemic species in the state. In the southern part of the peninsula, many plants, insects, and birds reveal the strong tropical influence of the West Indies. But in other groups, such as reptiles and amphibians, species diversity sharply decreases from the northern to the southern peninsula—a biogeographic phenomenon generally known as the *peninsular effect.* In recent years, an increasing number of exotic pet species, especially reptiles and birds, have escaped or been released from captivity in southern Florida.

Modern Florida is, in many ways, stuck in its own geological past. Present-day environments, geological processes, and some plant and animal life are not unlike those of even the distant past. Modern coastal dunes are the sedimentary descendants of giant interior Pleistocene-age dunes. Today's estuaries are like those that have deeply indented Florida's fluctuating shorelines since Miocene time, leaving fossil-rich clays and sands many miles inland from the modern shoreline. Sinkholes continue to open and collapse as they have since Oligocene time. And although significantly downsized from its heyday before the Ice Age, a limestone factory continues to manufacture coral and other limestone around the Florida Keys. The living coral reefs of the Florida Reef Tract are nearly identical to the reefs that formed the Florida Keys during Pleistocene time. Plants and animals that are more typical of the Appalachian Mountains still reside in Florida's cool, deep ravines in the panhandle, though they were more widespread during the cooler climate of Pleistocene time. Giant, docile manatees swim leisurely in coastal waters, as did their cousins the dugongs in Eocene, Oligocene,

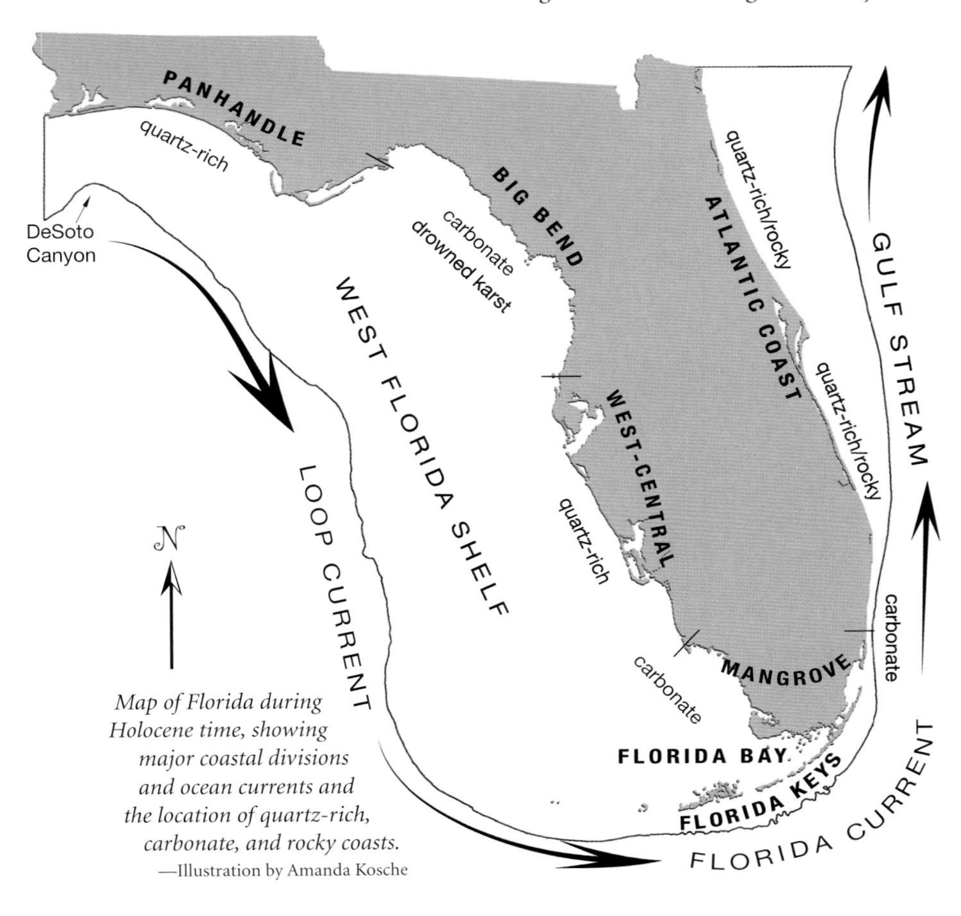

Map of Florida during Holocene time, showing major coastal divisions and ocean currents and the location of quartz-rich, carbonate, and rocky coasts.
—Illustration by Amanda Kosche

Miocene, and Pliocene time, and as the manatees have done since Pleistocene time. And finally—some things really never change—Florida has, since at least Miocene time, been infested with alligators!

The First Floridians—Geoarchaeology

Florida arguably has one of the richest archaeological records of any state, extending from late Pleistocene time to historic times. The Paleoindian Period (13,000 to 8,000 BC) is represented by Clovis, Suwannee, and Simpson artifacts, which are recognized in the archaeological record as large, lanceolate, chert projectile points. Paleoindian sites are uncommon, and finding dateable material in context is rare. Artifacts are commonly found in karst rivers and spring runs, where material concentrates in lag deposits of debris, but finding sites that still contain material in stratigraphic context is more important for archaeologists because it allows them to accurately date the material, and to collect associated artifacts. During the Paleoindian Period, sea level was 160 to 325 feet lower than today, so it is reasonable to assume that numerous sites exist offshore in the marine environment—and indeed they do. But most Paleoindian sites are found inland.

The Archaic Period (8,000 to 1,000 BC), which is well represented in Florida, is characterized by increasingly complex and sophisticated artifacts and homesites. As climate changed and large animals, or megafauna, of late Pleistocene time died out, cultures of this period adapted. They began growing crops, and the tools they used were different from those of the Paleoindians. As sea level stabilized during the late Archaic Period, populations began to exploit Florida's rich marine environment, as indicated by large shell middens. State archaeological sites preserve evidence from this time at Fort Walton Beach, Crystal River (the town), Fort Myers, and numerous places along the St. Johns River. Another important cultural development of this period was the use of clay pottery. This invention appeared about 1,000 BC in Florida, and it marked the transition to the Woodland Period.

During the Woodland Period (1,000 BC to AD 1,000), native Floridians utilized ceramics extensively, cultivated more crops, and became more sedentary. Woodland Period cultures were highly social, which is reflected in their extensive trade networks and mound complexes. Galena, copper, mica crystals, and greenstone, none of which naturally occur in Florida, are found associated with sites of this time. One of the best examples of mound building is near Tallahassee at the Lake Jackson Mounds. These mounds were built for powerful rulers who oversaw a complex agricultural society with ties to other villages throughout the southeastern United States.

The Mississippian Period (AD 1,000 to 1,600) was the last cultural period before European contact. Cultures of this period built extensive ceremonial mound complexes as their populations expanded across and settled Florida. Agriculture was an important aspect of these cultures, more so than with previous cultures.

It is not surprising that the cultural remains of Florida's earliest human inhabitants have geological contexts. Springs, coastal environments, and riverbanks were common homesites for prehistoric people, and geologically they are favorable locations for the preservation of artifacts because sediment accumulates in them. Karst has also proven to be an important archaeological environment. Karst springs were common sources of freshwater, food (they were watering holes for game), and stony raw materials, such as chert. Sinkholes, springs, and deeply incised rivers also act as natural traps for artifacts, animal remains, and other materials. Archaeologists have discovered intact sites in karst traps that date from Paleoindian time to after Columbus's contact.

Typical projectile points of early native Floridians. From left to right: Clovis waisted point, Paleoindian Period, Wacissa River; Bolen beveled point, early Archaic Period, Wacissa River; Kirk stemmed point, early Archaic Period, Wacissa River; Newnan point, middle Archaic Period, Ocklawaha River; Lafayette point, late Archaic Period, Apalachicola River; Hernando point, Woodland Period, Aucilla River; Pinellas point, Mississippian Period, Apalachicola River.

Two geological raw materials were especially important to Florida's first inhabitants: chert and clay. Chert is a fine-grained sedimentary rock composed of silica minerals that precipitate out of groundwater; it is commonly associated with limestone. Chert is exposed primarily in the northern and north-central portions of Florida where Tertiary-age limestones are at or near the surface. Most of the chert in Florida is related to weathering of the Hawthorn Group, which is composed of clays that, when weathered, provide a source of silica for chert formation. Native Americans learned to manipulate the conchoidal (circular) fracturing property of chert by percussion and pressure flaking. They produced highly effective tools for cutting, scraping, and piercing, including projectile points, scrapers, choppers, adzes, and small arrow points. Quarry sites were often utilized over several cultural time periods.

Clay also became an important geological resource. Archaic cultures were the first to manufacture pottery from this abundant natural resource found in formations of Miocene age and younger in Florida. There are different types of clay minerals in Florida, and they exhibit different properties when molded and fired. Early Floridians utilized a technique called *tempering* to make the clay more plastic. By adding materials such as sand, plant fiber, freshwater sponges, limestone, among others, the clay became easier to mold into pottery.

SCULPTING A LAND FROM THE SEA

Because of Florida's geologically recent emergence from the sea and lack of significant tectonic deformation, the origin of its modern surface topography must be understood in the context of processes related to sea level changes, coastal dynamics, and the exposure of ancient seafloors, coastal sands, and deltas to the terrestrial geologic forces of weathering and erosion.

Sea Level Change—Past and Present

Geologists have understood the general geologic history of global sea level change for many years, but this knowledge has been highly refined in recent decades thanks to exploratory work by petroleum geologists. In Jurassic and Cretaceous time, 200 to 65 million years ago, sea level was quite high, due in part to the absence of continental glaciers. Likewise, the Earth was mostly ice-free during Paleocene and Eocene time, 65 to 34 million years ago. Sea level was high and climates were warm. But by Oligocene time, there is clear evidence that the Antarctic ice sheet had grown substantially. In the middle of Oligocene time, global sea level fell dramatically due to this continental ice buildup. The high seas of Paleocene and Eocene time are well-documented in Florida's strata, as is the shallow water of Early Oligocene time and sea level drop of mid-Oligocene time. Since Oligocene time, Earth has effectively been in a glacial mode, and sea level has been relatively low compared with earlier times.

During the last 2 million years, Earth entered its most recent ice age (there were several earlier ice ages). Sea levels during this time, although low overall, have fluctuated dramatically as great continental glaciers periodically grew, causing a drop in sea level, and then partially melted, causing a rise in sea level. Scientists hypothesize that these glacial-interglacial cycles are caused by variations in Earth's orbital patterns, which can cause significant climate change once the planet has entered an ice age climate regime. For a continental ice sheet to grow the climate must be cool enough that polar winter snow does not melt appreciably during the polar summer, allowing the snow to continue to accumulate year after year.

At the end of Pleistocene time, from about 21,000 to 18,000 years ago, the great continental glaciers reached their maximum size, more than twice that of today's ice sheets, and, as mentioned above, sea level was as much as 390 feet below its present level. Large river systems spread across what is now the modern continental shelf, and deltas formed near the modern shelf edge. As the continental glaciers melted back to their current, more modest size (consisting of the Greenland and Antarctic ice sheets), sea level rose dramatically. This rise, called the Holocene

Time	Panhandle	N. Peninsula	S. Peninsula
Pleistocene	undifferentiated	undifferentiated	Anastasia Formation Miami Limestone Key Largo Limestone
Pliocene	Citronelle Formation	Cypresshead Fm. Miccosukee Fm.	Cypresshead Fm. Caloosahatchee Fm. Tamiami Fm.
Miocene	Pensacola Clay Alum Bluff Group	Hawthorn Group	Hawthorn Group Peace River Fm.
Oligocene	Chickasawhay Limestone Bridgeboro Limestone Bumpnose Limestone	Suwannee Limestone	Hawthorn Group Arcadia Formation (Tampa Member) Suwannee Limestone
Eocene	Ocala Limestone	Ocala Limestone	Ocala Limestone
	Claiborne Group	Avon Park Formation Oldsmar Formation	Avon Park Formation Oldsmar Formation
Paleocene	Wilcox Group	Cedar Keys Formation	Cedar Keys Formation Rebecca Shoals Dolomite
Late Cretaceous	Selma Group	Lawson Formation	Pine Key Formation

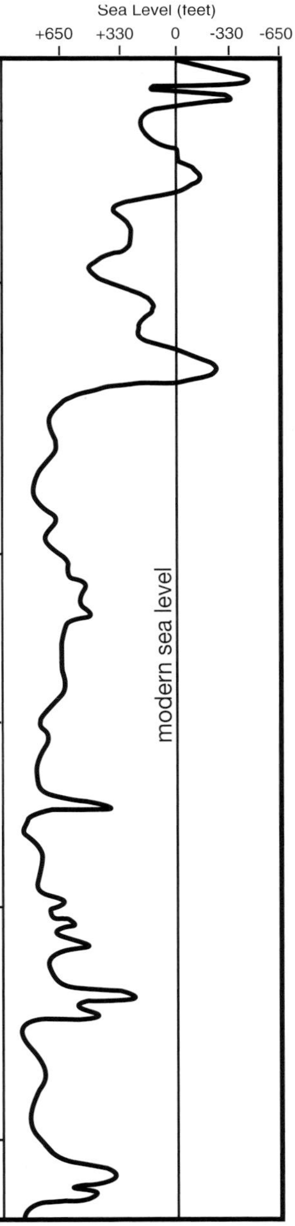

Sea Level (feet)

+650 +330 0 -330 -650

modern sea level

The sea level curve for Cretaceous through Holocene time, showing the major rock formations of Florida deposited during that time. The vertical line with the sea level curve represents modern sea level. When the curve is left of the line, sea level was higher than present. When it's right of the line, sea level was lower than present.

Transgression, occurred between about 19,000 and 6,000 years ago, with sea level rising to its present position. On average, this rise was very rapid at first, as much as 1 centimeter per year between 18,000 and 8,500 years ago, and at times sea level may have risen as much as 5 centimeters (about 2 inches) per year due to glacial meltwater pulses and the calving, or breaking off, of large icebergs. Eventually, the rate slowed to as little as 1 millimeter per year from about 8,500 to 6,500 years ago, and with the global deceleration in sea level rise, the dominant coastal process changed from marine transgression (shift of the coastline in a landward direction) and coastal erosion to delta growth and beach development. Humans began colonizing the modern coastline about 5,000 years ago.

Besides the growth and ablation of glaciers, there are other causes for sea level fluctuation. Sea level may rise due to the addition of water from volcanic emissions, an ongoing process throughout geologic time. Land may rise after glaciers have melted off them, resulting in a local drop in sea level. Called *isostasy*, this process is occurring along modern Scandinavian coasts and in northern Canada. Isostasy occurs in Florida, though not because of the removal of glaciers. As calcium carbonate dissolves from the Florida Platform, the platform rises and sea level falls relative to the land. Coastal areas may subside due to the addition of sediment, as is happening with the Mississippi Delta. Land may be tectonically lifted, resulting in a local drop in sea level. Inflation of the volcanic mid-ocean ridge during times of increased heat flow from Earth's mantle and lava extrusion may displace water over the continents. This occurred during Cretaceous time. Presently, sea level is rising at a rate of about 2 millimeters per year, and this rate has remained steady over the past several decades. Scientists think this rise is related to global climate change and its effects—namely, the warming and thermal expansion of ocean water and the melting of glacial ice, especially the smaller alpine glaciers, the melting of which may have resulted in up to 5 centimeters of rise in the past one hundred years.

Estimates for the total rise in sea level over the next one hundred years range from 8 inches to over 3 feet. Scientists base the low estimate on a continuation of present rates of sea level rise and the high estimate on more dramatic changes, including substantial contributions from continental glaciers. Scientists estimate that if the West Antarctic Ice Sheet collapsed and slid into the sea, sea level could rapidly rise over 16 feet globally. Indeed, from 130,000 to 100,000 years ago, in an interglacial stage of late Pleistocene time, sea level did rise about 16 feet higher than its present position. There is enough water locked up in the Greenland and Antarctic ice sheets to raise sea level some 250 feet, which would submerge most of Florida. Some scientists are concerned about the present stability of the Greenland ice sheet and the Arctic

Ocean pack ice. As indicated by these large masses of ice, we are still very much in an ice age, albeit in an interglacial period of warming.

Coastal Strandlines and Terraces

The rise and fall of the sea leaves a variety of geological high-water marks that show the former position of the coastline. Geologists can approximate the location of these earlier coastlines—called *strandlines*—in the stratigraphic record using many clues. The presence of certain fossils, such as land snails, sea grasses, or mangrove stumps, or fossil groupings where land animals are mixed with shallow marine fossils such as oysters, indicate the location of a former shoreline environment. Such fossil associations are extremely common in Florida. Certain types of sedimentary features, such as cross-stratification, can indicate very shallow water and proximity to an ancient shore. Coasts that are endowed with a rich supply of sand may build their beaches out into the sea by the accretion of beach ridges—berms of sand deposited by incoming waves. These ridges mark the location of successive beach scarps. Windblown sand dunes are another common coastal feature, and several ancient dunes and dune fields are found across the Florida interior.

As a rising sea floods a coastline, it will create a new, shallow sea bottom. After sea level drops again, the former seafloor is exposed as a broad, flat plain called a terrace. A scarp, or escarpment, may mark the landward edge of the terrace, where waves cut into upland sediments leaving

TERRACES OF FLORIDA

Terraces	Previous Names	Elevation (feet above mean sea level)	Probable Age
Hazelhurst	Brandywine	215–320	Miocene or Pliocene
Coharie		170–215	Pliocene
Sunderland	Okefenokee	100–170	Late Pliocene
Wicomico		70–100	early Pleistocene
Penholoway		42–70	mid- to late Pleistocene
Talbot		25–42	late Pleistocene
Pamlico	Pensacola	8–25	late Pleistocene (Sangamonian interglacial period)
Silver Bluff	Princess Anne	less than 1–10	latest Pleistocene or early Holocene

a steep face. Although rivers deeply erode old terraces and scarps in their gravitational drive to reach the sea, geologists have identified several relict terraces of Pleistocene, Pliocene, and even Miocene age around the state. Some terraces are quite obvious, while others are very subtle and disputed in the geologic community. Recognition of older terraces can be difficult due to erosion or uplift of the land after the terraces formed. Thus, a simple tracing of terraces across the landscape based on elevation above modern sea level can be misleading and inaccurate.

The modern shoreline, established 4,000 to 5,000 years ago, constitutes an escarpment (a beach scarp), and most of the continental shelf represents Florida's present flooded terrace. There are several other, older submarine terraces across the Florida Platform, some of which show evidence of karst development. Because the karst developed when these portions of the platform were exposed above sea level, they are strong evidence of recent rises in sea level.

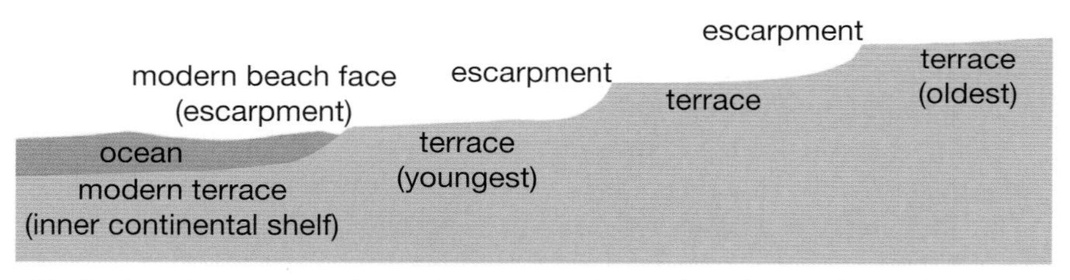

Idealized marine terraces and coastal escarpments in a setting where sea level is dropping.
—Illustration by Amanda Kosche

Coastal Environments

With 1,350 miles of oceanfront shoreline and some 5,000 miles of bay and estuarine shoreline, Florida is a natural geological laboratory for observing modern coastal environments and processes. The entire Florida coast is part of the larger eastern North American passive margin, so named because it does not border an active plate boundary and therefore exhibits minimal seismic and volcanic activity. The Florida coast is easily divided into seven segments, each with distinctive geological features.

The panhandle coast, from Perdido Key to Ochlockonee Bay, contains the whitest sandy beaches in the world. They are composed of almost pure quartz sand. The abundant sand constitutes the raw material for this coastline's numerous barrier islands and spits. Extensive sand dunes are also a notable feature of the panhandle beaches. The windblown sand of these dunes is noticeably finer than that on the beach. Dune fields are especially well developed along Santa Rosa Island and St.

Joseph Peninsula. The seaward-protruding Apalachicola area, a cuspate foreland, is an ancient delta that rivers have been building into the Gulf since Miocene time.

The Big Bend coast, from Ochlockonee Bay to Anclote Key, doesn't have barrier islands because quartz sand is not abundant. The floor of this coastline is limestone of Eocene, Oligocene, and Miocene age. The coast is rocky and marshy, and oyster reefs are common. Acidic groundwater has dissolved the limestone exposures here into jagged and porous karst. The global rise in sea level at the end of the Ice Age flooded much of this limestone surface, creating what is called a *drowned karst coastline*. Geologists also refer to this coastline as a *low-energy coastline* because of its very low wave energy. As waves travel across the broad and shallow limestone shelf, they continuously break and lose their energy, so they are greatly diminished by the time they reach the shore.

The west-central coast, from Anclote Key to Cape Romano, has numerous barrier islands—about twenty-nine—and is primarily composed of quartz sand. Unlike the panhandle coast, the sediment layer here is rather thin. This is a dynamic, shifting coastal system. In the past one hundred years alone, new barrier islands formed and tidal inlets opened. Two prominent estuaries in this segment, Tampa Bay and Charlotte Harbor, are very large river valleys that were drowned when global sea level rose at the end of Pleistocene time.

The southwestern mangrove coast, from Cape Romano to Cape Sable, is a continuous coastal forest of mangrove trees with abundant oyster reefs. With their specialized root systems, which can tolerate muddy water, and leaves that can excrete salt, mangroves are uniquely adapted to marine waters in tropical and subtropical climates. They develop a tangled network of roots that effectively traps sediment and provides some protection against coastal erosion. Ten Thousand Islands is an archipelago of mangrove islands with foundations of old oyster reefs and shell beds. Tides have strongly influenced the shape of the southwestern mangrove coast, and as a result there is an extensive network of tidal channels. This stretch of coast also represents an important sedimentary transition from quartz-rich sand at Cape Romano to carbonate sand farther south.

Florida Bay is a broad, shallow estuary on the northwest side of the Florida Keys. It is a carbonate-producing machine that cranks out a great deal of calcium carbonate mud. The bulk of this mud, which is called *micrite*, is created by calcareous green algae that produce very small, needlelike crystals of the mineral aragonite within their filaments. When the algae die, great volumes of these needles are released, forming the light-gray lime mud. Shells from marine mollusks are an additional source of calcium carbonate. Extensive mounds of micrite occur in the bay, and sea grasses and mangrove trees grow on and stabilize them, forming many small islands, or keys.

The Florida Keys of Monroe County consist of exposed limestone of Pleistocene age: fossilized coral reefs and oolitic limestone. The Keys represent merely the tip of an iceberg of limestone strata that is over 3.7 miles thick and reaches back some 100 million years, to Cretaceous time. A string of living reefs grows offshore of the Keys, and the abundant marine life associated with this tropical ecosystem and geological limestone factory constantly produces carbonate sand.

The Atlantic coast extends essentially from Miami to Amelia Island. This stretch of coastline is the longest barrier island chain in the United States, some 340 miles long. The northern portion has many tidal creeks and salt marsh wetlands. In the middle is the seaward-projecting promontory, or cuspate foreland, of Cape Canaveral. Both north and south of the cape are exposures of a shelly Pleistocene-age rock called *coquina*, which forms rocky coastlines in some areas and provides a foundation upon which the barrier islands grow. The quartz-sand barrier islands end at Key Biscayne; south of that point, biologically produced lime sand (calcium carbonate) becomes the dominant sediment type. The transition from quartz-rich to carbonate sediments is an important one in geology. It frequently indicates a climatic transition from temperate to subtropical or tropical conditions, and this is certainly the case in Florida.

Ocean Currents and Coastal Water Flow

In the middle of the Gulf of Mexico is a strong, clockwise-flowing current called the Loop Current. It enters the Gulf through the Straits of Yucatan, moves north toward the western Florida Panhandle, then turns right and flows along the edge of the continental shelf—the West Florida Shelf—down to the Florida Keys. West of the Dry Tortugas it turns east, forming the Florida Current, and flows swiftly through the Straits of Florida. Once through the straits, the current turns north and flows up the east coast, where it joins the Antilles Current to form the famous Gulf Stream. The flow pattern of the Loop Current and Gulf Stream have been in place since at least Cretaceous time, although from Cretaceous through Oligocene time the Loop Current took a shortcut to the Atlantic through the Suwannee Channel. The intensification of the Loop Current's flow around the peninsula during Miocene time, caused by falling sea level in Late Oligocene time, is well documented in continental shelf sediments of the western platform, and the current further intensified after the Central American Isthmus rose.

Wave and tidal energy along the Gulf Coast of Florida is generally low, with tidal ranges (the difference between high and low tide) of 2 feet or less. A coastal ocean setting with a mean tidal range of less than 2 meters (6.5 feet) is called *microtidal*. Much higher wave energy is found along Florida's Atlantic coast, and spring tides can exceed 2 meters there, placing it in the mesotidal range. Semidiurnal tides—two nearly equal high and low tides per day—occur along most of Florida's east coast. Mixed semidiurnal tides—two unequal high and low tides per day—characterize the southern tip of the peninsula and western peninsula up to Apalachicola. Diurnal tides—one high tide and one low tide per day—occur along most of the panhandle.

When waves enter shallow water, they rise and then break, releasing their energy on or near the shore. No matter what direction waves approach the beach from, they break nearly parallel to the shoreline due to a process called *wave refraction*. As a wave enters very shallow water and encounters the seafloor, its forward motion slows, allowing the rest of the wave to "catch up" and therefore strike the shore at an angle of 10 degrees or less. As incoming waves strike the shore at a slight angle, some of their energy may be deflected, generating a current parallel to the shore known as a *longshore current*. Longshore currents sweep sand along the shore in the shallow surf zone—and carry raft-floating sunbathers far away from the rest of their party on the beach. Breaking waves also generate swash (water rushing up the beach) and backwash,

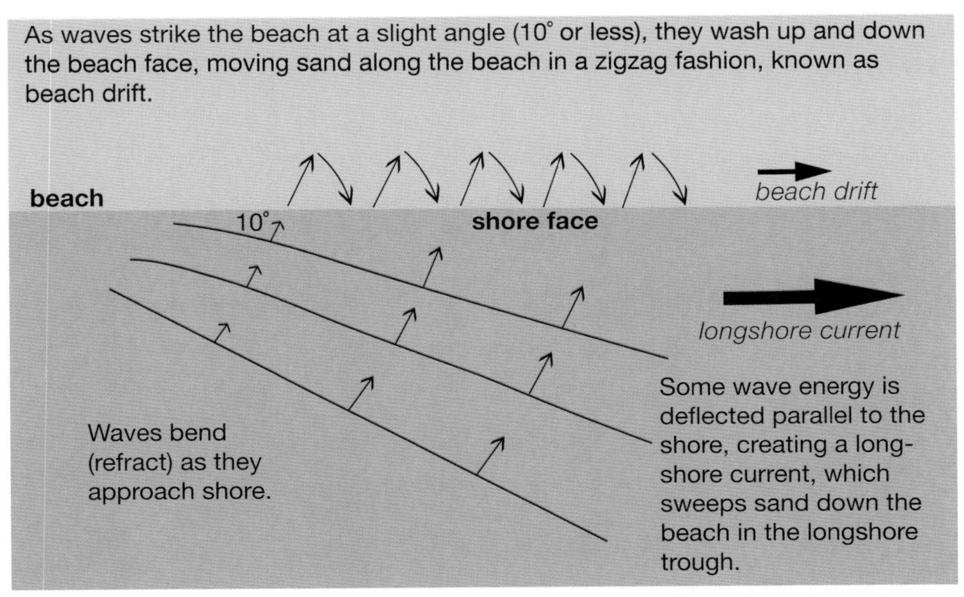

Longshore drift, a combination of a longshore current and beach drift, which together erode and construct beaches. —Illustration by Amanda Kosche

which create beach drift—a zigzagging movement of sand along the beach face. Collectively, a longshore current and beach drift are called *longshore drift*, or *littoral drift*, and their combined effect both erodes and constructs beaches.

If incoming waves push enough water forward, longshore currents that move in opposite directions can develop. This phenomenon usually develops when the waves are high and a lot of water is moved forward into the surf zone. When longshore currents converge from opposite directions, they create a coastal hazard called a rip current, also known colloquially as *rip tide* or *undertow*. Rip currents are narrow, powerful currents that flow back out to the sea, perpendicular to the shore. They usually occur in evenly spaced sets and only rarely occur singly. Sadly,

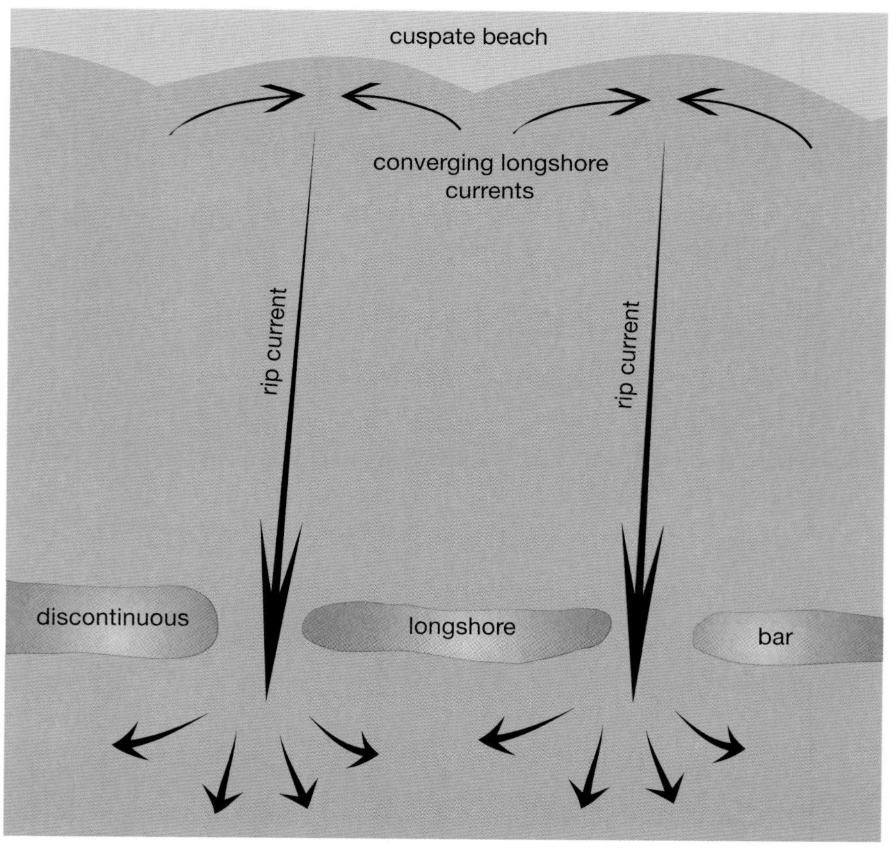

Cell circulation. As water moves shoreward and is impounded between longshore bars and the beach, longshore currents may begin to flow in opposite directions. When opposing currents meet, they form powerful rip currents. Cuspate (pointed) beaches and discontinuous, submerged longshore bars may occur when cell circulation is well-developed. —Illustration by Amanda Kosche

rip currents annually claim the lives of many unwary Florida swimmers. If you are ever caught in a rip current and feel unable to swim to shore, swim parallel to the beach. These currents are usually narrow, and several strokes should put you safely out of harm's way. The combination of incoming waves and water, the flow of longshore currents parallel to the shore, and rip currents is called *cell circulation*. It represents a coastal plumbing system of incoming and outgoing water flow.

Anatomy of a Beach

Most sandy shores have similar zones or environments that are created and modified by various coastal processes. The beach is a buildup of unconsolidated sand with a berm, or beach ridge, at its edge. The berm is a mound of sand that breaking waves deposit. There may be high and low tide berms, and perhaps a winter berm that marks the position of the shoreline during winter storms, which deeply erode the beach (beaches usually build back to their normal size during the calmer summer months). The beach face is the inclined ramp of the berm over which breakers rush up onto the shore. The beach may have a runnel, a shallow channel of flowing water that developed when the ocean breached the berm during a storm or high tide. The longshore trough is a channel through which the longshore current flows. There may be one or more longshore bars that temporarily hold some of the abundant coastal sand supply. Longshore bars may migrate landward and become welded, or accreted, to the beach, forming berms.

The area between average high tide and the first dune, or foredune, is the backshore zone. Water covers the backshore only during storms or especially high spring tides. The first permanent vegetation on a beach occurs at the toe of the foredune. The foreshore zone, also called the *intertidal* or *littoral zone*, is the area between average high and low tides, where most waves break. It's also called the *swash zone*. The nearshore zone, which extends from the low tide line to the last longshore bar, is very shallow. Also called the *breaker zone* or *surf zone*, it is where waves encounter the bottom and breakers begin to rise. The offshore zone is seaward of the last longshore bar.

Sand dunes are constructed of medium- to fine-grained sand (normal winds cannot easily lift and carry coarser sand). On coastal dunes, vegetation helps trap and secure the sand and greatly assists in dune buildup. Dune vegetation is salt-tolerant, and the plants are zoned according to the salinity of the dune soil. In the frontal zone, just landward of the backshore zone, the beach-facing dune (foredune) may support sea oats and other herbs that can withstand much salt spray. The back-dune zone supports the same, with the addition of some trees, vines, and shrubs.

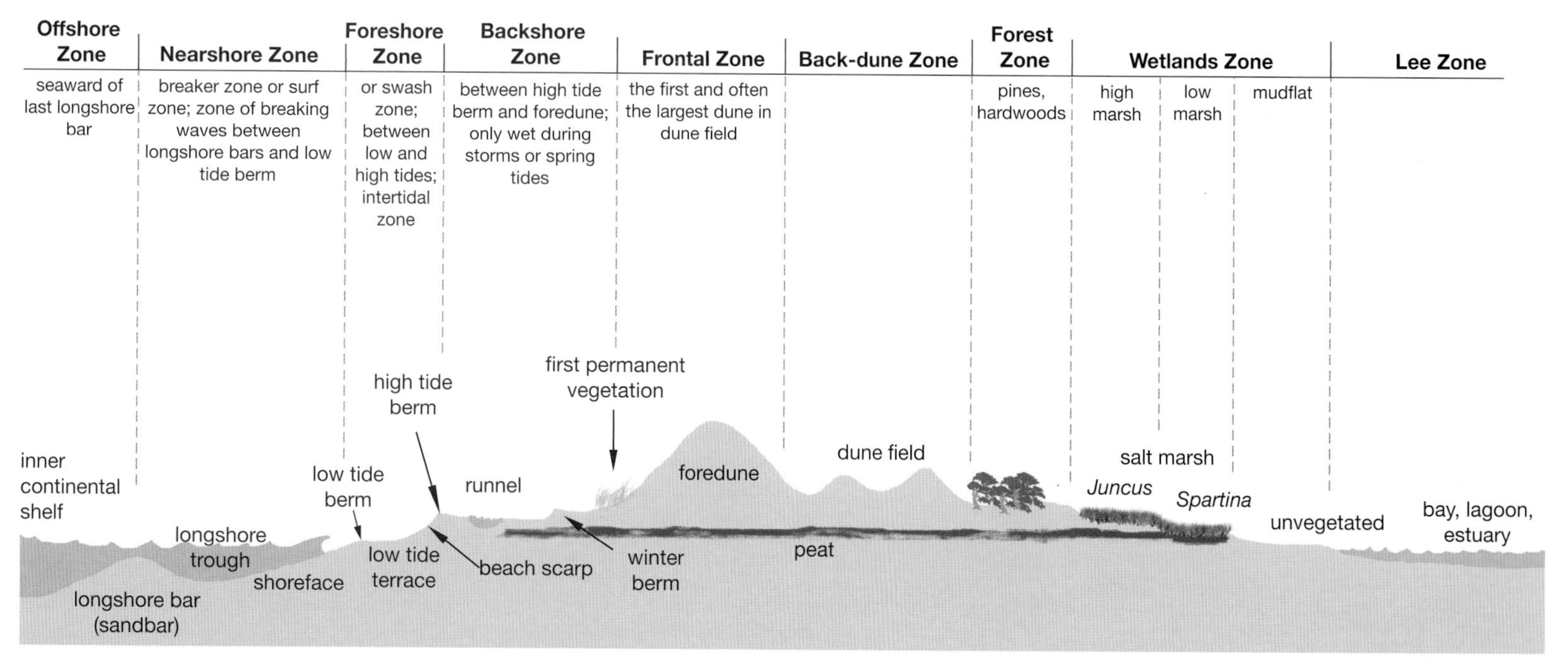

Generalized beach profile during low tide. —Illustration by Amanda Kosche

The forest zone supports pines and hardwoods. On barrier islands, next to the bay or lagoon side of the island is a marsh zone where sawgrasses such as *Spartina* and *Juncus* grow.

Origin of Barrier Islands

Long, narrow offshore strips of sand are called *barrier islands*. They form in several ways, most of which are exhibited along Florida's long coastline. All require an abundance of either quartz sand (most often) or, in some cases, carbonate sand. Commonly, barriers form when a spit, a sandy extension of a beach, grows and is modified by the ocean. A spit forms as longshore drift deposits sediment, and it becomes an island when a new tidal channel or storm channel cuts through it, isolating the spit from the mainland. The shoaling of longshore bars by the accumulation of sediment may also create a barrier island. Sea level rise may flood coastal berms or dunes, which can become the nucleus of new barriers. Or sea level can fall, exposing offshore sandbars and creating new islands. And finally, nearshore rocky shoals may form substrates on which sand accumulates, forming what are called *perched barriers*. All barrier islands in Florida are of Holocene age and are less than 6,000 years old. Some formed recently, within the past few hundred years or less.

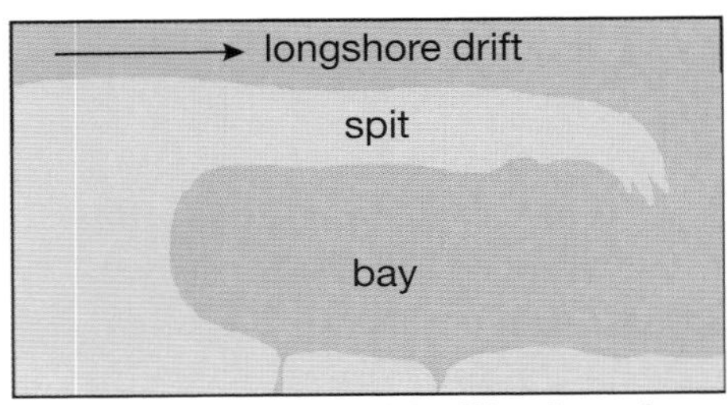

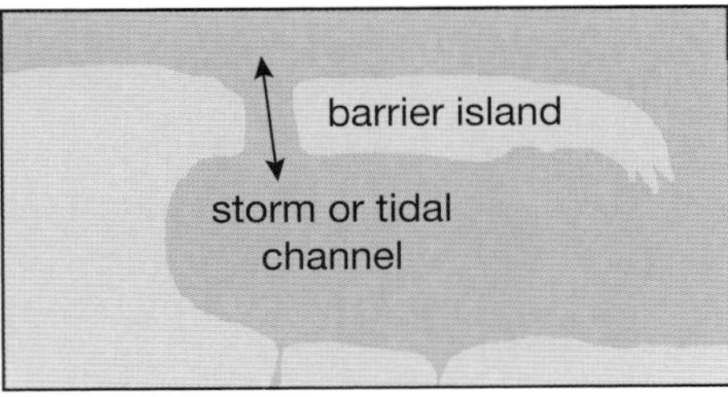

Modification of a sand spit into a barrier island.

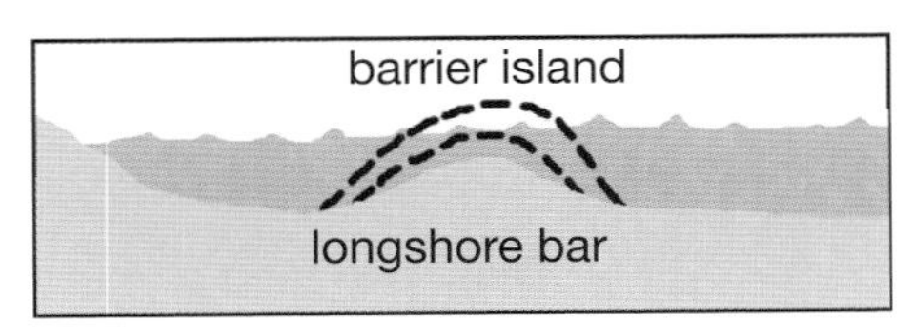

Shoaling of a longshore bar.

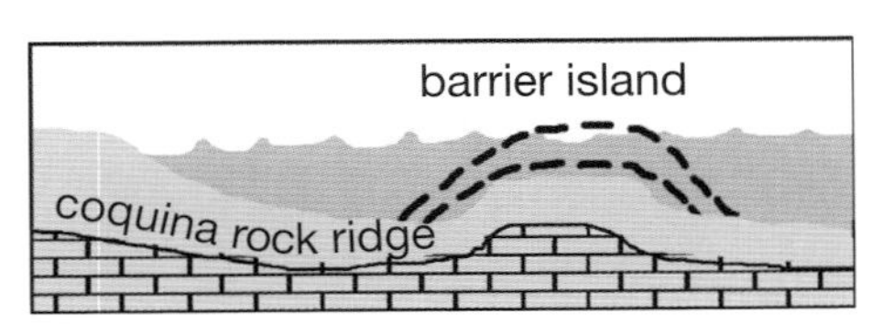

Buildup of sand over a rocky, offshore ridge. The barrier island is said to be "perched" on the ridge.

Three possible origins of barrier islands. —Illustration by Amanda Kosche

Coastal Erosion

According to coastal geologists, more than 75 percent of the U.S. coastline is presently eroding, meaning there is a net loss of sediment each year. Of the more than 800 miles of open, sandy beaches on the Florida coast, over 40 percent are eroding—more than 320 miles. The long-term consequences of beach erosion are evident as historic structures and various coastal industries are increasingly threatened by the landward retreat of the shore.

The causes for coastal erosion are varied and include sediment starvation because of dams, which trap sediment in reservoirs and deprive beaches of a primary source of sand—river sand; the "hardening" of the coast with seawalls and other structures, which tend to increase wave turbulence, deprive the coast of sediment, and increase erosion in the long run; land subsidence, sometimes caused by the withdrawal of groundwater; and sea level rise. These factors intensify the erosional effects of normal wave activity and tropical storms. Solutions to coastal erosion include the construction of jetties and breakwaters, and beach nourishment, in which sand dredged from nearshore areas is added to the beach to rebuild it. All are costly and provide only short-term gains. Prohibiting construction close to the shore may seem like a reasonable solution to coastal erosion, but the aesthetic and economic allure of the shore is perhaps too great to resist—especially in Florida.

Hurricanes

Hurricanes are tropical, cyclonic storms with sustained wind speeds of 74 miles per hour or more. Although they are not unique to Florida, there are few hurricanes that form in the Atlantic, Caribbean, or Gulf of Mexico that do not threaten the Sunshine State. The official hurricane season is June 1 to November 30, with most storms arriving in September or October. Their frequency varies. Several years may pass without a major landfall, or a single season may see several. Historically, the panhandle and southern peninsula have been the most likely targets of hurricanes in Florida. Hurricane intensity is ranked on a scale of 1 to 5, with 5 being the most extreme. In 2004, four hurricanes, a record number, struck the Florida mainland.

Hurricanes form over the ocean in areas of low atmospheric pressure, where warm, moist air rises to form thick clouds. As surrounding air begins to move toward the low-pressure cell (much like the air that rushes into a vacuum-sealed jar when it is opened), the spin of the Earth creates a phenomenon called the *coriolis effect*, which causes the air to rotate around the low-pressure cell, generating the familiar counterclockwise rotation of northern hemisphere hurricanes. For residents living on the coast, the greatest hurricane threats are from direct wind damage and storm surge. Storm surge, an exceptional rise in sea level as the storm makes landfall, is due to low atmospheric pressure and the "piling up" of water along a shallow coastline. Coastal flooding can be extreme, particularly if the storm hits during high tide. Farther inland, wind damage diminishes slightly, but tornadoes commonly spin off of hurricanes, and the impact of inland flooding can exceed that of the wind.

The geologic effects of a hurricane may be substantial. Coral reefs, barrier islands, and other coastal areas can be destroyed, and it may take many years for nature to reconstruct eroded beaches and coastal dunes. Many tidal inlets form during hurricanes as barrier islands and spits are breached by the ocean. But hurricanes are also constructional, meaning they can help build coastal landforms by transporting and depositing sand and shell debris. Geologists have located several hurricane deposits in Florida's strata, some reaching back several hundred

THE SAFFIR-SIMPSON HURRICANE INTENSITY SCALE

Category	Wind Speed (miles per hour)	Atmospheric Pressure	Storm Surge Potential (feet)	Potential Damage	Historic Florida Examples
1	74–95	more than 979	4–5	minimal	Floyd (1987).
2	96–110	965–979	6–8	moderate	Kate (1985).
3	111–130	945–964	9–12	extensive	Opal (1995) hit the western Florida Panhandle.
4	131–155	920–944	13–18	extreme	Donna (1960) hit southeastern Florida and Ivan (2004) devastated the Pensacola area.
5	more than 155	less than 920	more than 18	catastrophic	Andrew (1992) struck Homestead and the Labor Day Hurricane (1935) struck the Florida Keys and killed 588.

years. Layers of beach sand, for example, may wash into coastal lakes, forming obvious hurricane deposits. Carbon-14 dating can determine the age of the lake sediments, and thus the ancient hurricane. In other cases, the preserved geological record of a storm amounts to no more than a thin layer of sand or shells, usually deposited far inland from today's shoreline. However, some storm deposits in the Florida geologic record are several feet thick; these are known as *tempestites.*

Rivers

The shape of the modern land surface of Florida is a function of many variables, not least of which is the action of rivers and streams. Florida has more than 1,700 rivers. Almost all of them originate within the state, but a few begin in Alabama or Georgia. (The Apalachicola has its headwaters in the southern Appalachian Mountains of northern Georgia.) The location and drainage patterns of these rivers are controlled by underlying geology and water sources. Old coastal landforms, such as dunes, bars, spits, and former lagoons, can direct river flow, as can karst topography, and sea level can dictate water tables, spring locations, and stream gradients. The strange, meandering course of the Suwannee River, for example, is directed by underlying karst and the fracture patterns (jointing) in the rock. Several major rivers, such as the St. Johns, southern Withlacoochee, and Kissimmee, flow parallel to the axis of the peninsula and are largely controlled by elongate peninsular ridges and valleys. Although rivers respond to surrounding topography, they are also powerful agents of erosion, especially during flooding, and they can dramatically alter a landscape over time.

As rivers flow to the sea or a lake, seeking their lowest level, called the *base level,* they incise the land surface and create natural drainage basins. These basins, also called watersheds or catchments, are land areas that drain into a particular river. The separation between watersheds is usually a ridge of higher ground called a *divide.* The size of a watershed depends on scale—how large of an area you are looking at. A small stream has its own little basin, and a river—a large stream with many tributaries—has a larger basin that includes the basins of its tributaries. Understanding stream drainage patterns is especially useful in urban planning. Many roads, for example, are constructed on divides. Any deviation from this eventually requires the construction of a bridge. However, river basins may be poorly defined or disjunct over karst terrain, as is common in peninsular Florida. The state has numerous disappearing, or sinking, streams, which vanish underground through sinkholes called *swallets,* or *swallow holes.* Some resurface, while others remain underground, recharging the Floridan aquifer system.

Rivers and streams in Florida can be classified according to their hydrologic sources. However, individual rivers and streams are usually not exclusively one of the types outlined below; many have more than one source and may contain a variety of suspended sediments or dissolved materials.

Alluvial streams flow over clay-rich terrain, are supplied by surface water, and have many tributaries. They carry a great deal of detritus and sediment, so they are usually "muddy." Most alluvial streams originate in Alabama or Georgia. The Escambia, Choctawhatchee, and Apalachicola are classic examples.

Spring-fed streams flow over karst terrain and begin at springs. These waterways rise at a spring (springhead) and may flow a short distance, called a *spring run*, to a larger stream channel. Spring-fed streams are normally crystal clear. Many rivers of the northern peninsula are spring fed, such as the Wakulla, Santa Fe, and Ichetucknee.

Blackwater streams flow over sandy or clayey terrain and are supplied by surface water and seepage from swamps and freshwater wetlands. The "black" water, which is colored by dissolved organic materials and tannic acid, looks like iced tea. Blackwater River in the western panhandle and the St. Marys in the northeast corner of the state are blackwater rivers.

Seepage streams flow over sandy terrain and are supplied by groundwater that flows into them through their banks. Seepage flow is a component of many Florida streams, including the Ochlockonee and Suwannee rivers. Steephead streams are a specific type of seepage stream that erodes deep gullies in porous sand.

Wetlands, Lakes, and Soils

The water table is the upper limit of the ground that is saturated with water. Where the water table is at or above a land surface, standing water creates a wetland—and Florida has many wetlands! It is estimated that prior to settlement over half of Florida was freshwater wetlands, and about half of that original area has been destroyed by urban and agricultural development. Wetlands may be coastal, such as salt marshes, mudflats, and mangrove swamps; or freshwater, such as floodplains, marshes, swamps, and bogs. Marshes are grassy wetlands, such as the Everglades, and swamps are forested wetlands, such as Big Cypress Swamp. Bogs are soggy areas formed by the seepage of groundwater. They are also vegetated, sometimes with unusual plants. Swamps presently cover about 10 percent of the land surface in Florida. Many freshwater marshes in the peninsula are located within the valleys between the sandy ridges of the Central Highlands.

Although wetlands were once considered wastelands, we now know that they are extremely important and fragile ecosystems that support unique biological communities and endemic species. But wetlands are also distinct geological environments, and they leave a very legible sedimentary record. Many marshes and swamps, for example, accumulate extensive deposits of peat, a precursor of coal. Some peat deposits in Florida have been forming at an average rate of 3.6 inches per 100 years, and some deposits are more than 10 feet thick.

Florida has about 7,800 named lakes greater than 1 acre in size. Lakes also accumulate a distinctive sediment record, rich in organic material and pollen. Modern lakes may contain sediments reaching back to Pleistocene time, and a few Florida lakes have been accumulating sediments for more than 30,000 years, such as Lake Annie and Lake Tulane in Highlands County. Lakes are ubiquitous in the state and occur in many environments, but most are concentrated along the Central Highlands of the peninsula. More than half of Florida's lakes lie on this axis of peninsular highlands, most of which is covered by shallow marine sand of the Pliocene-age Cypresshead Formation. Geologists estimate that up to 70 percent of all Florida lakes are seepage lakes, meaning they derive their water directly from groundwater. This is a characteristic of karst terrain. It is a remarkable fact that almost all rainfall in Florida is absorbed by the ground or held in lakes or wetlands—almost none flows directly over the surface to the sea. This attests to the spongelike character of the Floridan surface and subsurface. Some Florida lakes, such as Okeechobee and Istokpoga, are actually the flooded depressions of former seafloors. In the Florida Panhandle there are many dune lakes, which form when coastal sand dunes impound freshwater. And, of course, there are many man-made lakes, such as reservoirs and those related to mining reclamation.

Just like wetlands and lakes, soils are intimately associated with surficial geology. The type of soil that develops is related to the rock type and surface features of the land on which it forms, and soil type, in turn, controls the nature and distribution of plant communities. Over three hundred soil types have been mapped in Florida. The variables that control the type and quality of a soil include seasonal water availability, proximity to water table, aeration, nutrients, acidity, and salinity.

Aquifer Systems

Florida is blessed with an abundance of fresh groundwater—water that flows through rocks and sediment. More than 90 percent of the state's drinking water—nearly 1 trillion gallons per year—is drawn from subterranean sources. Geologists estimate that there are more than 2 quadrillion gallons of freshwater in Florida's aquifer systems. But this figure may be misleading. For the resource to be sustained and renewable, no more water can be removed from the aquifers than recharges them. Approximately 4 trillion gallons of water recharge the aquifers each year, and about 3 trillion gallons naturally discharge through springs and seeps, leaving about 1 trillion gallons for human use. Some areas of the state have ample supplies of freshwater, while others, with rapidly increasing populations, are running out. In regions where groundwater supplies are stressed, cities and counties are obtaining freshwater by desalinizing brackish or saline groundwater and marine water and reclaiming wastewater.

Florida's aquifer systems occur in a complex sequence of sediments that range from limestone and dolostone to sand and shell. The aquifers are separated and often confined by carbonate rocks and clay-rich sediments with low permeability. The major aquifers are the Floridan aquifer system, the intermediate aquifer system, and the surficial aquifer system. The Floridan aquifer system is one of the world's most productive aquifers and provides most of the groundwater used in the state. In limited areas, the intermediate aquifer system is an important water source. The surficial aquifer system is the primary groundwater source for the western panhandle and southeastern peninsula.

Water within the Floridan and intermediate aquifer systems often occurs in confined conditions—trapped beneath or between low-permeability to impermeable sediments. Under these conditions, water is under pressure and will rise in a well to a point above the confining strata. Such water is said to be under *artesian* conditions, and if the pressure is great enough, water will flow to the surface without a pump. In some areas of the state, water can flow more than 10 feet above the land surface. However, artesian flow has stopped in some areas and declined in others as a result of increased withdrawal of groundwater by a burgeoning population.

The state's geological features affect the distribution of its aquifers. The strata of the intermediate and surficial aquifer systems were eroded from the Ocala Platform, Sanford High, and Chattahoochee Arch. In these areas, the underlying Floridan aquifer system is not confined by impermeable strata, so the water table can be at the surface. The vast majority of Florida's springs occur in areas where the Floridan aquifer system is not confined or where sinkholes have breached it. The intermediate and surficial aquifer systems thicken in the low or negative features, such as the Gulf of Mexico Basin, Okeechobee Basin, and Apalachicola Embayment. These hydrogeologic units become important sources of water where they are thickest.

The Floridan aquifer system underlies all of Florida. It consists of carbonate rocks of Eocene to Miocene age that can be thousands of feet thick. Some of these rocks are exposed at elevations of more than 100 feet above sea level. But in southern Florida and the panhandle, the top of this aquifer is more than 1,000 feet below sea level. The aquifer system occurs in part or all of the Oldsmar Formation, Avon Park Formation, Ocala Limestone, Marianna Limestone, Suwannee Limestone, St. Marks Formation, and Tampa Member of the Arcadia Formation. There are several aquifer zones within the system—permeable beds separated by low-permeability beds. The base of the Floridan aquifer system occurs in lower-permeability Paleocene- and Eocene-age carbonates. In the southern portion of the peninsula, the system contains saline water. West of Orlando, under the Green Swamp, the thickness of rocks containing freshwater exceeds 2,000 feet.

The intermediate aquifer system is most important as the unit that confines the Floridan aquifer system over much of the state, and it may be up to 1,000 feet thick in places. The top of it ranges between nearly 225 feet above sea level to more than 350 feet below sea level. It is found over much of the state but is absent on the Ocala Platform, Sanford High, and Chattahoochee Arch. This aquifer system is composed of sand, clay, and carbonate rocks of Late Oligocene and Pliocene age. These strata include formations of the Hawthorn Group, and at least part of the Tamiami Formation. In most areas where it is present, the intermediate aquifer system contains limited amounts of water due to the low permeability of its strata. However, in the southwestern peninsula, in Sarasota, Charlotte, and Lee counties, permeable zones provide the region's main water supply.

The surficial aquifer system occurs over much of the state but is also absent on the Ocala Platform and the Chattahoochee Arch. The system includes the Biscayne aquifer and the sand and gravel aquifer. It reaches a thickness of more than 350 feet in southern Florida and more than 500 feet in the western panhandle. The system comprises sand, shell, and carbonate strata of Miocene to Holocene age, including the Alum Bluff Group, Tamiami Formation, Caloosahatchee Formation, Citronelle Formation, Fort Thompson Formation, Miami Limestone, Key Largo Limestone, Anastasia Formation, and undifferentiated sand. The Biscayne aquifer is the principal aquifer for much of Miami-Dade, Broward, and Palm Beach counties in the southeastern peninsula. The sand and gravel aquifer provides water for Escambia and Santa Rosa counties in the panhandle.

Water quality varies between aquifer systems. Most groundwater in the state is hard water, meaning it contains high concentrations of dissolved ions. However, water from the sand and gravel aquifer is soft, meaning it has lower concentrations of dissolved ions. The other aquifers have hard water because the water dissolves calcite and dolomite in the carbonate and shell-bearing strata. Historically, water quality in Florida's potable aquifers was good. But, as the state has developed, water quality has declined in many areas due to septic tanks, fertilizers, runoff containing petroleum products and pesticides, and farming activities.

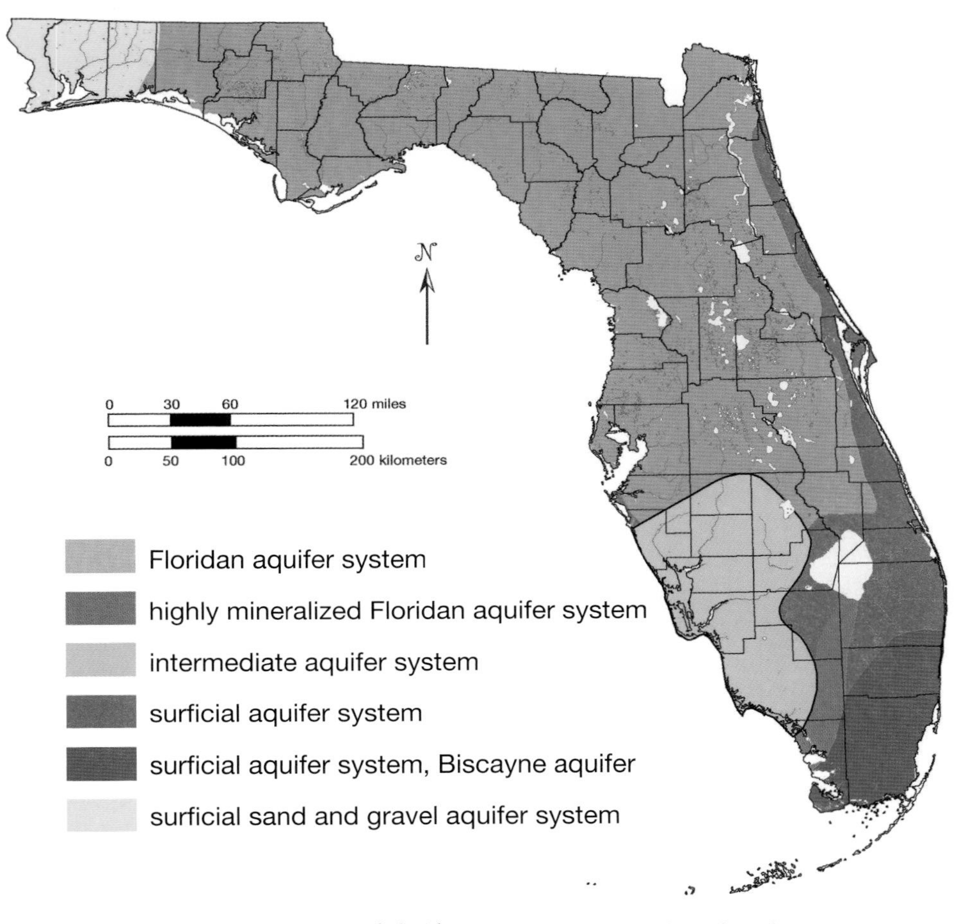

Aquifer distribution map of Florida. —Courtesy of the Florida Geological Survey

Another groundwater hazard of concern to Florida's coastal communities is saltwater intrusion. When fresh groundwater flows near a coastal area, it encounters saline groundwater from the ocean. The freshwater and saltwater tend to remain somewhat separate because seawater has a higher density, but if water wells remove potable groundwater faster than the aquifer is recharged with freshwater, saline groundwater may begin to move into the well and threaten the supply. Coastal water wells are carefully monitored for chloride, the primary dissolved solute in seawater.

Karstification—A Dissolving Landscape

Karst refers to an area that is characterized by sinkholes, depressions, cavities, caves, and related features. It forms where surface rock is composed of carbonate rocks. Karstification is a "native" geological process in Florida, and any discussion of karst processes should not fail to include Florida as a standard. Other famous karst regions around the world include areas in China, Jamaica, and Yugoslavia. Geologists estimate that as much as 5 to 10 percent of Earth's surface is karstic. Most of Florida's parks—Falling Waters, Florida Caverns, Wakulla Springs, Devil's Millhopper, and others—feature karst terrain. Florida sinkholes have been forming since Late Oligocene time and occur in strata that was deposited in the last 40 million years. Some of Florida's springs, such as Silver, Homosassa, and Wakulla, are sinkholes that filled with groundwater as water tables rose in response to sea level change.

Beneath Florida's varied terrain and lush vegetation, limestone and dolostone occur at or near the surface in many areas. These rocks are susceptible to dissolution by acidic groundwater. As rainfall absorbs atmospheric sulfates, nitrates, and carbon dioxide, it becomes slightly acidic. As the water seeps through the soil, it picks up more carbon dioxide and organic acids from decaying vegetation, making it more acidic. This relatively weak acidic solution reacts with carbonate rocks, partly dissolving them, enlarging pore spaces and fractures, and creating cavities, caves, and caverns.

One of the primary features of karst is the sinkhole, one of the most common landforms in Florida. Sinkholes have been variously classified, but three types are easily distinguishable. Solution sinkholes are simple surface depressions that form as limestone that is close to the surface dissolves. They form slowly over many decades or centuries and result in a gently rolling landscape. Cover subsidence sinkholes develop gradually as permeable, unconsolidated sediments slowly sink, like sand in an hourglass, into an underlying cavity that is enlarging. Terrains with these types of sinkholes are sometimes called *sand hill karst*. Cover collapse sinkholes form rapidly, in minutes to hours, as surface sediments fall into an underlying cavity that can no longer support the overburden.

Sinkholes are a natural phenomenon, but the excessive drawdown of groundwater for municipal, industrial, and agricultural uses, or drought conditions, may accelerate their formation because water supports rock. Excessive or rapid fluctuation of groundwater levels usually encourages sinkhole development. To our knowledge, no one has been killed or injured in a sinkhole collapse, although vehicles, homes, and businesses have been damaged or destroyed. The weight of a Florida Geological Survey drilling rig even caused a small sinkhole to develop—which the drillers discovered when they found the rig in the sinkhole upon returning to work one morning. Fortunately, the rig was not damaged, but other drillers have not been so lucky. Some rigs have been completely engulfed and lost in collapse sinkholes.

A cover collapse sinkhole.

Satellite and aerial photographs of Florida reveal the dramatic effect karst processes have had on the landscape. Large areas are pockmarked with circular lakes and coalescing circular lakes that are water-filled sinkholes. There are very few surface streams and rivers in these areas. Most streams flowing on karst are captured by sinkholes, or swallets, and drain into the subsurface. Some rivers, such as the Alapaha in Hamilton County or the Santa Fe on the Alachua-Columbia county line, disappear underground through a sink and reemerge some miles away with additional groundwater flow.

Another sinkhole-related phenomenon is periodic lake drainage. Sinkholes can develop in the beds of lakes, and with the right conditions, the entire lake can disappear underground. Lake Jackson in northern Florida is an excellent example. The lake is about 4,000 acres in area and drains, entirely, through sinkholes about every twenty-five years. Billions of gallons of lake water recharge the aquifer in a relatively short period of time, sometimes creating problems for people with water wells in the vicinity.

Where karst processes have been active for a long time, the landscape is relatively flat and has numerous sinkholes. In areas where karstification is relatively new, or in areas where sediments above the limestone are thick, sinkholes are less numerous and topographical relief is greater.

A smaller-scale karst, called *microkarst*, has formed shallow, small depressions in the Everglades District where the Pleistocene-age Miami Limestone has dissolved. Karst processes have also been active on the Atlantic Coastal Ridge, creating caves and large sinkholes.

Florida's carbonate rocks are riddled with cavities and caves, and spelunkers are constantly looking for new ones to explore. There are some 1,400 documented caves in the state. Approximately half are dry, or air-filled, caves, while the others are submerged. The number of caves that do not open to the surface or are filled with sediment is unknown. Based on drilling records, there are probably thousands of undocumented caves and cavities, most of which will never be seen. The best-known air-filled cave system in Florida is Marianna Caverns in Florida Caverns State Park. Wakulla Springs, south of Tallahassee, is an underwater cave system, one of the largest water-filled caves in the world. Divers have explored more than 28 miles of caves at Wakulla Springs. Florida's caves are scattered over a wide area from the central panhandle to the southern peninsula. There are even caves in the Miami area in the Miami Limestone, a Pleistocene-age deposit. Alachua County has the highest concentration of documented caves, with more than 175, including Warrens Cave, the longest air-filled cave in the state at more than 19,000 feet.

Fountains of Youth—Florida's Legendary Springs

The bank was dense with magnolia and loblolly bay, sweet gum and gray-barked ash. He went down to the spring in the cool darkness of their shadows. A sharp pleasure came over him. This was a secret and a lovely place.

—Marjorie Kinnan Rawlings, *The Yearling* (1938)

We may normally think of groundwater in only practical terms, as a natural resource that is necessary for our existence. And karst processes usually evoke images of collapsing sinkholes and other geological hazards that we must understand, manage, and live with. But the same groundwater that quenches our thirst, dissolves caves, and opens sinkholes also creates what are truly some of the most beautiful spots on Earth. Springs enchant us with their unusual beauty and mysterious depths.

Floridians have utilized springs for thousands of years. Any visit to the state of Florida would be incomplete without experiencing a spring, and we point out accessible spring localities throughout this guide. Most of the largest springs are located within state parks and are open to the public. Florida has more than seven hundred documented springs, which may be the largest concentration in the world. Florida also boasts thirty-three first magnitude springs—more than any other state. Spring magnitude is a measure of the volume of water a spring discharges over a given time. A spring that discharges 100 cubic feet per second (64.6 million gallons per day!) or more is called a *first magnitude* spring.

The vast majority of Florida's springs are located in the Ocala Karst, Central Lakes, and Dougherty Karst Plain districts. The cool, crystal clear water that flows from most of these is groundwater from the Floridan aquifer system. These submerged sinks and caves harbor native species of salamanders, crayfish, and isopods, some of which are completely adapted to the total darkness of the caves. Some spring runs are home to Florida manatees (*Trichechus manatus latirostris*) seeking water with a constant temperature during the cold winter months. Florida's springs provide life-giving water for many complex and diverse ecosystems.

Some of Florida's most popular attractions are centered around springs. Wakulla Springs, Silver Springs, Rainbow Springs, and Weeki

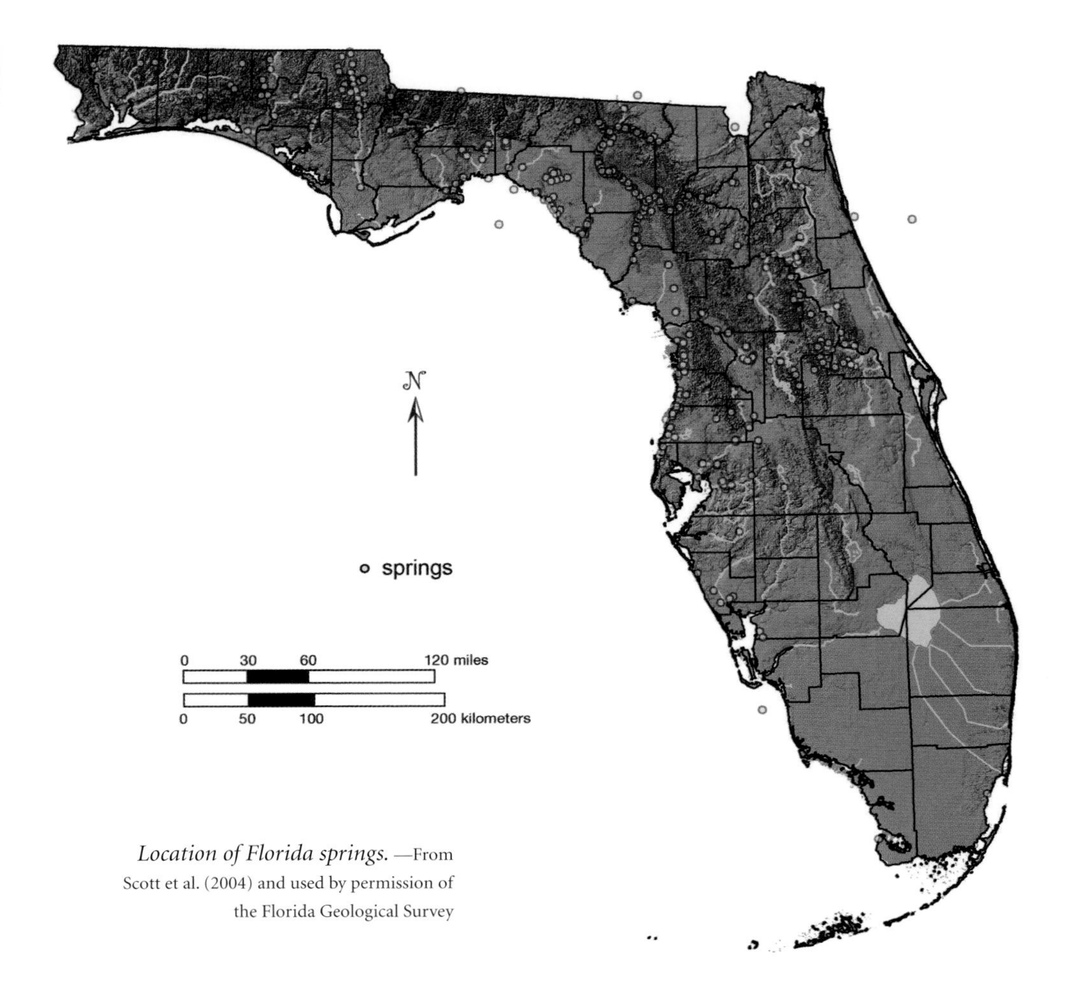

Location of Florida springs. —From Scott et al. (2004) and used by permission of the Florida Geological Survey

Williford Spring, on the Econfina River near Florida 20.

Wachee Springs are all popular tourist attractions. Many of Florida's springs occur along Florida's scenic rivers, including the Suwannee, St. Johns, and Wacissa rivers. Numerous outfitters specialize in boat rentals and portaging services for paddlers in areas where springs are numerous. Some springs, like Ichetucknee Springs, cater to visitors who are interested in floating down a spring run on an inner tube or other flotation device. Scuba diving is another popular way to view Florida's springs. Some spring parks are set up especially to cater to the diving community, like Ginnie Springs on the Santa Fe River in central Florida. Recreational opportunities abound in Florida's spring country.

Freshwater springs in Florida discharge into numerous environments, including spring runs, rivers, lakes, and the marine environment. Florida's largest spring, by volume of discharge, is Spring Creek Spring, located in southern Wakulla County. Water from this spring discharges directly into the ocean. Florida's shallow, submerged Gulf Coast harbors numerous undocumented springs. Fishermen and divers have reported springs discharging offshore, but scientists have documented very few of them. The total number of these features under Florida's coastal waters is not known.

FLOW-BASED CLASSIFICATION OF SPRINGS	
Magnitude	**Average Flow (discharge)**
1	100 cubic feet per second or more (64.6 million gallons per day or more)
2	10 to 100 cubic feet per second (6.46 to 64.6 million gallons per day)
3	1 to 10 cubic feet per second (0.646 to 6.46 million gallons per day)
4	100 gallons per minute to 1 cubic foot per second (448 gallons per minute)
5	10 to 100 gallons per minute
6	1 to 10 gallons per minute
7	1 pint to 1 gallon per minute
8	Less than 1 pint per minute

Florida's springs are valuable resources that provide economic and recreational benefits to the state, but they are useful in another way. They provide us a unique opportunity to investigate and monitor the aquifer systems from which they emerge. The water quality of spring-water is essentially the same as the water quality of the aquifer—the same aquifers that supply water to springs also supply drinking water to Floridians.

Over the past thirty years, Florida's population has expanded greatly. Land-use patterns have changed in response to this rapid growth. Unfortunately, this change has led to the degradation of water quality in many of Florida's treasured springs. Agriculture, sewage treatment facilities, landfills, and urbanization are impacting groundwater by introducing excessive chemicals and nutrients. Many of Florida's springs are facing threats caused by this increased nutrient loading in the springshed. Nutrients, like nitrate, naturally occur only in very low concentrations within these spring systems. Rising nitrate levels damage water quality by causing eutrophication, a phenomenon in which invasive, exotic plants and algae multiply quickly in the nutrient-rich water, crowding out native aquatic vegetation and using oxygen (by plant decay) that other aquatic organisms need to survive. Florida's springs are also threatened by the over-allocation of water resources within springsheds. Quenching the thirst of a rapidly growing state and protecting water flow to springs will continue to be an environmental challenge for Florida for the foreseeable future.

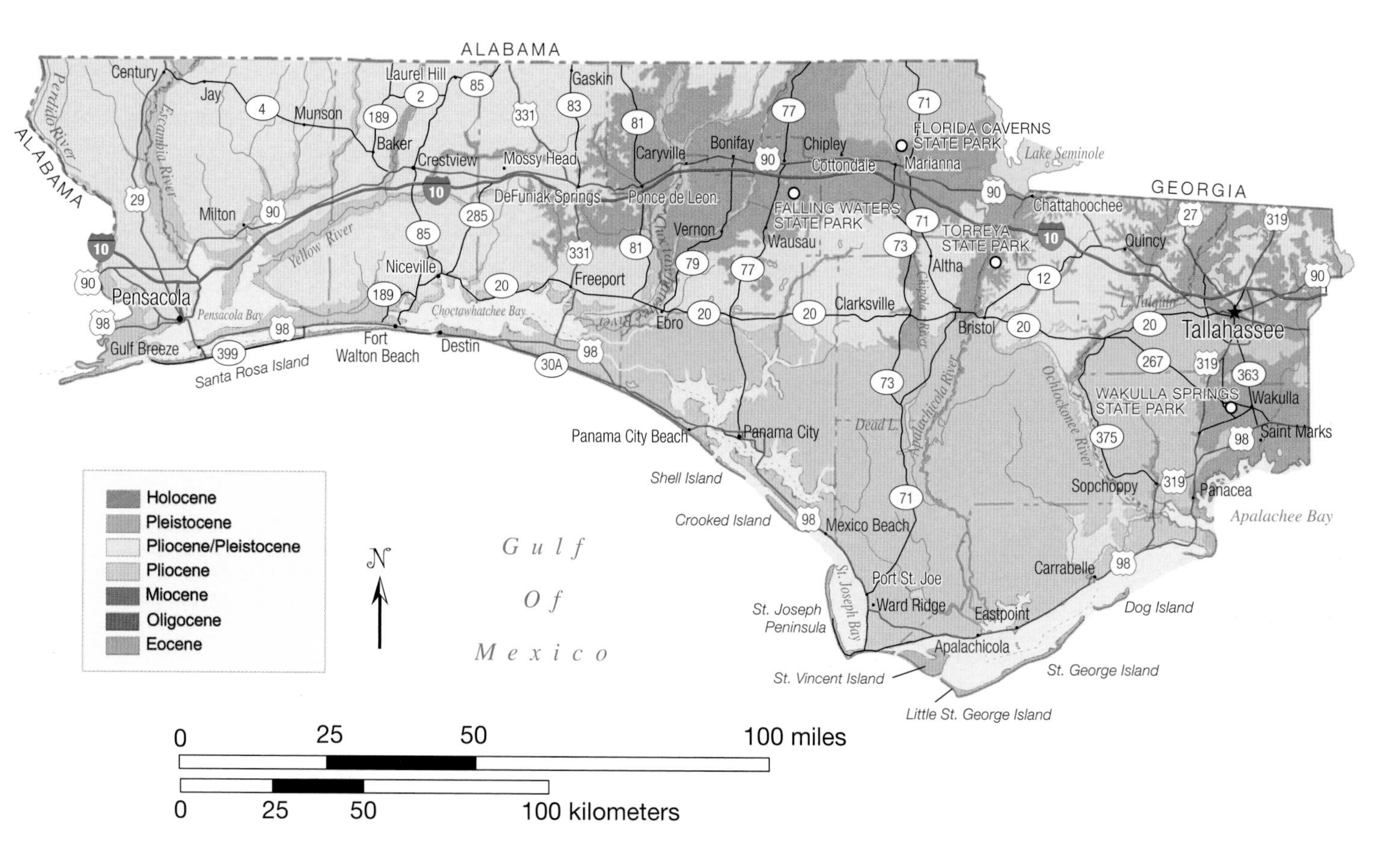

Road log map of the Florida Panhandle. —Map by Mountain Press

Panhandle
The Other Florida

Author Gloria Jahoda once referred to northern Florida as the "other Florida"—that part of the state frequently left off tourist destination maps and culturally far removed from Walt Disney World, Palm Beach, Miami, and the Florida Keys. Although rapid population growth and a mobile society have diluted many of its historic charms, northern Florida remains very much part of the Deep South. Indeed, Floridians frequently quip that, in Florida, "north is south."

There is a geological parallel to this cultural observation. The geology of the Florida Panhandle is transitional in nature, lying between the limestone-dominated platform of the Florida Peninsula and the more quartz-rich sediments of the Gulf Coastal Plain of Alabama and Mississippi. The geological development of this central part of the Gulf Coastal Plain has long been controlled by the Mississippi Embayment—a gigantic basin that, since Mesozoic time, has accommodated both rising seas and continental-scale rivers as it subsided to archive the sedimentary records of each. The influence of the embayment becomes increasingly evident westward across the Florida Panhandle.

Much of the panhandle is comprised of quartz-rich sediments of Miocene and Pliocene age that were deposited in rivers and deltas and now gently slope to the south. The deposits compose the Southern Pine Hills and Tifton Uplands districts. Terraced, quartz-rich strata of the Apalachicola Delta District, south of these other districts, are of Pliocene to Holocene age. Karst terrains are developed in the Dougherty Karst Plain District and in the western end of the Ocala Karst District, including the Woodville Karst Plain, but karst influence on surface features rapidly decreases in a westward direction because limestone is buried increasingly deeper beneath sand and clay. (See **Structural and Geomorphic Regions** in the **Introduction** for more information about these districts.) The surface topography of the panhandle has been deeply incised by erosion, resulting in some impressive relief that is uncharacteristic of most Floridan landscapes. Barrier islands and dunes of pure white quartz sand are prominent coastal features.

One peculiar geomorphic aspect of the panhandle is its pattern of stream drainage. Many of its streams form a trellis pattern over the surface, in which stream tributaries tend to flow at 90-degree angles to the main channel. What would be expected in this uniformly sandy terrain is a dendritic pattern, in which streams and their tributaries form a branching, bushlike pattern. The trellis pattern has led some geologists to speculate that relict beach ridges, dunes, or barrier islands of Pliocene and Pleistocene age may now be controlling some river and stream drainage. As sea level fell, these "fossil" coastal features were left high and dry, causing freshwater drainage to flow in the low swales between them. Such topographically low areas may have been former lagoons, embayments, or marshy wetlands.

But the trellis drainage pattern may also be the result of the formation of steephead ravines—deep, narrow, steep-walled "canyons" that form at the head of a ravine. Rainwater quickly seeps though porous sediment, becoming groundwater. When it reaches a less permeable, clay-rich layer, the water flows laterally until it emerges at the surface as a seep. The head of the seep enlarges quickly as loose sand slumps, forming the steep, amphitheater-like steephead. Steepheads in the panhandle form in the sandy, porous Citronelle Formation, where the clay-rich layers of the Alum Bluff Group inhibit the vertical flow of groundwater. Steepheads can be quite large and may contain unique assemblages of plants and animals. Steephead ravines may have inspired certain place-names, such as Mossy Head, Bear Head, and Deer Head.

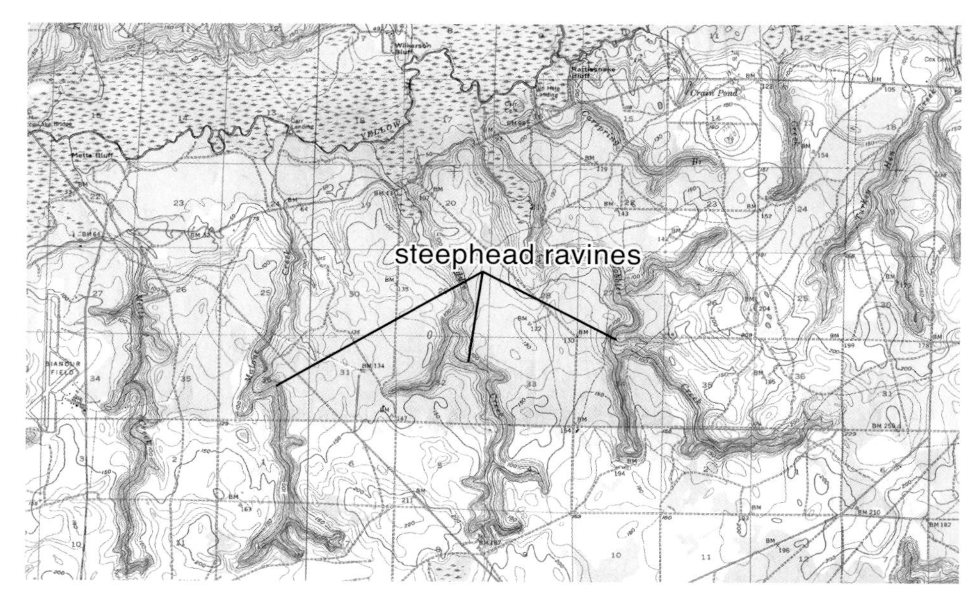

Portion of the Holt 15-Minute Quadrangle, showing trellis drainage pattern and location of steephead ravines. —Courtesy of the U.S. Geological Survey

US 29 and Florida 4

Pensacola—Century—Crestview

78 miles

The land surface in the northwestern corner of the state is developed on the Citronelle Formation, which in Florida can be river to shallow marine deposits of Pliocene age. The orange to rusty red sands of the Citronelle can be seen on many hillsides and in shallow roadcuts. When driving north on US 29 from Pensacola, it is easy to see the transition from the Pleistocene-age sands of the Gulf Coastal Lowlands to the Citronelle sands of the Western Highlands. This begins around Cantonment, where the topography becomes noticeably more pronounced and rolling. The Citronelle Formation also forms most of the sand and gravel aquifer, part of the surficial aquifer system, and supplies potable water for residents of westernmost Florida.

In Century, turn east on Florida 4. Just east of this historic logging town, the road crosses the 2-mile-wide floodplain of the Escambia River, one of the largest rivers in Florida, with a watershed that reaches far into Alabama. The Escambia ends in an extensive marsh and estuarine delta, or bayhead delta, in Escambia Bay. East of the river is the sleepy hamlet of Jay, which gained geologic notoriety in 1970 after the discovery of the Jay Field—said to be the largest oil field discovered in North America since the 1968 discovery of the Prudhoe Field of the Alaskan North Slope. Six additional fields have since been discovered in Santa Rosa and Escambia counties. The oil is recovered from the Smackover and Norphlet formations of Jurassic age at depths ranging between 14,200 and 16,200 feet. These fields account for over 80 percent of Florida's total production of oil, although production has declined sharply since a peak in 1979.

Scattered exposures of the Citronelle Formation are visible in Blackwater River State Forest, especially along some bluffs of the Blackwater River and its shallow tributaries—Coldwater, Sweetwater, and Juniper creeks. These are popular canoeing streams (the area was designated the Canoe Capital of Florida by the state legislature in 1981), and we highly recommend exploring them. They are outstanding natural laboratories for examining stream deposits such as gravel bars, point bars, cutoffs, oxbows, and cross-stratification. The Blackwater River and Juniper Creek erode into clay-rich sediments that can best be seen at Red Rock Bluff on Juniper Creek. The 40-foot-tall clay bluff is located about 100 yards downstream of the Red Rock Road bridge over Juniper Creek; it may be accessed by canoe or trail. To reach this bluff, turn right (south) on County Road 191 in Munson. Go 0.5 mile, then turn left on Sandy Forest Road. Continue 5.1 miles, then turn right (west) on Red Rock Road.

Continue 0.8 mile to the bridge. Fossilized oyster beds representing former estuarine environments have been found in this clay northwest of Baker, just east of the Blackwater River. The Citronelle Formation is exposed in various places between Munson and Crestview.

US 90 and Interstate 10

Pensacola—Crestview—DeFuniak Springs

80 miles

US 90 and I-10 parallel each other between Pensacola and Tallahassee. Though some sites in this road log are tied to one road, the geological narrative for both roads is essentially the same. On the west side of Escambia Bay are the Pensacola Bluffs, standing nearly 100 feet above present sea level. The bluffs were cut into the Citronelle Formation. The top of the bluff is the Wicomico Terrace, the oldest of the Pleistocene-age terraces. The bluffs are especially visible from I-10 as one travels west into Pensacola, but they can also be seen at Bay Bluffs Park. To reach the park, proceed south on US 90 from the US 90/I-10 intersection (the interstate exit on the west side of Escambia Bay) for about 5 miles to the park entrance. Here, the Citronelle Formation is visible from the boardwalk down to the bay.

While driving east out of Pensacola on either I-10 or US 90, you pass over Escambia Bay, the westernmost estuary that is entirely in the state (Perdido Bay straddles the Alabama-Florida border). US 90 passes directly over the bayhead delta of the Escambia River. Escambia Bay and East Bay, both part of the Pensacola Bay system, are classic examples of drowned river valleys—river mouths that were flooded during the Holocene Transgression, from 18,000 to 6,000 years ago. Estuaries such as these are geologically ephemeral. They usually last only a few thousand years because large volumes of river sediment eventually fill them in, turning them into marshes. The combination of sandy shoals and hurricanes has contributed to several historic shipwrecks in Pensacola Bay and surrounding areas, including the oldest known wreck in Florida waters—a 1559 vessel from the Spanish Luna Expedition, found off Emanuel Point.

US 90, Interstate 10, and the old Louisville and Nashville Railroad line (now a CSX Transportation line), are nearly parallel to one another along this section of the road log, and in places they are within a stone's throw. There was good reason to construct major transportation lines along this route, which is also part of the Old Spanish Trail. This is Florida high ground—the rolling hills of the Western Highlands. Scattered exposures of the Citronelle Formation are visible along this route and are especially evident along I-10 just west of the appropriately named town of Crestview. At 225 feet, the Hub City, as it is known, is a historic rail and trucking crossroads. The higher ground of the landscape seen along roads in the southwestern part of town is the Citronelle Formation. Excellent exposures can also be seen in the large sand pits near the US 90/Florida 285 intersection, west of Mossy Head. Along many rural roads in northern and central Okaloosa and Walton counties (Florida 189, County Road 2, US 331, Florida 85, Florida 285) the rolling terrain of the Western Highlands is beautifully displayed and exposures of the Citronelle Formation are common. The Western Highlands are evident as far south as Niceville on Florida 85 and Florida 285.

A rather famous fossil-rich exposure of the Alum Bluff Group is located near Mossy Head, north of US 90, on the slopes overlooking the Shoal River. Under the luxuriant vegetation of a small steephead ravine, a small waterfall spills over a densely fossiliferous shell bed of the Shoal River Formation, a Miocene-age deposit dated at just over 12 million years old. Crates of fossils from this site were once packed by mule down to the railroad and transported to the U.S. Geological Survey in Washington, where they were described by paleontologist Julia Gardner between 1926 and 1950. This site is now protected and part of the Shoal Sanctuary, a small, private nature preserve. Visitors are welcome with advance notice. By canoe, additional outcrops of the Shoal River Formation are visible along the Shoal River from Dorcas, accessed from County Road 393, to south of Crestview off Florida 85.

Fossil bivalve mollusks form the Shoal River Formation of Early Miocene age, Walton County. Ruler is in centimeters.

SIDE TRIP ON FLORIDA 85

Rock exposures are not plentiful in Florida. If you have the time, there are some interesting things to see north of Crestview. From US 90 in Crestview, turn north on Florida 85. Drive 13 miles and turn west on County Road 2. Proceed 3.7 miles to an abandoned sand pit in the Citronelle Formation on the north side of the road. Cross-stratification is very evident, and iron-cemented quartz sandstone can be collected here. West of this exposure, County Road 2 descends rapidly to the floodplain of the Yellow River. About 0.6 mile west of the Citronelle outcrop, a dirt road on the east side of the bridge takes you to a bend in the river and boat landing on the north side of the bridge. Along the riverbank, the dark gray, fossiliferous Oak Grove Sand is exposed. It is a deposit of the Alum Bluff Group of Early Miocene age with abundant mollusks and occasionally shark teeth. This outcrop is the westernmost surface exposure of richly fossiliferous strata in the state, but samples can be collected only when the river is low. By canoe, tall bluffs of the Citronelle Formation can be seen along the east side of the Yellow River between County Road 2 and US 90.

To see another good exposure of the Citronelle Formation, from the Yellow River bridge head east on County Road 2 about 2.3 miles. Turn south on Senterfitt Road (a small power station is on the north side of County Road 2 at this intersection). Proceed about 1.1 miles to Senterfitt Road (dirt road) and turn east. At 0.6 mile, interbedded clays and cross-stratified sands of the Citronelle are well exposed, especially along the south side of the road. Small pieces of petrified wood can occasionally be found in the sandstone rubble piles along this exposure.

Numerous exposures of the Citronelle Formation can be observed in hillsides, roadcuts, and borrow pits located around the area of Laurel Hill

A large scallop shell and other mollusks in fossiliferous sediments of the Oak Grove Sand of Early Miocene age along the Yellow River.

Fossilized wood (a cast) from the Citronelle Formation, found at the Perdido Landfill, Pensacola. Ruler is in centimeters.

on Florida 85 and most of northern Okaloosa County. Fossilized wood is regularly found in the Citronelle here, and west as far as Pensacola. Extensive networks of shrimp burrows, a trace fossil named *Ophiomorpha nodosa*, are very abundant in coarse Citronelle sands that also contain an abundance of the white clay mineral kaolinite. These burrows were made by blind shrimp of the genus *Callianassa*, and their presence indicates that the Citronelle Formation throughout much of the Florida Panhandle was deposited in a nearshore marine environment. Sandstone casts of several species of marine bivalve mollusks are associated with the burrows in some areas.

The trace fossil Ophiomorpha nodosa, *made by a nearshore burrowing blind shrimp (genus* Callianassa*), from the Citronelle Formation of Pliocene age at Eglin Air Force Base, near Niceville.*

SIDE TRIP TO THE HIGHEST POINT IN FLORIDA

In northern Walton County, just east of Paxton, is the highest point in Florida. From US 90, exit onto US 331 west of DeFuniak Springs (from I-10 in DeFuniak Springs, take US 331 north for 2.3 miles and US 331/90 west for 1.9 miles). Follow US 331 north for 18.2 miles to County Road 285. Proceed north on County Road 285 for 2.8 miles. Lakewood Park is on the west side of the road at 345 feet above mean sea level. The park is situated on the surface of the Citronelle Formation, and there is a small roadside stop with a monument for Florida's highest Point, Britton Hill. The pastoral, panoramic view here will convince you of your comparatively high elevation—not bad for Florida, which has the "distinction" of having the lowest highest point of any of the fifty states.

Just east of Florida's highest point is the westernmost exposure of limestone in Florida at Natural Bridge. Ten feet or more of the Marianna Limestone is exposed in this remote outcrop, which is rich in larger foraminifera, bryozoans, and mollusk molds. (Some geologists classify this outcrop as the Byram Formation of Alabama.) Take Florida 83 north from DeFuniak Springs to Gaskin. Turn left (west) onto County Road 181 and proceed 4.1 miles to a dirt road on the left (south). Turn left and proceed 0.5 mile to the outcrop. Access may be limited.

Marianna Limestone of Early Oligocene age exposed at Natural Bridge. The car is parked over the "bridge," which partially separates two spring-fed ponds.

DeFuniak Springs was built as a railroad town around Lake DeFuniak, a spring-fed seepage lake of karst origin. Around DeFuniak Springs, exposures of Alum Bluff Group sands and clays of Miocene age are visible in some large roadcuts and riverbanks. North of DeFuniak Springs, mollusk-rich sands are exposed in small creek beds, some of which have fragments of vertebrate fossils of Miocene age, including sharks, dugongs, early horses, and even camels. Southeast of DeFuniak Springs, near Redbay, baleen whale bones have been recovered. Other scattered exposures of the Alum Bluff Group in this area have yielded petrified wood (hardwood, palm, and cypress).

Fossilized cypress wood (top) and palm wood (note the pithy texture) from the Alum Bluff Group of Miocene age near DeFuniak Springs. They were collected by the late Mr. Harold Gillis of DeFuniak Springs. Ruler is in centimeters.

US 90 and Interstate 10

DeFuniak Springs—Marianna

50 miles

At the surface between DeFuniak Springs and Marianna, the Citronelle Formation becomes less common than it is in the western panhandle, and Alum Bluff Group sediments of Miocene age are prominent. Many hills in the area are capped by the Citronelle Formation, while most valleys, marshes, and other low areas are floored by clay-rich Alum Bluff sediments. Some of the Citronelle outcrops show clear evidence of slumping or other movement. It is very likely that the weight of the Citronelle Formation compacted and dewatered the Alum Bluff clays underneath it, creating much of the rolling topography of the area. As older limestones of Oligocene age appear closer to the surface in the Dougherty Karst Plain, around Chipley and Marianna, karst processes become increasingly important in shaping the landscape.

Several springs are located in this area. These are popular swimming and diving spots. While they may not be fountains of youth after all, a dip in the chilly (68 degrees Fahrenheit) waters of Morrison, Ponce de Leon, and Vortex springs will certainly restore your blood circulation

The Citronelle Formation of Pliocene age exposed in a sand pit along County Road 279 southeast of Caryville.

to a youthful rate! These springs issue from limestones of Oligocene or Miocene age that can be viewed with a snorkel and mask. Ponce de Leon Springs State Park and Morrison Spring are southeast of the town of Ponce de Leon. To reach the park, follow Florida 81 north from I-10 for 1 mile, then US 90 east for 0.6 mile, and then turn right (south) on County Road 181A and follow it for 0.6 mile. To reach Morrison Spring, take Florida 81 south for 3.7 miles, turn left (east) on County Road 181 and follow it for 1.7 miles, and then turn right (south) on County Road 181A and follow it for 0.5 mile. Vortex Springs is to the north. From I-10 follow Florida 81 north for 4.8 miles.

The Choctawhatchee River flows through this region, and both routes cross it at Caryville. The headwaters of the Choctawhatchee are in southern Alabama, and the river mouth forms a bayhead delta at the eastern end of Choctawhatchee Bay in southern Walton County. The river has a substantial load of suspended silt and clay (probably because of increased soil erosion due to tree harvesting) and is considered the muddiest river in Florida. This sediment load has resulted in a wide floodplain, and the river is especially prone to flooding. Although difficult to collect, many vertebrate fossils have been recovered from the lower end of the river, including those of Pleistocene-age mammoths, bison, and horses. Miocene-age shark and ray teeth and dugong ribs, vertebrae, and skulls have also been found. Rivers are common locations of vertebrate fossils in Florida because the fossils erode out of riverbanks and become concentrated in channels. The northern reaches of the Choctawhatchee, in Holmes County up to the Alabama line, expose the westernmost outcrops of the Ocala and Bumpnose limestones in rocky shoals and banks.

Holmes Blue Spring is a remote and beautiful little spring on the Choctawhatchee River. You can access this spring by boat about 1.5 miles north of the public boat landing near Cerrogordo. Exit I-10 at County Road 279 and drive north to Caryville and the County Road 279/US 90 intersection. Turn west (left) on US 90 and drive 2.5 miles. Proceed north on County Road 179A/181 for 5 miles, and then turn east (right) at Cerrogordo and drive 0.5 mile to the river. The spring is easy to find on the western side of the river, where its crystal clear water flows into the brown, muddy river. From the confluence of the spring and river, you can walk along the spring run to the springhead, which gushes from the Ocala Limestone of Eocene age.

There are many "blue" springs in Florida, so it has become a convention to use the county name before the title Blue Spring. If there is more than one Blue Spring in the county, then an additional location designation is added to its name. Washington Blue Spring Choctawhatchee and Washington Blue Springs Econfina are examples.

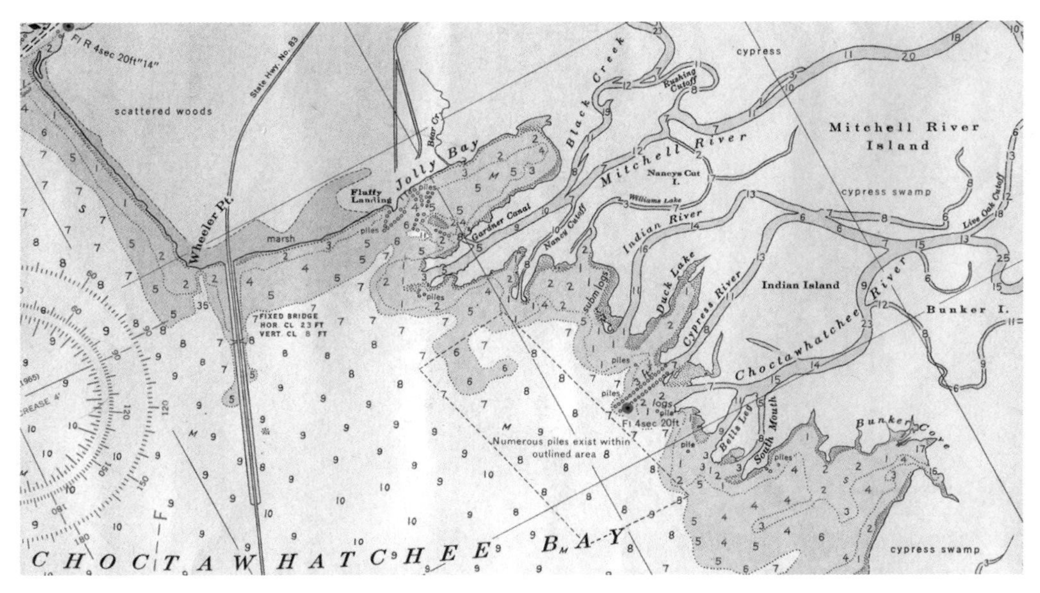

The bayhead delta of the Choctawhatchee River where it joins Choctawhatchee Bay. The numerous branches in the river, called distributaries, *form in response to the enormous sediment load the river delivers to the delta.* —Courtesy of National Oceanic and Atmospheric Administration

Holmes Creek, which US 90 and I-10 cross just east of Bonifay, is a major tributary of the Choctawhatchee. The creek is fed by many springs. On Florida 79, south of Bonifay and north of Vernon, Cypress Spring issues out of the Bridgeboro Limestone of Oligocene age. Nearby Beckton Spring is located approximately 2 miles northeast of Vernon. From the intersection at Florida 79 and Florida 277 in Vernon, turn east and then northeast on Florida 277. Drive 2.2 miles to Culpepper Lane. Turn left and go 0.3 mile to Big Pine Lane. Turn left and go 0.2 mile to the boat landing. The spring flows into Holmes Creek from the northwest, approximately 0.4 mile downstream from Big Pine Lane boat ramp.

Cypress and Beckton springs are the largest on Holmes Creek—both are second magnitude. Numerous other springs and seepages are visible from a canoe along the length of Holmes Creek, and, depending on water level, fossiliferous exposures of the Alum Bluff Group are common along the southern end of Holmes Creek.

The Marianna Limestone of Oligocene age was obviously named for exposures in the town of Marianna, but this formation extends west continuously as far as the Mississippi River, and erosional outliers are found far into south-central Georgia. The limestone forms an east-west ridge, or escarpment, that extends through the town of Marianna, from Jackson Blue Spring, northeast of town, to Cottondale.

SIDE TRIP TO FALLING WATERS STATE PARK

Falling Waters State Park, located south of Chipley, is a must-see for understanding the geology of the Florida Panhandle. The park entrance is 2.5 miles from I-10. From Chipley head south on Florida 77 for 0.75 mile and then east on County Road 77A (State Park Road). Travel east on County Road 77A for 1.7 miles to the park entrance. You will cross Falling Waters Road. The park is located on Falling Waters Hill, which is one of several large hills in the region, some of which reach elevations of over 300 feet above sea level. These sandy hills are erosional outliers—all that remain of a highland landscape that was formerly more extensive. Interestingly, there is an erosional gap in the sedimentary record in this area, essentially between the Choctawhatchee and Apalachicola rivers, with the Alum Bluff Group and Citronelle Formation removed. This gap, or eroded landscape, interrupts the Western Highlands. Karst development on the limestone here, which is close to the surface, has dissolved and lowered the surface, while large rivers and their tributaries have removed sandy strata as well as limestone.

The main attraction of Falling Waters State Park is a vertical, cylindrical sinkhole, called a *solution pipe*, 20 feet in diameter and 100 feet

The falls of Falling Waters sinkhole.

deep. During rainy seasons, a spectacular waterfall drops 67 feet into the pipe. The water flows only a very short distance before it quickly disappears into a swallet to enter a cave system. All the surface water of this park flows unimpeded into the limestone of the Floridan aquifer system. The Bridgeboro Limestone of Oligocene age and the Chattahoochee Formation of Miocene age are exposed at the sinkhole. These formations are extremely fossiliferous and represent shallow marine deposits. Alum Bluff Group sands and clays of Miocene age and sands of Pliocene and Pleistocene age overlie the Chattahoochee Formation elsewhere in the park, but they are not exposed at the waterfall.

A boardwalk in the park will take you around the margins of some impressive and spectacular karst terrain, with sinkholes, caves, and limestone arches. One of Florida's first oil wells was drilled on Falling Water Hill around 1920. The well reached a depth of 4,912 feet, into strata of Cretaceous age, but it was a dry hole.

Falling Waters is one of the few areas in the panhandle where the Bridgeboro Limestone is exposed. Geologists have referred to this formation as the Suwannee Limestone, and in older literature it was called

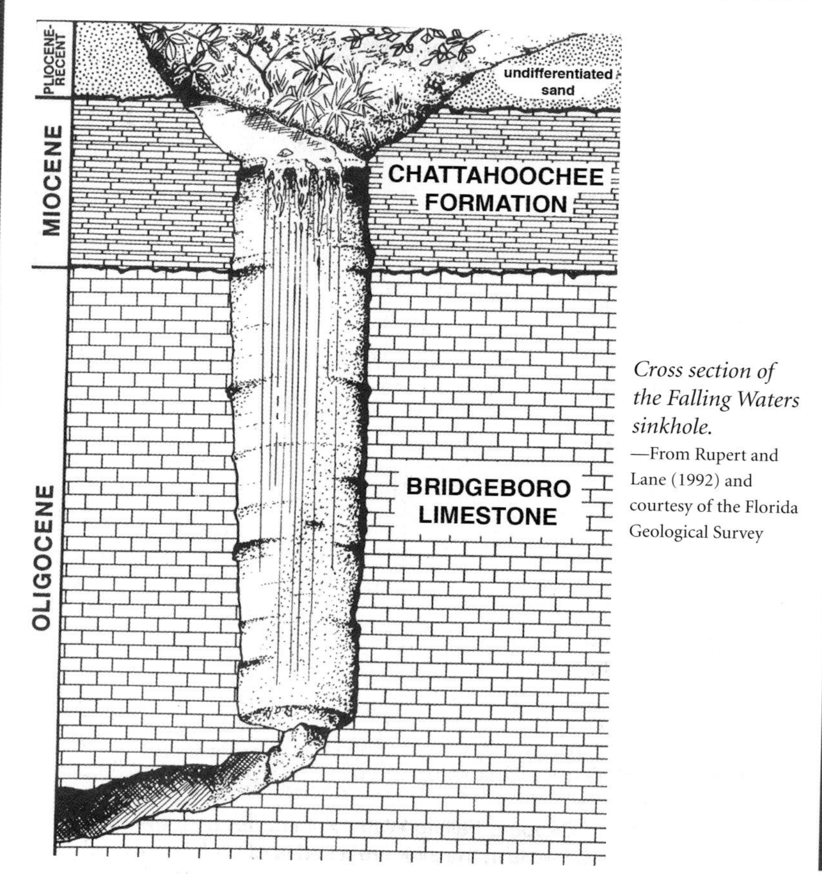

Cross section of the Falling Waters sinkhole.
—From Rupert and Lane (1992) and courtesy of the Florida Geological Survey

A sinkhole along Sinks Trail at Falling Waters State Park.

the Duncan Church Beds of the Suwannee Limestone. Named for exposures near Bridgeboro in southwestern Georgia, it is an entirely unique limestone. The Bridgeboro Limestone is a coarse grainstone, meaning it consists of large grains with almost no mud in the rock. This composition indicates it was deposited in high-energy waters that removed muddy sediments. The sediments are largely composed of coralline red algae, a type of algae that precipitates calcium carbonate and superficially resembles coral. Most of the grains are rhodoliths—concentrically laminated algal balls—that range from about 0.5 to 4 inches in diameter. The Bridgeboro also has abundant larger foraminifera, especially those of the genus *Lepidocyclina*, and reef corals.

Rhodoliths commonly grow in association with coral reefs, but they are also found in cooler waters. They have been found near Chipley but are actually more common in southwestern Georgia. Interestingly, the

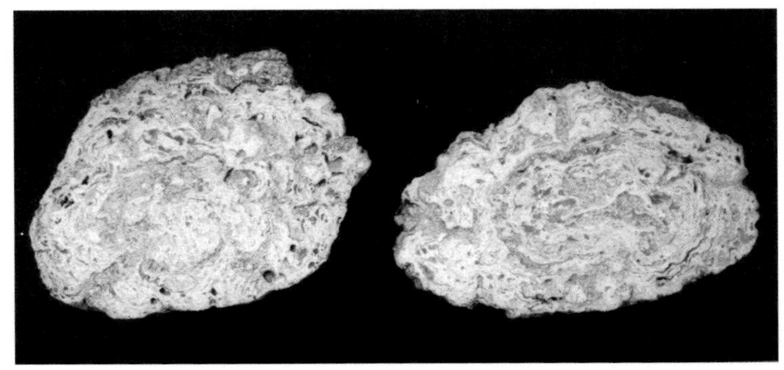

Cut rhodoliths from the Bridgeboro Limestone of Oligocene age, showing concentric layering of coralline algae.

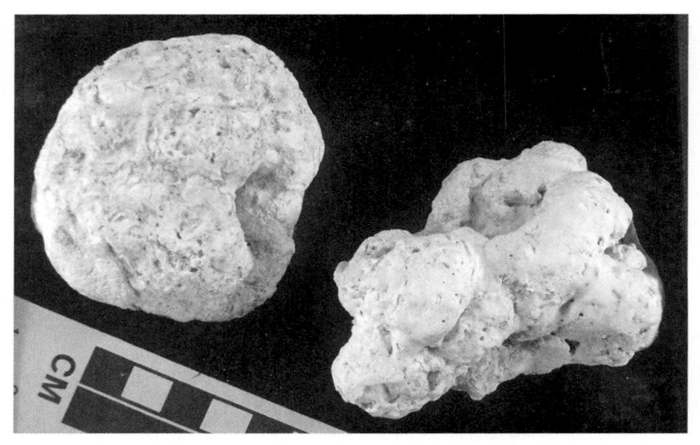

Modern rhodoliths from the Caribbean Sea.

Bridgeboro Limestone, with its rhodoliths and reef corals, was restricted in its development and only occurs along the flanks of the Gulf Trough. This is very similar to development of modern reefs and rhodoliths along the southern Florida continental shelf, where they grow along the flanks of the current-swept Straits of Florida.

The Bridgeboro Limestone is also exposed with the Marianna Limestone in the Duncan Church Quarry near Wausau, which is southwest of Falling Waters State Park on Florida 77. Follow Florida 77 to County Road 276 and turn right (west). Go 1.8 miles and turn right again (you are still on County Road 276). Go 0.3 mile to Piney Grove Baptist Church and turn right onto a dirt road. The quarry is 0.1 mile ahead on the left, behind the church. Permission to enter the quarry is required.

The Bridgeboro Limestone of Oligocene age at Duncan Church Quarry in Wausau. The pebblelike nodules are rhodoliths.

The pure, fine-grained limestone continues to be quarried for agricultural lime and other uses. The rock is so soft that it was once sawed into blocks with standard wood saws and used as bricks (over time, the surface of the limestone becomes cemented as a hard, exterior crust). This softness earned the Marianna Limestone the name "chimney rock" for its ease of cutting and use in former years as a building stone. Exposures are visible in various places around Marianna, but it is most easily seen in a hillside bluff on the north side of US 90 just west of the Chipola River. You can observe countless fossils of the very large foraminiferan *Lepidocyclina mantelli* and other small fossils in this bluff. This is also the type section of the Marianna Limestone—the outcrop that geologists use as a comparative standard to describe the formation.

The Marianna Limestone has long been quarried in Florida, and abandoned mines are common in Jackson County. Perhaps the best exposure of the Marianna is the active Marianna Lime Products quarry, located on the west side of Marianna. Geologists know it as the Brooks Quarry. Over 70 feet of limestone is exposed in the quarry, including the Ocala, Bumpnose, and Marianna limestones. From US 90 on the west side of Marianna, take Florida 73 for 0.4 mile to a partly paved road on the right. Proceed about 1 mile to the quarry entrance. You must obtain permission before entering the quarry.

The Brooks Quarry is one of the most important stratigraphic sections of rock exposed in the state for several reasons. First, it is one of the best and most accessible areas to see the boundary between Eocene and Oligocene time in Florida, which represents a critically important time of climatic and biotic change in Earth's history. Second, its geographic location in the eastern Florida Panhandle is between time-equivalent but deeper-water formations in Alabama and the rest of the Gulf Coastal Plain and the shallower-water deposits of the Florida Platform, thus providing a crucial link for the accurate correlation of rocks between these two regions. Third, the limestones in this quarry were deposited near the northern flank of the Gulf Trough and include unique and rare fossil assemblages, such as beds in the Ocala Limestone that are rich with large foraminifera of the genus *Asterocyclina.*

Several springs located near Marianna offer excellent opportunities to see—and swim in—the water of the Floridan aquifer system. Most springs issue out of the Ocala Limestone, including the popular Jackson Blue Spring. This first-magnitude spring discharges water at an incredible rate—over 100 cubic feet per second (or 64.6 million gallons per day). Cave divers have mapped more than 18,000 feet of underwater passages here. General Andrew Jackson and his forces camped here in 1818, during an expedition through west Florida. To visit the spring, from US 90 just east of Marianna go 1.1 miles north on Florida 71.

The Brooks Quarry in Marianna.

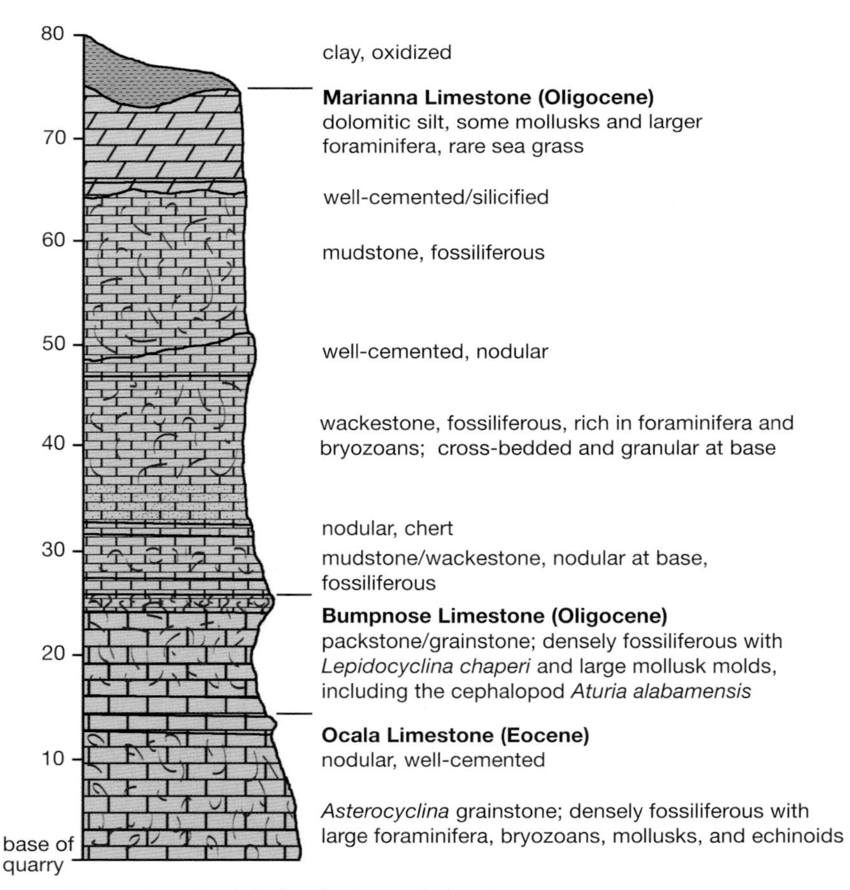

The stratigraphy of the Brooks Quarry in Marianna. —Illustration by Amanda Kosche

Proceed 3.3 miles east on County Road 164 (Blue Springs Road) to the entrance to the recreation area.

By canoe you can see many limestone exposures along the Chipola River. A beautiful riverside cave is exposed on the west side of the Chipola River about 0.5 mile south of US 90. Locally called The Ovens or Alamo Cave, it developed in the Marianna Limestone. The Marianna is exposed along the Chipola as far as Hollis Branch, west of Altha, and the Chattahoochee and Chipola formations can be seen along the riverbanks

A tooth of the extinct great white shark Carcharodon sokolowi, *from the Bumpnose Limestone of Early Oligocene age at Brooks Quarry in Marianna.*

Divers in Jackson Blue Spring.

farther south. South of Marianna, the Marianna Limestone is a dolostone and is exposed in three quarries along Florida 71 between I-10 and County Road 278. A coral- and algae-bearing formation, probably the Suwannee Limestone, is found in the southernmost of these quarries; it indicates that the region was near the northern, reef-bearing margin of the Gulf Trough during Oligocene time.

The Ovens, a cave within the Marianna Limestone of Early Oligocene age along the Chipola River.

Mermaid's Pennies—Larger Foraminifera

The most conspicuous and abundant fossils in Florida's limestones of Eocene and Oligocene age are the larger foraminifera, a group of marine protists that are closely related to the familiar freshwater protozoa of the genus *Amoeba.* Most foraminifera, or forams, are housed in a small shell called a test, which is made of calcium carbonate. They are helpful fossils to geologists, who use them to identify specific geological time intervals, reconstruct past environments, decipher ancient climate changes, and document evolutionary patterns. The small size of forams, usually less than 1 to 2 millimeters, ensures that they are recovered in rock cores that geologists collect. They have been used extensively in subsurface geologic investigations, such as oil exploration. The study of fossil foraminifera falls under the branch of paleontology called micropaleontology.

Larger foraminifera are characterized by their relatively large and complex tests. In fact, some hardly seem to qualify as microfossils at all. There are records of extinct foram species that reached 6 inches or more in diameter. In Florida, some fragments of the genus *Lepidocyclina* exceed 2.5 inches in diameter. Considering that these creatures are single-celled protozoa, these are extraordinary sizes. Most fossil foram species routinely found in Florida's strata are 0.25 inch or less in diameter, with larger species averaging 0.5 to 1.25 inches in size.

By studying modern species, scientists can infer much about fossil species. With few exceptions, the relatively large and complex tests of living species of larger foraminifera are related to algal symbiosis. The highly compartmentalized tests of these species contain multiple "greenhouses" that shelter single-celled algae. These symbiotic algae provide photosynthetic nutriment to the foram, and the foram provides shelter for the algae. Large forams are frequently associated with coral reefs and other shallow-water, tropical or subtropical limestone deposits. They can be so abundant that they compose

Common larger foraminifera of Florida. Special techniques are required to study larger foraminifera. To identify forams at the species level, paleontologists often need to split the test open or partially grind it away to see the complex internal chambers. Rulers are in millimeters.

Nummulites floridensis, *Ocala Limestone, Late Eocene age. Internal view (left) of test showing chambers and whole test (right).*

Nummulites willcoxi, *Ocala Limestone, Late Eocene age. Whole test and internal view of test showing chambers.*

Asterocyclina americana *in* Asterocyclina-*rich Ocala Limestone, Late Eocene age.*

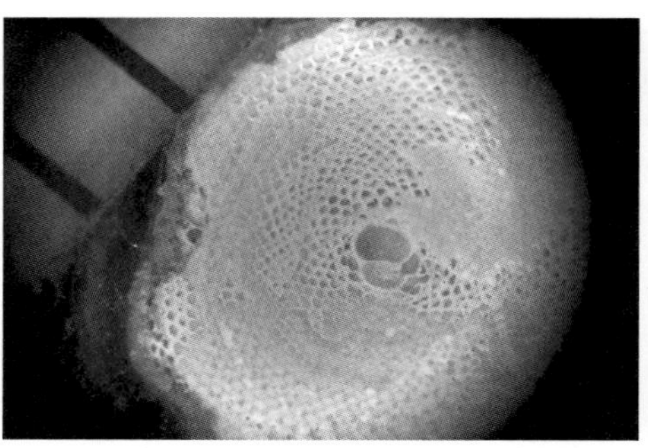

Lepidocyclina mantelli, *Marianna Limestone, Early Oligocene age. Internal view of test showing chambers.*

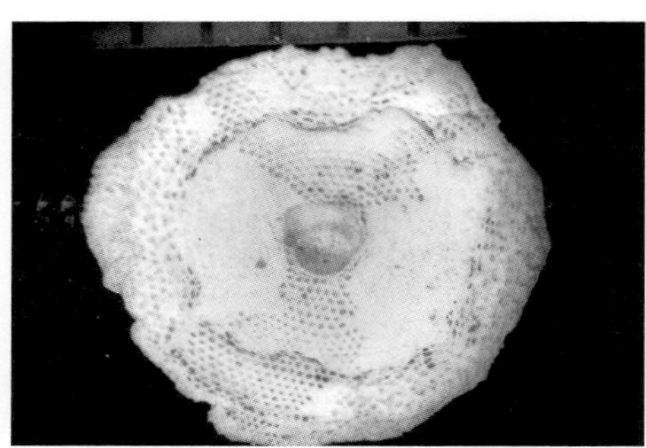

Lepidocyclina chaperi, *Bumpnose Limestone, Early Oligocene age. Internal view of test showing chambers.*

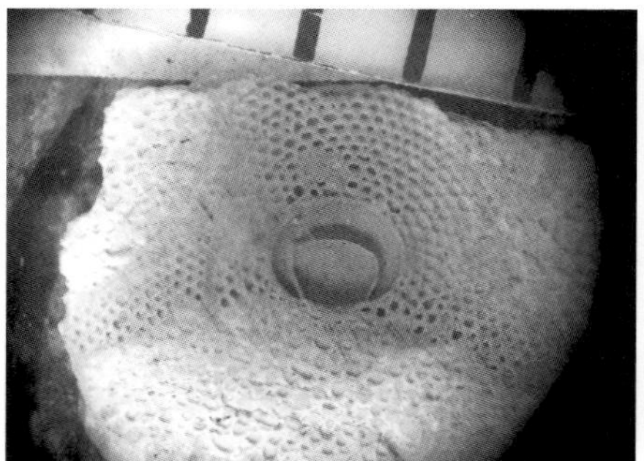

Lepidocyclina undosa, *Bridgeboro Limestone, Early Oligocene age. Internal view of test showing chambers.*

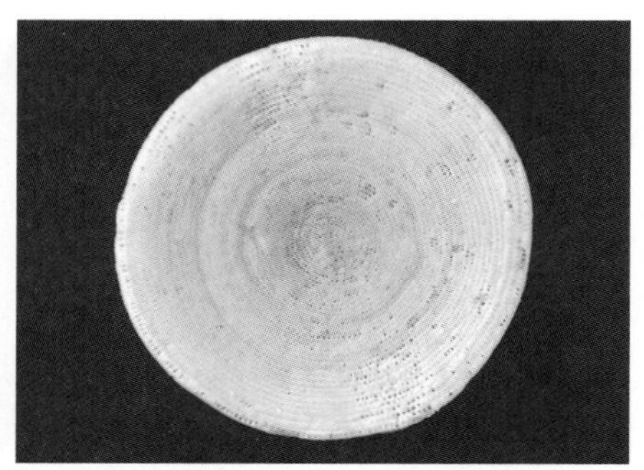

Miosorites americanus, *Chipola Formation, Early Miocene age.*

Archaias angulatus, *Florida Reef Tract, modern.*

Laevipeneroplis proteus, *Florida Reef Tract, modern.*

the majority of particles in some limestones—even 100 percent. Such a limestone is called a coquina. Coquinas composed of larger foraminifera of the genera *Lepidocyclina* or *Nummulites* are common in the Ocala Limestone. The Eocene-age Giza Limestone of Egypt, also a nummulitic limestone, was used as building stone for the great pyramids. Florida has been home to larger forams from Cretaceous time to the present.

Pearly Nautilus in Florida

Probably no other modern animal has captured the attention of paleontologists more than the pearly nautilus. A classic living fossil, the nautilus does indeed have a long geological pedigree, with ancestors reaching far back into Paleozoic time, hundreds of millions of years ago. One of the last of the cephalopods with a coiled shell, the nautilus provides scientists with much information on the biology of extinct nautiloids and their extinct cousins, the ammonites. Today, there are only three definite living species: *Nautilus pompilius, Nautilus macromphalus,* and *Nautilus scrobiculatus.* Scientists dispute the legitimacy of two or three other possible modern species. All are found only in the South Pacific and perhaps part of the Indian Ocean. They are primarily scavenger-predators, usually foraging on the sea bottom for small crustaceans. The unique and beautiful chambered shell acts as a protective submarine for the animal. Behind the large body chamber are many smaller chambers, all connected by a tubular siphuncle. Through the siphuncle, the animal can either fill or empty the chambers with gas or water. This allows the nautilus to control its buoyancy and float at whatever water level it desires. A nautiloid swims like a squid

The extinct nautiloid Aturia alabamensis. *This specimen was recovered from the Bumpnose Limestone of Early Oligocene age in Marianna and is about 8 inches in diameter.* —Courtesy of the Invertebrate Paleontology Division, Florida Museum of Natural History, University of Florida

and octopus, by jet propulsion; it ejects water from its mantle cavity through a tube called the *hyponome.*

Nautiloids were common in the early part of Cenozoic time in the continental shelf waters of the Gulf of Mexico. They are particularly common in Paleocene-age rocks of Alabama. Globally, they reached their peak diversity in Eocene time with over one hundred species; the most prominent and diverse genus was *Aturia.* Paleontologists have described over forty *Aturia* species in rocks of Paleocene to Miocene age. In Florida, the species *Aturia alabamensis* is found in the Ocala Limestone (upper strata of Eocene age) and the Bumpnose Limestone (lower strata of Oligocene age). On rare occasions, *Aturia* species are found in the Chipola Formation (lower strata of Miocene age). They are usually preserved as an internal mold, and because they're never found in abundance, the discovery of an *Aturia* fossil is a real prize. Some individuals reached extreme sizes, up to 2 feet or more in diameter. A few specimens with original, mother-of-pearl shell have been recovered from the Chipola Formation at Tenmile Creek in Calhoun County, and at Alum Bluff in Liberty County. Although they once thrived in the Gulf of Mexico, the elegant nautiluses were extinct in this region by Middle Miocene time.

*The modern chambered nautilus (*Nautilus pompilius*).*

Florida Caverns State Park and Speleothems

Florida Caverns State Park, located 2.7 miles north of Marianna on Florida 166 (Caverns Road), is a geological must-see. The caverns, which developed in the Ocala Limestone of Eocene age and the Bumpnose Limestone of Oligocene age, are 20 to 50 feet below the surface. More than 2,900 feet of passageways have been mapped in this cavern system. A 0.75-mile guided tour will take you through some rather spectacular cave development, but many caves in the park have been placed off-limits and can only be explored by qualified spelunkers with special permission. The Florida Caverns certainly do not have the grandeur of Kentucky's Mammoth Cave or New Mexico's Carlsbad Caverns, but this park is nonetheless a fantastic natural laboratory for observing karst processes and cave development. Florida Caverns is one of several designated state geological sites.

Scattered artifacts indicate that humans have known about the caverns for at least 1,000 years, and probably much longer. The area around the caverns was mentioned in 1693 by Friar Barreda, who was part of the first Spanish overland expedition to Pensacola Bay. In 1818, Native Americans successfully hid from General Andrew Jackson in the caverns, and local citizens took refuge here during a Civil War skirmish. A surveyor rediscovered the caverns in 1937, and the trails and lighting were developed by the Civilian Conservation Corps between 1938 and 1942. The first public tour of the caverns was in 1941.

The "wedding cake," a column in Florida Caverns.

The building stone used to construct the visitor center has an abundance of the larger foram *Lepidocyclina chaperi* and mollusk molds. These blocks are primarily Bumpnose Limestone.

Geologists do not know the precise timing of cave formation in this region. Karst terrain developed in Florida as early as Oligocene time, about 30 million years ago. But it is likely that the caverns developed more recently, during Pliocene or Pleistocene time, as fluctuating sea levels altered the position of the water table. Caves often form near the water table, where old groundwater is regularly replaced by new acidic groundwater, which dissolves the limestone. It is probable that earlier caverns became dry as the water table dropped, and then were destroyed by surface erosion. Cave formation proceeded along rock fractures that trend in northeast and northwest orientations, almost at right angles to each other. Caves have also developed preferentially along the contact of the Ocala and Bumpnose limestones, where water can easily flow.
Fossils are abundant in the Florida Caverns but can only be seen with a bright flashlight. There are countless large foraminifera, scallop shells, and echinoids (also called "sea biscuits"), as well as an ancient nautilus (*Aturia alabamensis*) and even shark teeth preserved in the ceiling. Ask a ranger to point them out on the tour. You can also observe many karst-related features along the nature trails in the park. The road to Blue Hole Spring crosses a natural bridge where the Chipola River disappears into a swallet and reemerges about 1,500 feet to the south. Be sure to walk Bluff Trail, which leads from the parking lot to the caverns entrance and visitor center. The trail features many additional cave openings (most are barred), a rock shelter, sinkholes, and a tunnel cave—a long solution tunnel with a conspicuous fracture along the ceiling. This tunnel perfectly illustrates how fractures channel water, which in turn facilitates the dissolution of limestone. Along the park's high limestone cliffs next to the Chipola River are the remains of several other caves, some of which have

Speleothems in Florida Caverns.

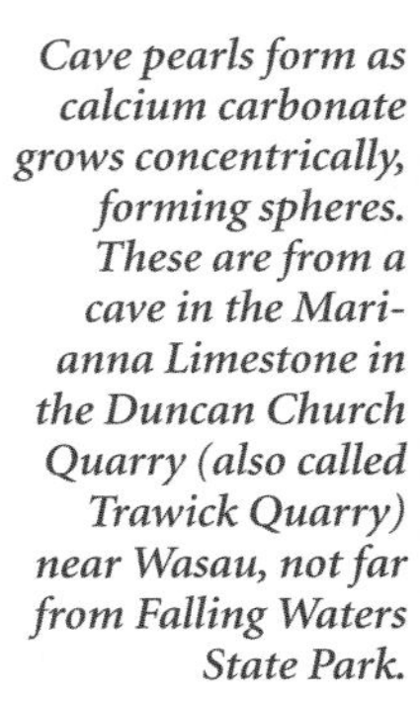

Cave pearls form as calcium carbonate grows concentrically, forming spheres. These are from a cave in the Marianna Limestone in the Duncan Church Quarry (also called Trawick Quarry) near Wasau, not far from Falling Waters State Park.

speleothems. The river has almost entirely eroded these caves. Many more limestone exposures are visible along the nature trail. All of the exposures are the densely fossiliferous Ocala Limestone.

About 20 feet or more of the upper part of the Marianna Limestone is exposed on a hilltop on County Road 167 located 0.8 mile south of Florida Caverns (or 1.9 miles north of US 90 in Marianna). Molds of mollusks and larger foraminifera are evident. Approaching this outcrop from the south, it is very obvious that it is the top of an elongated hill. This hill is an erosional remnant of Marianna Limestone sitting next to the floodplain of the Chipola River. About 0.3 mile north of this outcrop is another natural exposure of the Marianna that someone has evidently sawed into a blocklike wall.

Cave formations, or speleothems, are abundant and well-developed in the Florida Caverns and other Florida caves. Speleothems form as water saturated with calcium carbonate passes through the cave ceiling and precipitates thin layers of calcite in the process of dripping and evaporating. Thin formations of calcium carbonate called *soda straws* hang from ceilings and thicken into stalactites. Stalagmites rise from the ground beneath most stalactites and may eventually connect with them to form a column. Thin, wavy draperies may grow through cracks in the ceiling, and as iron and other minerals enter the water, red-banded drapes, locally called "bacon," may form. Sheetlike layers of calcite called *flowstone* may grow over some speleothems or limestone surfaces, creating the appearance of white cake-icing over the rocks. Helictites are a type of speleothem that forms when there is a persistent wind through a cave system. The wind causes the water on speleothems to drip at an angle, creating uniquely shaped stalactites. Another type of speleothem forms when calcite precipitates around the rims of standing pools of water, producing a bathtub-like wall called *rimstone*. Because most speleothems grow at the expense of the cave ceiling, which dissolves but only partially remineralizes in speleothems, caves eventually cave in as their roofs thin to the point of collapse. In places the Florida Caverns are not far below the surface, and at times of high rainfall water may rush in at certain points and cover much of the cave floor.

US 90 and Interstate 10

Marianna—Chattahoochee—Tallahassee

55 miles

Most of the region between Marianna and Tallahassee is part of the Apalachicola River basin, and it has some spectacular topography. High, rolling terrain of bluffs and ravines, part of the Tifton Uplands District, is very evident on the east side of the Apalachicola River from I-10 and US 90 (particularly from the westbound lane). The overlook to the Apalachicola River from US 90 is stunning—you would never guess you were in Florida! The high bluffs along the eastern bank of the Apalachicola, extending from Chattahoochee to just south of Bristol, are the southern extension of a prominent erosional cliff, or cuesta, called the Pelham Escarpment, which extends far into southwestern Georgia. Some of Florida's best geologic exposures are found along this cuesta, along the eastern banks of the upper Apalachicola River. These outcrops include Aspalaga Landing (just south of the I-10 bridge), Rock Bluff at Torreya State Park, Alum Bluff at Apalachicola Bluffs and Ravines Preserve, and Estiffanulga Bluff south of Bristol.

Why is the topographic relief so pronounced here? Geologists believe that the Apalachicola, Chattahoochee, and Flint rivers shifted their courses and merged at some point after Pleistocene time, and their combined erosional force cut deeply into the Miocene- and Pliocene-age sands of the Tallahassee Hills. This, coupled with the lowering of terrain (through karstic processes) of the Dougherty Karst Plain District west of the river, has accentuated the relief. Isolated hills west of the river (such as Falling Waters Hill at Falling Waters State Park) were once continuous with the Southern Pine Hills and Tifton Uplands districts to the north and west. The hills that remain today testify to this area's erosional past.

Just west of Chattahoochee, Lake Seminole, impounded by Jim Woodruff Dam, is visible north of US 90. The Chattahoochee Formation of Early Miocene time can be seen along the west side of Jim Woodruff Dam, near the town of Chattahoochee. In Chattahoochee, take US 90 west over the Apalachicola River. Only 0.1 mile west of the bridge, turn right (north) on an unmarked, paved road and proceed about 0.4 mile. You will pass through a gate. About 20 feet of the Chattahoochee Formation is well exposed for nearly 0.25 mile along the west side of the road from the gate to the guardhouse on the dam. There is a scenic overlook on the loop road at the top of the outcrop. The Chattahoochee is a sandy and clay-rich limestone and dolostone with many mollusk molds, echinoid casts, and some larger foraminifera, including those of the genera *Archaias* and *Sorites*. It also has many beds of intraclasts, flat limestone pebbles that form a rock called an *intraformational conglomerate*. Such a rock is indicative of deposition in very shallow marine water and even in the intertidal zone of a shoreline.

Another excellent exposure of the Chattahoochee Formation is on the east side of the dam, although this outcrop is actually in Georgia. Go north on Bolivar Street at the first light east of the US 90 bridge for about 0.5 mile, then veer left on the fork to Woodruff Dam and proceed another 0.5 mile. About 40 or more feet of the Chattahoochee crops out here along a lengthy exposure leading down to the boat ramp. And an access road by the dam leads down to the eastern shoreline on the Florida side, where there are many boulders of the Chattahoochee Formation. Additional exposures can be seen in a city nature park on the southeastern corner of the US 90 bridge and the river, reached by an access road on the east end of the bridge.

Florida 65 north out of Quincy passes through a region where clay deposits called *fuller's earth* are mined in the Torreya Formation of Miocene age. Fuller's earth consists of clay minerals that have many useful qualities, including being especially good absorbent and decoloring agents. One common type of this clay mined near Quincy is called *palygorskite*; it is also called *attapulgite*, named for the clay-producing Georgia town Attapulgus just a few miles north of Quincy. Some of these mines have

An exposure of the Chattahoochee Formation of Early Miocene age along the west side of the Jim Woodruff Dam near Chattahoochee.

*A petrified cypress tree (*Taxodium*), 18 million years old (Early Miocene age), collected in 1963 in the Englehard Company's Midway Fuller's Earth Mine near Quincy. Internal growth rings and bark were preserved by petrification. This specimen is on display at the Florida Museum of Natural History, in Gainesville.*

Modern cypress trees, Suwannee River.

yielded spectacular Miocene-age fossils, including petrified cypress logs and the complete skeletal remains of dugongs. The fuller's earth of this area was deposited in the Gulf Trough, which was probably an estuarine system at the time.

SIDE TRIP TO TORREYA STATE PARK

From I-10, take Exit 166, about 4.5 miles east of the Apalachicola River bridge, onto County Road 270A north. Proceed about 1.7 miles and turn left (south) onto County Road 269 (from US 90, exit onto County Road 269 in Chattahoochee). Continue for about 1.5 miles to the I-10 overpass. The rolling terrain of County Road 269 is very pronounced and includes some exposures of the Citronelle Formation. Just before reaching the I-10 overpass, an abandoned sand pit cuts deeply into the Citronelle. From the I-10 overpass, proceed 2.3 miles south on County Road 269 to County Road 270 and turn right. Continue 0.3 mile, then turn right again and continue on County Road 270 for 1.7 miles to where the road descends into the very deep ravine of Crooked Creek. This rather spectacular drop is typical of the deeply incised and eroded terrain common in this region.

Continue on County Road 270 to Torreya State Park. When the river is low, exposures of the Torreya Formation of Early Miocene age are visible below the River Bluff Loop Trail and the Confederate gun pits along

Deeply incised terrain through the Citronelle Formation of Pliocene age along County Road 270.

the banks of the Apalachicola River. You can access this trail from the Gregory House. Below the trail but above the river is a river terrace and former floodplain. The limy sands of the Torreya Formation contain invertebrate fossils of species that lived in shallow marine conditions as well as vertebrate animals, including several species of early horses and other mammals. The mix of shallow-marine and terrestrial fossils indicates that the Torreya was deposited in a nearshore area, probably in an estuary. The presence of certain foraminifera, those of the genera *Ammonia* and *Elphidium*, indicate there were also brackish water conditions while this formation was deposited.

The Torreya Formation exposed in the park along the river at Rock Bluff is the type section. Rock Bluff can be reached from the River Bluff Loop Trail. The trailhead is located immediately inside the park entrance. Along the trail not far from the trailhead is an exposure of the Citronelle Formation at the "grand canyon," which has been revealed by deep erosion and steephead formation. Note the numerous and lush steephead ravines along the trail as you descend to the river. When you reach the river at Rock Bluff, note the impressive bluffs of the Torreya Formation and overlying sediments that rise up to 150 feet above river level. You can also access Rock Bluff from the Weeping Ridge Trail, which begins in the park's main campground. Weeping Ridge is an intermittent waterfall flowing over the Torreya Formation. Near the beginning of the Weeping Ridge Trail, a connecting trail leads down to River Bluff Loop Trail; heading left will take you to Rock Bluff.

The Torreya Formation of Miocene age at Rock Bluff in Torreya State Park.

Tallahassee

Both I-10 and US 90 pass through Tallahassee, as does Florida 20. There is much interesting geology in this area; unfortunately, it is not easily visible from these roads. You'll have to leave them—and your car in some instances—to see it.

Lake Jackson Mounds and Lake Jackson

Located on the north end of Tallahassee is Lake Jackson Mounds Archaeological State Park, a Mississippian Period ceremonial site of six earthen temple mounds of the Fort Walton culture. This mound-building people lived between AD 1200 and 1500. From I–10 in Tallahassee, turn north on US 27 and drive 2 miles. Turn right (east) on Crowder Road and proceed to Indian Mounds Road and turn right (south). Continue to the park's entrance. This site was a political and religious center and part of what is known as the Southeastern Ceremonial Complex, a series of sites with artifacts that indicate the existence of powerful chiefs who wielded extensive political, military, and religious power. They were buried at such mounds with great ceremony and adorned with a wealth of diverse artifacts, such as pearls, embossed copper plates, and nonnative mineral artifacts. The largest mound is 278 feet by 312 feet at its base and about 36 feet high. Numerous cultural artifacts have been recovered from the area.

Archaeological sites of this particular farming culture are commonly found in the loamy soils that developed over the Pliocene-age Miccosukee Formation of the Tifton Uplands District north of Tallahassee (the Miccosukee composes the many orange-red bluffs around Tallahassee). The Lake Jackson Mounds are located on the shore of Lake Jackson, a 4,000-acre karst lake that periodically loses its water through sinkholes. The Native Americans who occupied this area referred to the lake as *Okeehepee* or *Okeehepkee*, which mean "disappearing waters." The lake last drained completely in September 1999 at the beginning of a five-year drought in northern Florida. By August 2005, the lake had refilled and reached the low end of its normal water levels. However, the sinkhole into which the lake drained, called Porter Hole, was still open and draining 6 to 8 million gallons of water per day. The lake drained into Porter Hole again in early June 2006 in response to continued drought conditions. Porter Hole filled with sediment following the draining and became plugged. A week later, a sinkhole about 8 feet in diameter opened 25 feet north of the plugged Porter Hole. Then, in August 2006, another sink, about 6 to 7 feet in diameter, opened 25 feet south of the plugged hole. This, in conjunction with numerous smaller holes that have opened since 1999 and the abundance of old, filled sinks in the area, attests to the complexity of the karst system responsible for the creation of Lake Jackson and other lakes in the vicinity.

Florida Geological Survey

While in Tallahassee, be sure to visit the Florida Geological Survey, located at 903 W. Tennessee Street (US 90) on the campus of Florida State University.

Lake Jackson sinkhole (Porter Hole) after the lake drained in 1999.

Since 1907, the survey has been conducting basic and applied research and disseminating crucial information on the state's mineral and environmental resources. The survey has displays of Florida's minerals and fossils and a geological research library. Survey publications are available in the library for a nominal cost.

Cody Scarp

While driving south of Tallahassee on Monroe Street, as you pass in front of the state capital building, you quickly leave behind the sands and clays of the Tallahassee Hills, which were deposited in deltas in Pliocene time, and pass into the Ocala Karst District, a nearly flat terrace that was flooded by a sea in Pleistocene time. Separating these two regions is the Cody Scarp, an obvious escarpment that rises 100 to 150 feet above the lowlands. The scarp marks the landward extent of the highest sea level of Pleistocene time, the Wicomico Sea and Wicomico Terrace of the early Pleistocene. The base, or toe, of this scarp is located at the southern end of town and nearly coincides with Tram Road (County Road 259), which runs southeast of the Leon County Fairgrounds. The Cody Scarp is especially pronounced near the small town for which it is named, Cody, which is in Jefferson County on County Road 259.

South of the scarp, the topography is noticeably flat, unlike the rolling hills of most of Tallahassee. This part of the Ocala Karst District, including much of Wakulla, Leon, Jefferson, and Taylor counties, is called the Woodville Karst Plain. Its gently rolling surface ranges between sea level and 50 feet in elevation and is characterized by a number of karst features, such as springs, sinkholes, disappearing streams, and caves.

Leon Sinks Geological Area

From Monroe Street in front of the state capitol, proceed south to US 319 (Crawfordville Road). About 5.8 miles south of SR 263 on US 319 is Leon Sinks Geological Area. A 3.1-mile walking trail takes you past numerous sinkholes, natural bridges, disappearing streams (swallets), karst windows, and karst swamps. Many of the sinks are dry, but some are karst lakes. The largest sinkhole is Big Dismal Sink, with a diameter of 250 feet and a depth of 50 feet. At water level there are dolostones of the St. Marks and Chattahoochee formations overlain by the fossiliferous Torreya Formation, all of Miocene age, capped by dune and terrace sands of Pleistocene age. Nearby Hammock Sink, or Little Dismal Sink, leads to an extensive underwater cave network—in fact, it is one of the longest underwater cave systems in the world, extending more than 18 miles with depths exceeding 200 feet. Using underwater dye tracing and cave diving, this cave system has been linked with the cave system of Wakulla Springs, to the southeast in Wakulla County.

Natural Bridge Battlefield Historic State Park

South of Tallahassee is Natural Bridge Battlefield Historic State Park. Here, in March 1864, a group of Confederates from Quincy and Tallahassee successfully repelled a Union attempt to sever rail and bridge connections to the St. Marks area, which would have isolated Tallahassee. Natural Bridge is so named because the St. Marks River disappears here, flowing underground through karst cavities in limestones of Oligocene and Miocene age and reemerging several times before flowing continuously again at the surface roughly 1 mile to the south. While underground, the St. Marks gains a significant amount of flow from groundwater and emerges as a much larger river than it was where it first disappeared. The ground between where the river disappears and reappears is the "natural bridge."

Florida 20

NICEVILLE—TALLAHASSEE

135 MILES

Between I-10 and US 98, lonely Florida 20 runs east-west through a nearly forgotten portion of the panhandle between Niceville and Tallahassee. Between Niceville and Freeport, the route skirts the northern coast of Choctawhatchee Bay. There are many underwater caves near Ebro, in which divers have recovered numerous fossils of Oligocene age, including reef corals. In one cave, divers encountered the skeleton of a modern alligator. The poor reptile apparently went too far back into the submerged cave and was unable to find the way back to the surface.

Near the intersection with Florida 77 (especially east of it), lies a field of very large dunes and lakes called the Sandhill Lakes. This area resembles the sand hill karst that is so common in the central peninsula of Florida. Large sand hills, up to 190 feet in elevation, and round lakes are evident for a few miles east of Florida 77. About 8.6 miles east of Florida 77, Florida 20 crosses Pitt Spring Recreation Area on Econfina Creek (pronounced, "E-con-*fine*-a," and not to be confused with a river of the same name in Taylor County pronounced "E-con-*feen*-a"). A canoe can be launched here for a 3-mile run south to County Road 388. Alum Bluff Group limestones are spectacularly exposed along this part of the Econfina, and a number of beautiful springs (the Gainer Springs Group) can be visited. Approximately 5 miles upstream of Florida 20 are exposures of the mollusk-rich fossil beds of the Jackson Bluff Formation of Pliocene age.

SIDE TRIP ON FLORIDA 73

One of the most well-known fossil localities in Florida is located 4.7 miles north of Clarksville (20.7 miles south of Marianna) on Florida 73, where the road crosses Tenmile Creek, a tributary of the Chipola River. Exposed along this otherwise inconspicuous creek is one of the most species-rich fossil deposits in the world—the Chipola Formation of Early Miocene time, dated at over 18 million years old. The Chipola Formation contains an astounding 1,100 or more species of mollusks. Furthermore, these fossils are preserved in pristine condition in the stiff, clay-rich sands. The original color markings of some shells have even been preserved. This site was first described in 1889, and paleontologists have collected fossils here ever since.

The Tenmile Creek site preserves a phase of the Chipola Formation that is rich in quartz sand and clay. In nearby Farley Creek, to the east, the Chipola is a limestone, and the formation is rich in corals, coralline algae, and other tropical species. Chipola sediments were deposited in coral patch reef, back-reef lagoon, oyster reef, and shoreline environments.

Chipola Formation mollusks represent the last of the truly tropical marine faunas to extend as far north as the panhandle. Many of these species have strong Tethyan affinities, and today closely related species exist in the South Pacific. The successive, younger faunas of the Oak Grove, Shoal River, Intracoastal, and Choctawhatchee formations of Miocene and Pliocene age clearly represent cooler-water molluscan species, reflecting the well-documented global cooling that occurred in both marine and terrestrial environments during Miocene time.

About 2.5 miles north of the Tenmile Creek locality on Florida 73 (or 18.4 miles south of US 90 in Marianna), take County Road 274 east for 1.6 miles to the Chipola River. The descent to the river valley is pronounced. Turn right (south) on a dirt road along the west side of the river and proceed 0.8 mile to

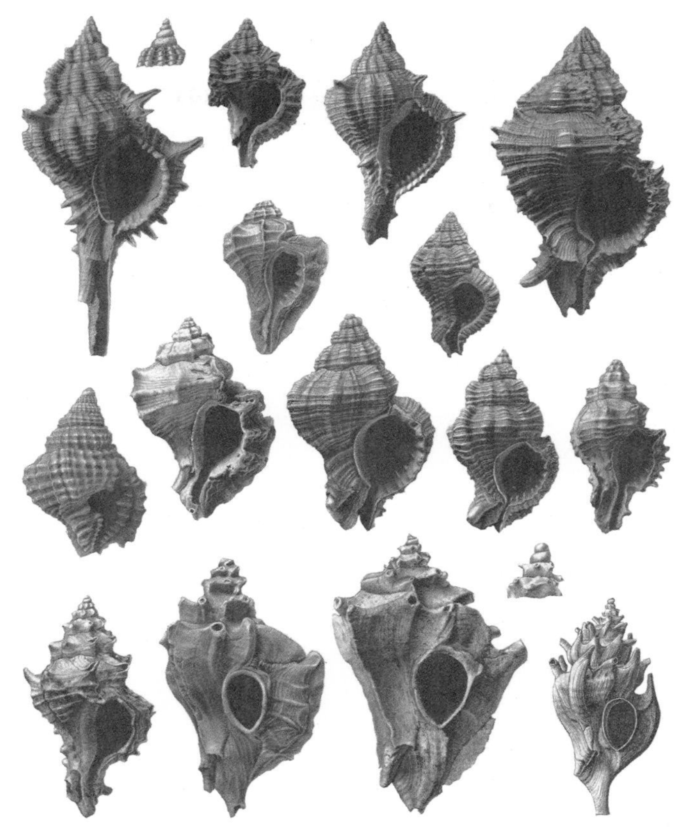

Various gastropod mollusks of the Chipola Formation, Shoal River Formation, and Oak Grove Sand of the Alum Bluff Group of Early Miocene age from the Florida Panhandle. —From Gardner (1947) and courtesy of the U.S. Geological Survey

Look and Tremble Shoals. At low water, exposures of the Chattahoochee Formation form rapids in the river here. The Chipola Formation is exposed along the riverbanks for several miles below this point. This area is a popular canoe run and picnic area. The Chattahoochee Formation is also exposed beneath the County Road 274 bridge (Willis Bridge), and nearly continuously from the bridge to Look and Tremble Shoals.

The Chipola River eventually enters the larger Apalachicola River, but not before flowing through Dead Lakes in the muddy floodplain of the Apalachicola. Stands of dead bald cypress trees emerge from the lake, testifying to its recent origin. It is thought that excess erosion resulting from the clearing of forests in the early nineteenth century may have drastically increased the sediment load of the Apalachicola, producing the excess sediments that dammed the Chipola, creating Dead Lakes. The lake can be accessed at Dead Lakes State Recreation Area, just north of Wewahitchka in Gulf County.

SIDE TRIP ON FLORIDA 12

Probably the most spectacular natural geologic exposure in Florida is what geologists have traditionally called Alum Bluff. This exposure is owned and managed by the Nature Conservancy as part of the Apalachicola Bluffs and Ravines Preserve. First described in 1889, the 130-foot bluff exposes formations of Miocene, Pliocene, and Pleistocene age, some of which are densely fossiliferous. This is also the location of E. E. Callaway's Garden of Eden (see **The Foundations of Eden** in the **Introduction**). The preserve entrance is 1.4 miles north of Bristol off Florida 12. From the entrance to the preserve, it's just 0.2 mile to the trailhead for the Garden of Eden Trail, a beautiful and educational geo-ecologic trail to the river. The 3-mile (round-trip) trail to the scenic Alum Bluff crosses two steephead ravines, which preserve cooler microclimates and provide habitat for endemic flora and fauna. You can also reach Alum Bluff by boat. It is about 3 miles upriver of a public boat launch that is about 1 mile north of both Bristol and the Florida 20 bridge.

The strata of Alum Bluff include the Chipola Formation of Miocene age, which is rich in well-preserved mollusks, corals, and other invertebrates and visible at water level; undifferentiated Alum Bluff Group sands and clays of Miocene age with well-preserved fossil leaves and wood and mammal bones; the Jackson Bluff Formation of Pliocene age, which has abundant fossil mollusks, overlain by unfossiliferous clays and sands (these clays are the source of the aluminous clays for which the Alum Bluff exposure was first named); and at the top, the Citronelle Formation of Pliocene age and wind-deposited sands of Pleistocene age. The plant fossils of the Alum Bluff Group include a mixture of tropical and more temperate plant species representing mesic

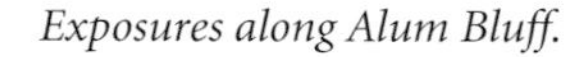

Exposures along Alum Bluff.

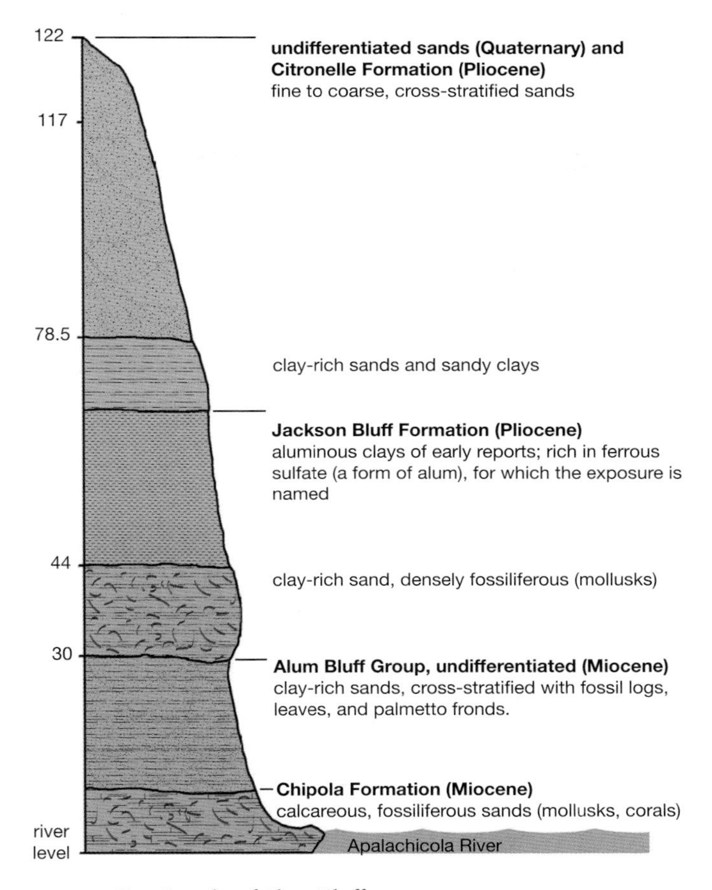

Stratigraphy of Alum Bluff. —Illustration by Amanda Kosche

(moist) climatic conditions. Mesic forests have persisted in Florida ever since Miocene time. The most impressive of the plant fossils are entire fronds of the palm *Sabalites apalachicolensis*, which is very similar to the modern sabal palm (*Sabal palmetto*), the state tree of Florida.

For over 125 years botanists have made excursions to the ravines at and near Alum Bluff to study their many exotic plants. The Apalachicola is the only Florida river with headwaters that reach to the southern Appalachian Mountains, and many Appalachian plant species can be found in the cool, deep steephead ravines that retain a favorable environment for these otherwise more northern species. It is believed that during the cooler climates of Pleistocene time, northern plant species extended their ranges southward. With the warming that accompanied the end of the Ice Age, these species contracted their ranges northward. However, a few were able to maintain their southern range in favorable microclimates, such as the ravines and bluffs along the Apalachicola. These sorts of isolated habitats are referred to as *refugia.*

*The fossiliferous Jackson Bluff Formation (*inset*) of Pliocene age at Alum Bluff, with the gastropod* Ecphora quadricostata.

Among these species is a well-known conifer, the gopherwood or Torreya tree (*Torreya taxifolia*). Although gopherwood trees in this area were once quite large, reaching up to 60 feet high, the only remaining examples of these rare trees are spindly—apparently the result of a fungal blight that has decimated them. Fewer than two thousand saplings are all that remain. Even so, the ravines are hot spots of biodiversity, harboring more than one hundred rare and endangered species.

South of Bristol there is a public boat ramp area and extensive riverside exposure of the Citronelle Formation at Estiffanulga Bluff, the southernmost of the tall bluffs along the Apalachicola. From Bristol, drive about 8.4 miles on County Road 12, and then proceed west on County Road 333 for 2.4 miles. Because Estiffanulga Bluff is located about 9 miles south of Alum Bluff, only the Citronelle Formation is exposed. Because of the gentle dip of strata toward the Gulf of Mexico, younger formations usually appear at the surface in more southern exposures and older formations appear at the surface in more northern exposures. Thus, Alum Bluff has many formations exposed, reaching as low as the Chipola Formation of Miocene age. On the nearby Chipola River, the older Chattahoochee Formation and Marianna and Ocala limestones progressively appear at the surface as one travels upstream.

Just south of Estiffanulga, on Forest Road105, is Camel Lake, one of the few lakes in Florida that preserves a long sediment and pollen record, reaching back into Pleistocene time. The nearly continuous sediment record, from 44,000 years ago to the present, reveals major climate-related vegetation changes. At the very end of the Ice Age, from 14,000 to 12,000 years ago, a spruce-hickory forest was present at the lake, exhibiting conditions similar to those of southern Canada today. Continued changes in pollen fossils document warming throughout most of Holocene time, until the modern turkey oak and longleaf pine forest was established 7,600 years ago.

Miocene-age sabal palm frond from Alum Bluff Group sands at Alum Bluff.

About 8 miles east of Hosford, Florida 20 crosses the Ochlockonee River. On the east side of the river and downstream of the hydroelectric dam is Jackson Bluff, an outcrop famous for its rich deposits of mollusk fossils. The Torreya Formation of Miocene age and the Jackson Bluff (the densely fossiliferous shell bed) and Miccosukee formations of Pliocene age are exposed in this outcrop. Well over two hundred molluscan species occur in the Jackson Bluff Formation, as well as many corals and other invertebrates, representing shallow marine and brackish depositional environments. The mollusks of the Jackson Bluff Formation have proven to be very important to paleontologists in their reconstruction of the geographical distribution of sea life in Pliocene time in the Gulf

Coast and Caribbean regions. You need permission to access the exposure at the hydroelectric dam, but the Torreya Formation is well exposed along the east bank of the river below the dam. You can easily access the Ochlockonee River by canoe from Florida 20.

About 0.5 mile east of the Florida 20 bridge, take Florida 375 south for 3.3 miles to Forest Road 320, a dirt road. Turn west (right) on Forest Road 320 and proceed 1 mile to Rock Bluff Scenic Area. This popular canoe landing is located about 2 miles downstream of Florida 20 on the Ochlockonee River. Both the Torreya Formation and the shelly Jackson Bluff Formation are exposed here, and there are large, sandy point bars inside the river bend.

Although no longer accessible, another important fossil site in the Torreya Formation—the Seaboard Airline Railroad Site—occurs on the south side of Tallahassee, not far from the Florida State University football stadium, between Jackson Bluff Road and Roberts Avenue. In the early 1960s, the construction of a spur in a railroad switchyard exposed a fossil oyster bed with abundant vertebrate remains, including fossil horses (*Parahippus leonensis*) and camels (*Oxydactylus* species), as well as many rodents, birds, and fish (sharks, rays, and bony fish). It is sites like this, with both marine and terrestrial fossil remains, that precisely locate former shorelines—in this case, a shoreline of Early Miocene time.

US 98

Pensacola—Panama City

95 miles

Welcome to the most beautiful beaches in the world—the Emerald Coast of the Florida Panhandle. The water is indeed green, especially on a sunny day. Its color is related to the preferential scattering and reflecting of green wavelengths of light over white sand in shallow water. Visitors regularly comment, "The beach looks like snow!" and "It's like sugar!" Why are the beaches so white? They are almost pure quartz sand, the residual mineral crumbs of untold volumes of decomposed igneous and metamorphic rock of Paleozoic age. The sand has been eroded, washed, deposited, re-eroded, sorted, and redeposited countless times in rivers and along shorelines. The great hardness of the quartz sand grains makes them nearly immortal at the Earth's surface, while other softer, rarer minerals are removed through chemical and physical weathering. The Florida Panhandle's beaches are regularly ranked among America's top ten beaches because of their sand. Visitors from around the world testify that there's nothing quite like the panhandle's beaches anywhere.

It Came from the Bog— Carnivorous Plants of the Panhandle

A unique geobotanical environment found across much of the highlands and sandy lowlands of northern Florida, and along the bases of sand ridges in central Florida, is what is called a *seepage wetland*—where cool, moist soil is continually dampened by the seepage of groundwater. A marsh is similar, but it may periodically dry out during times of low rainfall. Seepage wetlands develop on slopes that dip below the local water table, allowing a constant flow of groundwater to reach the surface. Many seepage wetlands are also called *bogs*. These areas are usually heavily vegetated, and peat may accumulate in their soggy bottoms.

Among the great variety of bog vegetation is a group of fascinating plants that thrive in the highly acidic bog soils—carnivorous plants! Sundews, butterworts, bladderworts, and the famous pitcher plants are among the flora that have devised a variety of ways to entrap insects and other prey on sticky, fuzzy leaves. Carnivory allows these plants to secure a much-needed supplemental food source in the acidic, nutrient-poor, soggy soils. The large pitcher plants compose a primitive group of plants with origins that extend back to Cretaceous time, some 80 million years ago. Seepage bogs with pitcher plants are found in the Florida Panhandle as far east as the Apalachicola National Forest. Sites of interest with pitcher plants include Tarkiln Bayou Preserve State Park, southwest of Pensacola on Florida 293, and Blackwater River State Forest, near Holt. Yellow River Marsh Preserve State Park, which boasts the largest community of pitcher plants in the state, is located in Santa Rosa County on Garcon Point. This park is between Escambia Bay and Blackwater Bay on County Road 191, about 1 mile north of the intersection with County Road 281.

The unusual, carnivorous pitcher plants from the Florida Panhandle.

Perdido Key, southwest of Pensacola, is the last bit of Florida's northwestern coast, and it actually lies partly in Baldwin County, Alabama. It is part of a deeply eroded beach ridge plain that is more than 2,000 years old. Extensive salt marsh environments are common on the landward side of this barrier island. Big Lagoon State Park is an especially good place to observe salt marsh habitat and tidal creeks. Beach ridge topography is very evident along County Road 292A, just east of the park entrance.

As you cross the Pensacola Bay Bridge on US 98, be sure to look southeastward to the southern shore of Pensacola Bay. The tall, unvegetated bluff along the southern shore is a sand dune that formed in latest Pleistocene time. This dune reaches up to 50 feet in elevation and is composed of a windblown cap of sand over an older Pleistocene-age barrier island. This impressive dune is made of very fine quartz sand and is only a remnant of what was probably a much larger dune.

If you have time, the dune is worth visiting. From US 98, take the northbound access road within the Naval Live Oaks Reservation (located 1.3 miles from the eastern boundary, and 0.8 mile from the western boundary of the reservation) about 0.4 mile to a parking area and picnic pavilion. There is a trailhead to the left of the parking lot. About 50 yards up this trail is another trailhead, and about 250 yards farther is yet another trail. The first trail leads to a beach along the southern end of Pensacola Bay. You can walk westward along the beach to the dune. The second trail leads to the top of the dune.

A dune of late Pleistocene age at Naval Live Oaks Reservation, east of Gulf Breeze.

Indian Temple Mound, a pyramidal ceremonial mound dating to around AD 1200, is off US 98 in Ft. Walton Beach. It is part of a late pre-Columbian (prior to Spanish contact) coastal culture that thrived in the Ft. Walton and Pensacola areas and westward to Mobile Bay (Alabama) and as far as Louisiana. The mound was strategically located for optimal access to the gulf, narrow Santa Rosa Sound, and the larger Choctawhatchee Bay.

Like several other estuaries in the panhandle, Choctawhatchee Bay is essentially a drowned river valley, but it was modified by the westward growth of the Moreno Point sand spit during late Pleistocene or early Holocene time. Such a sand ridge is called a *baymouth bar.* A thick bed of oyster and other mollusk shells is located just a few feet below the surface sediments in the bay, and radiocarbon dates indicate the bed is approximately 3,000 years old. It may have been at this time

SIDE TRIP TO SANTA ROSA ISLAND

County Road 399, which branches off US 98 at Gulf Breeze, traverses Santa Rosa Island, part of Gulf Islands National Seashore. At about 46 miles long, this is one of the longest barrier islands along the Gulf Coast, second only to the famous Padre Island of Texas. Santa Rosa Island is Holocene in age, although it formed on an older Pleistocene-age barrier island. Westward-flowing longshore drift delivered sands that accreted to the older island. On the far western end of the island is historic Fort Pickens, built in 1834, which was held by Union forces throughout the Civil War but saw minimal action. The Apache chief Geronimo was imprisoned here between 1886 and 1888.

Some of Santa Rosa Island's dunes reach 40 feet above sea level. Pristine dune fields are especially well developed along the 7-mile stretch of County Road 399 through the protected Gulf Islands National Seashore, between Pensacola Beach and Navarre Beach. Hurricanes periodically interrupt dune growth on Santa Rosa and similar barrier islands. They generate storm surges and waves that erode through or pass over the island, creating lobes of sand called *washover fans.*

The beaches of Santa Rosa Island are a popular tourist destination, but they are also known for their deadly rip currents. Nearly every year, several unsuspecting swimmers are tragically drowned by these powerful currents. Never swim without a dependable flotation device, and be aware of local surf conditions and warning flags.

that Moreno Point and Santa Rosa Island joined one another, isolating Choctawhatchee Bay and transforming it into a brackish or freshwater dune lake. In 1928, Destin (East) Pass between the east end of Santa Rosa Island and Destin, at the western tip of Moreno Point, was opened, allowing more saltwater exchange between the bay and the gulf.

At several places around the shore of the bay, and also along the entire panhandle coast, you may see sand that is cemented with brown or black material. It's called *humate sand.* The humate is an organic material derived from the decay of plant debris and transported by groundwater that percolates through soil and sediment. As the groundwater moves through the porous coastal sands, the carbon-rich humate comes out of solution and cements quartz sand into a crumbly sandstone. There are other dark sands along the shore that should not be confused with the black humate sands. Thin layers of heavy-mineral sands are also common among the otherwise white quartz sands. Heavy-mineral sands

DUNES AND DUNE LAKES—A SIDE TRIP ON FLORIDA 30A

Florida 30A parallels US 98 for almost 20 miles and offers access to state parks with many sand dunes and dune-impounded lakes. A beautiful dune field and two large dune lakes may be seen at Topsail Hill Preserve State Park, which is near the intersection of US 98 and Florida 30A, just east of Sandestin. Perhaps the best view of the park's geology is at a secondary entrance, located on US 98 about 1.4 miles west of the Florida 30A/US 98 intersection (eastbound access only). Here, take Topsail Road about 0.75 mile down to a picnic area and an entrance to a dune field and beach. The park's main entrance is on Florida 30A, immediately after you turn south off US 98. A tram at the park will take you down to a boardwalk that crosses over a dune field to the beach.

There are two relatively short nature trails (accessible only from the secondary entrance) to Morris Lake and Campbell Lake, the two large dune lakes in the park. These coastal lakes are flanked on their southern sides by dunes. A blackwater creek connecting Morris Lake to the gulf—and seeming so out of place on a beach—flows intermittently. East of Topsail Hill on Florida 30A, additional dune lakes and relict dune fields may be seen at Dune Allen Beach and Blue Mountain Beach, and in their vicinity. Nearby, the north-south county roads 393 and 83 pass over dunes of Pleistocene age.

Additional outstanding dune lakes and dune fields are at Grayton Beach State Park, located about 0.7 mile east of the Florida 30A/County Road 283 intersection. Western Lake is the large dune lake in the park. Perhaps the best view of the lake, however, is from US 98 in front of the park. Here, the tall,

Cross-stratified dune sands along Barrier Dunes Nature Trail at Grayton Beach State Park.

confining dunes are especially visible in the distance along the southern shore of the lake. Western Lake, which maintains a strong hydraulic connection with the gulf, is brackish and has a well-developed and extensive salt marsh. The barrier dune nature trail in the park is an especially good geological tour through a coastal dune environment. Here, you can see cross-stratification (excellent dune cross-strata are visible near trail marker 7), wind ripple marks, and migrating dunes. The unique and diverse flora of the dunes consists of many salt-tolerant species, including the common sea oats (*Uniola paniculata*). Because they trap wind-borne sand and hold it in place, sea oats and other salt-tolerant species are important for allowing dunes to build and in controlling dune erosion. The sand at Grayton Beach is especially rich in heavy minerals, giving the beach a salt-and-pepper appearance. These heavy minerals have been exposed by erosion along this beach and were most likely deposited during a pulse of heavy mineral deposition in the gulf in Pleistocene or early Holocene time.

Eastern Lake, another dune lake, lies 4.5 miles east of Grayton Beach on Florida 30A, and 5.8 miles east of Grayton Beach, Florida 30A crosses yet another coastal dune and dune lake environment at Deer Lake State Park. Deer Lake has perhaps the most extensive dune field of any of the coastal lake areas. A long boardwalk extends through the wide dune field. It is a barren yet beautiful relict landscape. The boardwalk ends at the crest of an eroded dune, which descends perhaps 40 or 50 feet to the beach, offering an impressive view. Cross-strata are very evident in the dunes and the beach is rich in heavy-mineral sands. On a blustery, overcast winter day, the waves in the gulf can be high, the sea covered with mist, and the winds strong across the high dunes. Little imagination is needed to transport yourself back to cooler late Pleistocene time, when these dunes first formed. The beach has been heavily eroded in recent years, but it represents one of the few unspoiled stretches of coastline in the panhandle. About 4.5 miles east of Deer Lake, Florida 30A rejoins US 98.

Dunes of Holocene age at Deer Lake State Park.

Dark, heavy-mineral sands along the coast at Deer Lake State Park. This layer of heavy minerals has been exposed by beach erosion.

consist of a variety of minerals that are denser than quartz. Examples are ilmenite, rutile, zircon, staurolite, garnet, tourmaline, corundum, spinel, and topaz. These sands are black or brown, and when mixed with white quartz sand, they create a salt-and-pepper appearance. Nearshore currents concentrate them in low areas.

From Destin to Panama City, there are no barrier islands but rather a smooth, concave, coastal strip of nearly pure quartz sand with numerous large dune lakes that are absolutely unique to this coastline. They formed as coastal creeks became impounded behind spits, baymouth bars, beach ridges, and especially large dunes, all of which are very common here because of the great abundance of quartz sand. Examples include Fuller, Morris, and Campbell lakes. The dune lakes may also have formed as a result of humate-cemented sand beneath the lakes. Humate sand is common here and may create a relatively impermeable layer that holds the water at the surface. Geologists call such water *perched* because it rests above the normal water table.

The dune lakes are home to some unique biological communities, and periodically they discharge freshwater into the gulf through narrow blackwater creeks. These dune lakes appear to primarily be associated with older, mature dune fields, some of which are Pleistocene in age. Some of the dunes rise to heights of 70 feet or more and originally formed at the very end of the Ice Age, when sea level was far lower and a drier climate prevailed. Some of the lakes can be accessed off US 98 east of Destin, but the majority can be easily seen along coastal Florida 30A to Panama City (see **Dunes and Dune Lakes—A Side Trip on Florida 30A**). After you pass County Road 83, look south for a view of Western Lake. The tall, confining dunes that impound the lake are especially visible in the distance along the southern shore of the lake.

Offshore of the panhandle, at depths up to 100 feet, are many limestone ledges or outcrops, some of which extend up to 200 feet in length and rise 10 or more feet off the seafloor. These ledges, often called "reefs" by divers, provide a hard substrate for encrusting sponges, hard and soft corals, and other marine life. These ledges are popular scuba diving locations, and the encrusted rock was once popular as the "Destin live rock" of saltwater-aquarium hobbyists; however, harvesting of live rock is now prohibited. Limestone exposures such as these have not been adequately mapped, so geologists do not know what rock formation they represent.

Camp Helen State Park, which has an extensive dune field, is just on the west side of Powell Lake, the largest of the dune lakes in this region. There is an exceptionally wide beach from the base of the dunes to the shore, and a long nature trail extends through the dune field. Powell Lake has the distinction of having its own lake monster—a small Loch

Ness–type creature, complete with a long neck, large head, rounded flippers, and a strange dorsal fin. Last seen in 1959 and associated with sightings of sharks and other animals, the "creature of Powell Lake" fortunately lives on in our imaginations, if nowhere else.

In the Panama City Beach area, St. Andrews Bay, along with West, North, and East bays, are drowned river valleys. The Econfina River is the primary source of freshwater. Shell Island, which in part forms St. Andrews Bay and is accessible only by boat, is composed of three sets of shore-parallel ridges of sand called *beach ridges.* Beach ridges usually form by the action of wave swash, which transports sand up the beach face to make a berm. The oldest of the beach ridges on Shell Island, located on the north side of the island, are separated by wide swales containing peat deposits, which formed in marshes that were behind the beach ridges. Shell Island was formerly part of a sand spit, but construction of Land's End Canal on its western side separated it from the mainland in the 1930s. The western end of the former spit is now St. Andrews State Park, which you can reach from Florida 392. The park is about 3 miles east of the town of Panama City Beach. Ridge-and-swale and dune topography is well exposed here.

The coastline in this area has been especially dynamic in recent centuries. Modern dunes are migrating over the ancient ridges on Shell Island, draping them with a layer of fine windblown sand. Most of Crooked Island, offshore of Tyndall Air Force Base and northwest of Mexico Beach, has emerged from the sea only since 1779. It too is composed of beach ridge deposits.

US 98

Panama City—Port St. Joe—St. Marks

126 miles

The very distinctive portion of the Florida Panhandle between Panama City and St. Marks traverses the impressive Apalachicola delta. The delta has been building seaward into the Gulf of Mexico since as early as Late Miocene time, and relict coastal features are found far upriver. This long history of delta advance has resulted in a protruding, pointed coastal landform called a *cuspate foreland.* Coastal processes modified the large cusp in Holocene time, forming many barrier islands and spits. An extensive spit-barrier complex surrounds the ancient delta. Today, the Apalachicola River terminates as a bayhead delta in Apalachicola Bay.

US 98 between Mexico Beach and Port St. Joe parallels several Holocene-age beach ridges, many of which are visible along the highway, especially around St. Joe Beach. Wetlands develop in the low swales between the ridges. Because of its high, white dunes, the northern end of St. Joseph spit is remarkably visible from US 98 between St. Joe Beach and Port St. Joe. Also called St. Joseph Peninsula, the spit was two separate barrier islands about 5,000 years ago. They merged some 1,000 years ago at Eagle Harbor, near the middle of the peninsula, to form the modern 11-mile-long sand spit. Sediment transport along this spit by longshore drift continues today, primarily in a northward direction, so the spit is still growing. It is constructed of a series of beach ridges and has a large dune field. Some dunes are over 30 feet tall.

Reachable only by boat, the unusual St. Vincent Island, home to a national wildlife refuge, is probably the oldest barrier island in the state. The oldest part of the island is some 4,400 years old. It has grown in size by the regular accretion of beach ridges, giving the island an appearance similar to a washboard or a ruffled potato chip. There are twelve sets of ridges and up to 180 individual ridges on this island. Each set of beach

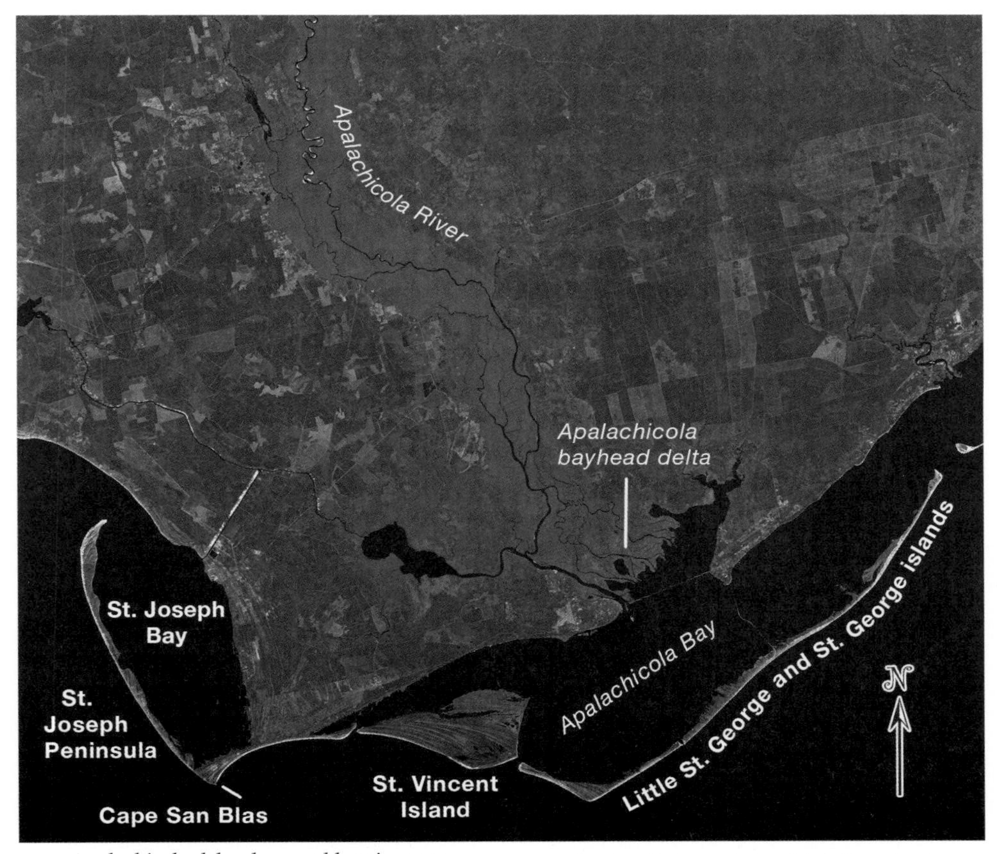

Apalachicola delta, bay, and barrier system. —Courtesy of the U.S. Geological Survey (labeled by authors)

ridges records a time of beach growth along a particular area of the island, and perhaps fluctuating sea levels over the past several thousand years. The beach at St. Vincent Island is densely covered with seashells.

The Holocene-age St. George Island is accessible by taking Florida 300 from Eastpoint. St. George Island State Park is at the northeastern end of this barrier island. Beach ridges are well developed on the eastern end of the island, especially around Gap Point. Migrating sand dunes (for example, Sugar Hill) are covering some of these ridges. Little St. George Island, or Cape St. George Island, which is west of St. George Island and accessible only by boat, is constructed of parallel beach

SIDE TRIP TO ST. JOSEPH PENINSULA

To reach St. Joseph Peninsula, follow Florida 30A approximately 6.6 miles south of its intersection with US 98 in Ward Ridge. Turn west on Florida 30E. St. Joseph Peninsula makes a dogleg at Cape San Blas, which, like the peninsula, is constructed of beach ridges and dunes. Geologists used radiocarbon dating to determine that a peat layer under the oldest set of ridges is 750 years old. The southern end of the peninsula is eroding badly, largely a result of hurricane activity. A large pile of boulders has been placed along the beach to prevent erosion, which, if allowed to progress, would soon turn the peninsula into a barrier island with no road access.

Tree stumps exposed along the southern end of St. Joseph Peninsula—the result of extensive beach erosion.

Peat exposed by erosion along the southern end of St. Joseph Peninsula. The peat layer extends under the beach to the salt marsh on the bay side of the island.

Along the gulf side of the southern end there are many large tree stumps on the beach that were once part of a coastal marsh. A thick layer of peat—the organic soil and muck of the same salt marsh—has also been exposed on the beach by erosion. The salt marsh existed behind the beach in the past, but erosion has exposed it to the open ocean.

With the exception of the stable northern tip of St. Joseph Peninsula, this barrier complex has been subject to much erosion in historic times, particularly around Cape San Blas, parts of which have been eroding at rates of 15 to 25 feet per year over the past 150 years or so. These are the highest coastal erosion rates in the entire state. Consequently, the lighthouse on Cape San Blas has been relocated six times. Several distinct offshore sand shoals are located southwest of Cape San Blas in depths of less than 50 feet. These shoals are most likely former spits, barriers, or other coastal features that have been eroded or submerged.

Behind the spit and cape is the aquatic preserve of St. Joseph Bay, a shallow, non-estuarine lagoon—one that is not diluted by significant river input. The bay is surrounded by coastal salt marsh wetlands in the intertidal zone, and there are extensive sea grass communities in the shallow subtidal areas. Both of these environments are critically important to the marine ecosystem of this region. Eagle Harbor cove, on the southern end of the peninsula, is quite shallow and affords an excellent opportunity for snorkeling to observe the abundant marine life. St. Joseph Peninsula State Park is located near the middle of the peninsula. The remote, pristine beach and high dunes in the park are worth the trip.

Beach ridge sets of Holocene age on St. Vincent Island. There are twelve distinct sets of ridges and at least 180 individual ridges, each recording a former position of the shoreline. —Courtesy of the U.S. Geological Survey

ridges. It was separated from the main island in 1954 when a channel was dredged to facilitate boat traffic. A kitchen midden with Fort Walton culture pot shards up to 800 years old was discovered atop a peat layer on this island. Radiocarbon dates indicate the peat is 480 to 1,545 years old.

Until recently, Cape St. George Lighthouse stood on the southern tip of Little St. George Island. Although it was built about 1,500 feet from the shore in 1852, the eroding coastline brought the old lighthouse increasingly closer to the surf. In 1992, Hurricane Andrew deeply eroded the island's beach, and the lighthouse was deactivated in 1994. After Hurricane Opal in 1995, the lighthouse began to lean significantly. Further beach erosion from Hurricane Earl in 1998 exposed the lighthouse to the surf zone. A heroic local effort restored the lighthouse to an upright position, but erosion continued to undermine the foundation. The lighthouse finally collapsed around 11:45 a.m. on October 21, 2005. Having watched over the gulf for 153 years, Cape St. George Lighthouse became a testimonial to the rates and inexorability of geological processes. This lighthouse was, in fact, the third on St. George. The first, built in 1833, was destroyed by a hurricane in 1846. The second fell after taking a beating from two storms, one in 1850 and the other in 1851. But as one part of the coast erodes, another emerges. The isolated Dog Island, a barrier island east of St. George Island, made its appearance between 1951 and 1952 due to shoaling of a submarine sandbar.

Between Little St. George and St. George islands and the mainland is Apalachicola Bay, a bar-built estuary. Estuaries like this form when barrier islands or spits isolate a body of river-fed coastal water. The Apalachicola River discharges into the bay, which is one of the most productive fisheries in the country, particularly for oysters—90 percent of the state's oyster harvest, and 10 percent of the national harvest, comes from Apalachicola Bay.

East of the Apalachicola River is an extensive wetland called Tates Hell Swamp, which formed on delta plain sediments of Pleistocene age and the Intracoastal Formation of Pliocene age. This swamp is named for Cebe Tate, who in 1875, armed with shotgun, knife, and dogs, set out to kill a livestock-raiding panther. Tate got lost in the swamp, lost his dogs to the panther, and was bitten by a rattlesnake. Drinking swamp water to curb his thirst and dragging a swollen, poisoned leg behind him, he managed to follow a creek 8 miles to the coastal town of Carrabelle.

From Apalachicola to St. Marks, US 98 traverses an ancient shore with many relict beach ridges and sand dunes. Royal Bluff, in the town of the same name, is a dune that formed at approximately the same time—late Pleistocene time—as the relict dunes between Destin and Grayton Beach, such as Topsail Hill and the dune field of the town of Blue Mountain Beach. Large-scale cross-stratification related to wind processes can be seen at Royal Bluff.

Twelve miles northeast of Carrabelle on US 98 is the small cusp of Turkey Point, which is the location of the Florida State University Marine Laboratory. You can visit extensive mudflats and oyster beds at low tide on the west side of the facility. You can also collect dolomitic rocks and large fossil molds of bivalve mollusks from boulders of the St. Marks Formation of Early Miocene age that have been dredged from the nearby channel and placed here. The Ochlockonee and Sopchoppy rivers enter nearby Ochlockonee Bay, another drowned river valley. The Miocene-age Torreya Formation is exposed along some high banks of the Sopchoppy River and is visible near the bridge north of Oak Park if you make a short side trip. About 4.5 miles north of Sopchoppy on Forest Road 365, head west on Forest Road 346/343 to the bridge over the river.

At the Franklin-Wakulla county line is the mouth of the Ochlockonee River. There is a very significant change in coastal geology after you cross the bridge over Ochlockonee Bay. The entire panhandle coast is mantled by an abundance of quartz sand. But this changes abruptly around the town of Panacea, where the surface rock is the limestone and dolostone of the Woodville Karst Plain, a part of the Ocala Karst District. There is a thin veneer of quartz sand over the limestone, but not enough to construct barrier islands or sandbars. The absence of barrier islands and sand spits is obvious on any map. The broad, arcuate coastline here, called the Big Bend, continues south to Pasco County, north of Tampa.

Wakulla Springs State Park

Ten miles north-northwest of St. Marks, on Florida 363 and Florida 267, or 13.5 miles south of Tallahassee, is the famous Edward Ball Wakulla Springs State Park, one of Florida's greatest natural attractions. The main spring pool, combined with nearby Sally Ward Spring, forms the headwaters of the Wakulla River. Combined, the two springs gush nearly 175,000 gallons of freshwater every minute, an average 250 million gallons per day, making Wakulla Springs one of the largest spring systems in Florida and the world. The entrance to the underwater cave from which the spring emanates is over 100 feet in diameter. Divers have mapped more than 148,000 feet of passages, including two passages that are each more than 14,000 feet long. Wakulla is supposedly a Seminole word meaning "strange and mysterious waters."

Divers prepare to recover mastodon remains in the fall of 1930 at Wakulla Springs. —Courtesy of the Florida Geological Survey Photo Archives

A hard-hat diver with a mastodon mandible at Wakulla Springs.
—Courtesy of the Florida Geological Survey Photo Archives

Between 1955 and 1958, scuba divers first explored the cave, or main spring vent. They traveled 1,100 feet from the main spring entrance and reached depths of over 200 feet. A massive exploration effort was mounted in 1987 by the U.S. Deep Caving Team. The Wakulla Springs Project, as it was called, determined that the spring's primary conduit extends some 900 feet and then splits into five separate tunnels that are mostly elliptical in shape. One of the tunnels extends an additional 4,500 feet from the main spring vent and is 360 feet deep. The walls show much scalloping, indicating turbulent water flow in the past. The conduits connect with the Leon Sinks Geological Area to the northwest. Before the 1987 exploration, divers had mapped conduits in these two systems to within 3 miles of each other. Dye tracing tests in 2004 confirmed that the sinkholes, swallets, and cave systems of Leon Sinks and the springs at Wakulla Springs are connected, and in July 2007, scuba divers found a connection between the two systems. The combined cave system is thought to be one of the longest underwater cave systems in the world. Efforts are now focused on finding a link between the Wakulla system and the Spring Creek cave system just offshore in the Gulf of Mexico.

The upper 90 feet of the cave at Edward Ball Wakulla Springs State Park, and the ledge of the main spring pool, consist of the St. Marks Formation of Early Miocene age, a limestone with abundant foraminifera, mollusks, and colonial corals. But the majority of the cave, including all of the mapped conduits, developed in the Suwannee Limestone of Early Oligocene time, which also has abundant foraminifera, mollusks, and colonial corals.

On the spring and cave floor there are fossil remains of many mammals of late Pleistocene age. In 1930, a diver recovered a nearly complete American mastodon (*Mammut americanum*) at a depth of about 25 feet. This mastodon, later affectionately named Herman, is on display at the Museum of Florida History in Tallahassee. Also recovered were remains of mammoths, ground sloths, tapirs, deer, camels, horses, bison, giant armadillos, and charred wood. Divers have discovered three so-called bone beds, one as far back as 1,250 feet from the cave entrance and up to 285 feet deep. The dive team named the deepest bone bed the Mega-Fauna Mausoleum. There is ample evidence that humans occupied the site from the time of the Paleoindians, about 12,000 years ago. Over six hundred spear points made of bone were recovered in the cave.

The geologic history of this cave is fascinating. First, the Suwannee and St. Marks limestones formed in tropical seas during Oligocene and Miocene time. These limestones were subjected to chemical dissolution (karstification) and other erosive forces, probably during Pleistocene time (the Pliocene record has been eroded here but is found farther north). Curiously, the cave has no speleothems, suggesting that the karst may have developed underwater. Throughout much of late Pleistocene time, during low sea levels, large mammals either wandered into the cave (then dry), fell into sinkholes or sinkhole lakes, were washed into the cave and spring, were trapped there, or were taken there in some fashion by Paleoindians. With global warming and rising sea level after the Ice Age, the water table rose, submerging the cavern and creating the modern springs.

Cross-sectional diagram of the cave entrance at Wakulla Springs.
—Courtesy of the Florida Geological Survey

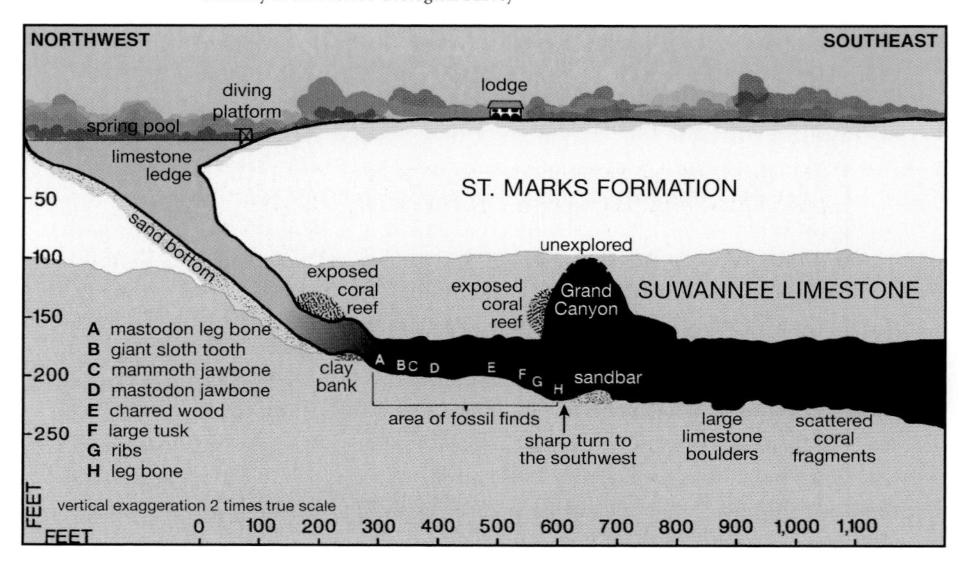

The Gill Man, also known as the Creature from the Black Lagoon, *a late Holocene inhabitant of Wakulla Springs.* —Courtesy of the State Library and Archives of Florida

Along with the Ice Age megafauna, there have been more-recent but no less "prehistoric" denizens of Wakulla Springs that are no longer with us. This list includes Old Joe, the docile, 11-foot alligator that served as mascot and sentinel of the springs for untold years (and now resides, permanently, in the lobby of the historic park lodge); Tarzan (several films were made here in the 1940s); and the Gill Man from the 1954 horror flick *Creature from the Black Lagoon* (sequels were shot at both Wakulla Springs and Silver Springs, near Ocala).

San Marcos de Apalache Historic State Park

San Marcos de Apalache Historic State Park at St. Marks is 2 miles south of US 98 on Florida 363. Located at the confluence of the Wakulla and St. Marks rivers, this strategic site has been fortified, occupied, attacked, captured, burned, and reoccupied numerous times between 1528 and 1861 by such notables as Panfilo de Narvaez, Hernando de Soto, William Augustus Bowles, and General Andrew Jackson, as well as assorted pirates and Confederates. All that remains of earlier fortifications, including the Spanish fort constructed between 1750 and 1785, are large blocks of the Miocene-age St. Marks Limestone. The St. Marks is rich in mollusk fossil molds and contains fossilized corals in exposures near the town of St. Marks.

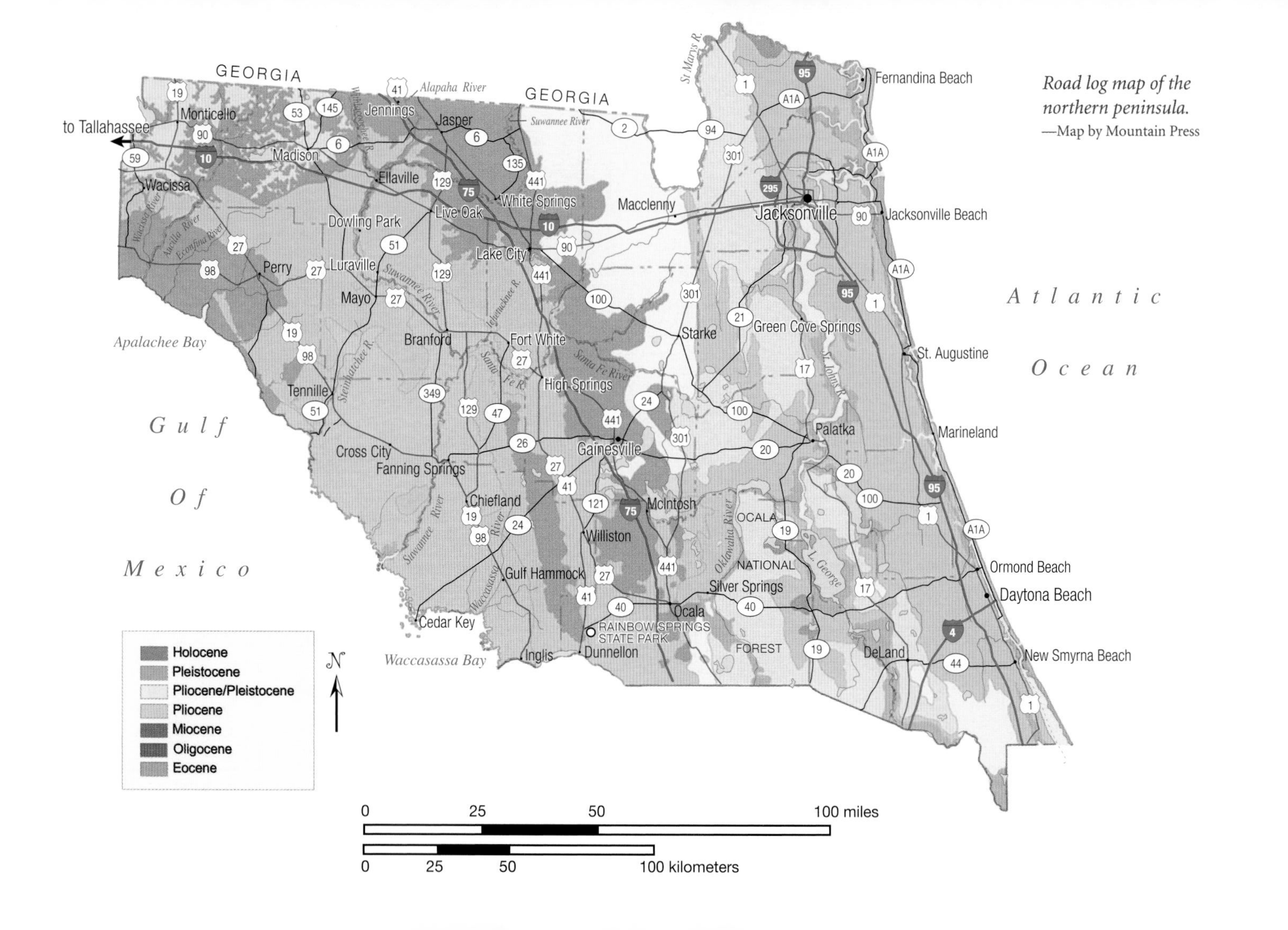

Road log map of the northern peninsula.
—Map by Mountain Press

Northern Peninsula

Upon the Suwannee River

It is not clear whether Stephen Foster ever saw the Suwannee River, but his melancholy 1851 folk tune about Southern plantation life, "Old Folks at Home" ("Way down upon the Suwannee River . . ."), is the state song of Florida, and it has forever endeared this river in the American consciousness. For many Floridians, the Suwannee River is nothing less than a state icon of old Florida. Just attend the Florida Folk Festival at the Stephen Foster Folk Culture Center State Park in White Springs, along the banks of the Suwannee, and you will soon realize that the idea of the Suwannee River has reached almost mythic proportions. But the river's namesakes carry geological meaning as well—Suwannee Suture, Suwannee Basin, Suwannee Channel, Suwannee Limestone. These too have significance for Florida—very old Florida.

The origin of the name "Suwannee" is uncertain, but two ideas have been circulated. Some claim that the river was named after an early mission, San Juan de Guacara, which was later corrupted to San Juanee, then to Suwannee. Others contend that the name is derived from a Native American word, *Suwani*, meaning "echo river." Whatever the case may be, both theories reflect that the region of northern Florida around the Suwannee River has a long and colorful history. Humans have inhabited the Suwannee River basin for at least 11,000 years. Evidence of past cultures is found along and in the river and is concentrated near the many springs along the Suwannee. Projectile points and other stone tools are abundant; the chert from which they were made is found along the river. Paleoindian, Archaic, Woodland, and Mississippian cultures left behind artifacts from the headwaters of the Suwannee to its mouth. Hernando de Soto's expedition in 1539 as well as seventeenth-century Spanish missions represent the early European influence in this region. Naturalist William Bartram visited the Seminole Indians along the Suwannee in 1773, and General Andrew Jackson led raids here to recover runaway slaves at a Seminole Indian village. Then there was the golden age of riverboats in the late 1800s. The remains of several of these steam paddleboats now repose on the river bottom. The youngest of these, the 141-foot-long *City of Hawkinsville*, was in service as late at 1922. It now lies south of the railroad trestle at Old Town.

With headwaters in Georgia's Okefenokee Swamp and its mouth at the Gulf of Mexico, the Suwannee River flows for over 200 miles across much of the northern end of the Florida Peninsula. This northern part of the peninsula offers a diverse sampling of the state's geologic record, including ample exposures of the Eocene-age Ocala Limestone, the Oligocene-age Suwannee Limestone, fossiliferous and phosphatic sediments of the Miocene-age Hawthorn Group, Pliocene-age sands, Pleistocene-age dunes and coquina rock, and Holocene-age beach ridges and barrier islands.

US 90 and Interstate 10

Tallahassee—Lake City

104 miles

East of Tallahassee to Madison, US 90 and I-10 follow the southern margin of the Tallahassee Hills and Madison Hills, part of the Tifton Uplands District—an extensive delta plain complex composed of the Torreya Formation of the Hawthorn Group of Miocene age and the Miccosukee Formation of Pliocene age. Farther east, to Lake City, these uplands comprise the Okefenokee Basin District, the Ocala Karst District, and the Central Lakes District. Undifferentiated sediments of Quaternary age and the Statenville Formation of the Hawthorn Group occur closest to the surface in the eastern segment of this route.

A Lonely Trilobite in Florida

In 1944, the Hunt Oil Company drilled a deep well in Madison County, and between 5,154 and 5,162 feet, drillers encountered a trilobite, *Plaesiacomia exsul*, from Middle Ordovician time (470 to 458 million years ago) in a dark gray shale. This modest little trilobite, the only one ever found in Florida, was preserved as an impression (external mold). Though it is only about 0.625 inch long, it is of enormous geologic significance. It belongs to a group of trilobites of Ordovician age that are primarily found in northern Africa and central and southern Europe, not North America. *Plaesiacomia exsul*, along with many other fossil invertebrates found in Florida's subsurface, is strong evidence that during Paleozoic time, the bedrock of Florida was not part of the North American continent but part of the megacontinent Gondwana, from which Africa, South America, Antarctica, Australia, India, and the Middle East were born.

The Gondwanan trilobite Plaesiacomia exsul *of Ordovician age.* —Modified from Lane (1994) and courtesy of the Florida Geological Survey

About 15 miles east of Tallahassee and south of US 90 (about 6 miles west of Monticello) is Letchworth Mounds Archaeological State Park. This Woodland Period site has four mounds of the Weeden Island culture (about AD 200 to 800), including burial sites. The largest, Letchworth Mound, is the highest ceremonial mound in Florida at 46 feet. From I-10, head north on Florida 59 2.5 miles to US 90, then go 2 miles east on US 90 to Sunray Road. Turn south to the mounds.

Withlacoochee and Alapaha Rivers

North of Madison, an easily accessible exposure of the Suwannee Limestone can be observed along the Florida side of the Withlacoochee River, beneath the Florida 145 bridge at the Florida-Georgia line. (There are two Withlacoochee Rivers in Florida, the southern Withlacoochee is in the west-central peninsula.) From 5 to 10 feet of rubbly limestone is continuously exposed along the river at this bridge. This is the put-in point to begin a canoe trip on the Florida portion of the Withlacoochee. Some canoeists consider it to be one of the finest paddles in the state. Another good Suwannee Limestone locality—and a popular swimming hole—is Madison Blue Spring, a first magnitude spring at the Madison-Hamilton county line. It's on Florida 6, about 8 miles east of the US 90/Florida 6 intersection in Madison. Phosphatic sands of the Statenville Formation of the Hawthorn Group are also exposed here along the west bank of the river near the bridge, where scattered vertebrate fossils can be found. The Suwannee Limestone is also visible from Interstate 10 when crossing the Suwannee River.

The section of the Withlacoochee River between Madison Blue Spring and the Georgia-Florida state line (and farther north, up to US 84 in Brooks and Lowndes counties, Georgia), is a special locality for rock hounds. This stretch of river is one of the best collecting grounds for Florida's state stone—agatized coral—and gem enthusiasts have amassed impressive collections from this area. The fossilized coral colonies are primarily eroding out of the Suwannee Limestone along the banks. Over time, they become concentrated in the river channel. In some areas the corals form rocky shoals, and some fossilized coral colonies are several feet in diameter. Corals of this size and abundance must have formed patch reefs in this region during Early Oligocene time. When cut with a diamond-bladed rock saw and polished, the corals form beautiful geodes with a deep brown to burgundy, bulbous, layered chalcedony called *agate*. Some specimens have small quartz crystals.

The Alapaha River, east of the Withlacoochee, is also a tributary of the Suwannee River. The Suwannee Limestone is exposed along most of the river's

Agatized coral from the Withlacoochee River, from the collection of Mr. Jerry Giles of Perry.

southern portion to its confluence with the Suwannee River northeast of Ellaville. The Hawthorn Group crops out along the river to the north up to the state line. Along that northern section, shoals composed of nonprecious opalized clays and carbonates are exposed when water levels are low.

The Alapaha is an intermittent stream. It may flow violently during flood stage or be completely dry and sandy during dry seasons. When water levels are low, parts of the river disappear through swallets. Two prominent swallets occur between Jennings and Jennings Bluff near the state line. The disappearing stream reemerges through the Suwannee Limestone at the Alapaha Rise, a first magnitude spring about 0.5 mile above the Alapaha's confluence with the Suwannee River, and flows 300 feet to the Alapaha channel. There are canoe access points along County Road 150, at Jennings Bluff, and along US 41 between Jennings and Jasper. The lower Alapaha is best accessed from the boat launch at Gibson Park on the Suwannee, just upstream of the confluence of the two rivers.

The Upper Suwannee River

Rivers are perhaps the best natural ally of the field geologist. Rivers cut deep valleys, exposing vertical cliffs of strata, and keep them swept clean of debris and vegetation with every flood. Of all of Florida's rivers, the Suwannee probably has the greatest variety of rock exposures of various ages along its banks. Numerous canoe guidebooks are available for the Suwannee. These are extremely useful, and we highly recommend them. The Suwannee River Water Management District also produces a very convenient map and guide to boat ramps and canoe launches on the Suwannee and its major tributaries, the Withlacoochee, Alapaha, and Santa Fe rivers. Several local outfitters provide canoe and kayak rentals.

No one road follows the Suwannee, so discussing this river in the Roadside Geology format is difficult. The following discussion is essentially a Riverside Geology of the Suwannee from the Georgia-Florida border to Suwannee River State Park, north of Ellaville on US 90. Since there are many good exposures of bedrock—and interesting historic locales—along the river that are visible or accessible from various roads, we mention them in this section as well. The southern portion of the Florida Suwannee River is discussed later in this chapter in **Exposures and Springs on the Lower Suwannee, Santa Fe, and Ichetucknee Rivers.**

The upper reaches of the Suwannee River, near the Georgia border, emerge from the Okefenokee Swamp. The few geologic exposures it offers are mostly of Quaternary-age sands. In northernmost Florida, the river is rather narrow, with high, sandy banks of recent origin. A few miles downstream of the Florida-Georgia border, exposures of Hawthorn Group sediments appear. The Hawthorn Group is well exposed along the section from Florida 6 south to near White Springs. The Statenville Formation of this group is exposed from near Florida 6 to south of Cone Bridge. It is of Middle Miocene age, 14 to 12 million years old, and consists of interbedded phosphatic sands, clays, and dolostones with abundant phosphate and dolostone pebbles and, commonly, dugong ribs. Since the 1960s, the Statenville Formation has been mined for phosphate southeast of Jasper off County Road 137. Vertebrate fossils are abundant in the mine, especially sharks and dugongs. Downstream of Cone Bridge, the Marks Head, Penney Farms, and Parachucla formations of the Hawthorn Group are exposed in various places. However, the formational contacts have not been adequately identified. The best places to see the Hawthorn Group along the river are from Florida 6, just north of the bridge, and from Cone Bridge Road, located off County Road 135 about 9 miles northeast of White Springs.

In places, the trace fossil shrimp burrow *Ophiomorpha nodosa* is abundant in the Hawthorn Group, indicating former nearshore depositional environments. Downstream from Cone Bridge to Big Shoals (5 miles upstream of White Springs), there are spectacular Hawthorn Group outcrops, including impressive cross-stratified phosphatic sandstones and conglomerates, with many rip-up clasts of rock from lower strata. Cross-bedding is a common and distinctive feature of the Statenville Formation, and in places trough cross-stratification is present, indicating deposition in nearshore channels called *longshore troughs.* At Big Shoals, rocky, rounded, and well-cemented exposures of the lower Hawthorn Group are visible. For about half of the year, approximately 1 mile of the Suwannee in this area becomes class III white water because of these rocks. A smaller set of rapids, Little Shoals, occurs downstream. Big Shoals is a favorite canoe or kayak run when water levels are appropriate, but when levels are too low, boaters must portage across the shoals. The Big Shoals Public Lands north entrance has a canoe launch and is easily accessible off County Road 135, about 3.7 miles north of US 41 in White Springs.

On the east side of White Springs, US 41 crosses the Suwannee River. Here, a small park on the west side of the river provides easy access. Little Shoals, visible about 100 to 200 yards upstream of the bridge, is accessible by foot. Strata at this location, as well as equivalent strata near White Springs, have yielded vertebrate fossils, both marine and terrestrial, of earliest Miocene (and possibly latest Oligocene) time: several sharks and other fish; reptiles, including the estuarine crocodile *Gavialosuchus americanus*; rodents; three species of dugong; the primitive horse *Miohippus*; a small camel; and, of great interest to paleontologists, the best-preserved fossil oreodont found in Florida. Oreodonts were a group of short-legged artiodactyls, even-toed ungulates (hoofed mammals) that lived in North America from Late Eocene to Miocene time. They are more commonly found in the Badlands of South Dakota and in Nebraska than in Florida. The White Springs oreodont was a new species, *Mesoreodon floridensis.*

Just west of White Springs, Florida 136 crosses the Suwannee River. Beneath the bridge (accessible from the west side of the river) and upstream of the bridge, there are excellent fossiliferous exposures of the Parachucla Formation, which is rich in mollusks and other invertebrates. North of the Florida 136 bridge (downstream), near the old White Springs bathhouse, specimens of the stony starlet coral (*Siderastrea siderea*) are common. These coral colonies have eroded out of the lower part of the Parachucla Formation, or perhaps the Suwannee Limestone, and have been concentrated in the river. Some are quite large, indicating that small patch reefs developed in this area during Early Miocene time. The corals are silicified but generally do not produce the agate that is so common in many Oligocene-age corals. *Siderastrea* coral heads are also found in southern Columbia County and northwestern Alachua County, near the Santa Fe River. Today, starlet coral is a hardy, sediment-tolerant species commonly found in nearshore patch reefs.

The old bathhouse at White Springs, constructed in 1908, encloses what was formerly called White Sulphur Springs. This was a popular health resort some one hundred years ago and was active as late as the 1960s. An earlier springhouse was constructed here as early as 1835. Many people thought the aromatic, sulfur-rich waters had great healing qualities. Similar bathhouses surrounding other springs are found at several places along the Suwannee, and most are constructed of native Suwannee Limestone.

Just downstream of the White Springs bathhouse, the Suwannee Limestone begins to crop out with regularity along the banks of the river. Exposures continue to Suwannee River State Park near Ellaville and several miles downstream. In some areas, high vertical banks of 20 feet or more enclose the river and exposures are nearly continuous for long stretches. Access points along this segment of the river are the Florida 136 bridge, the Stephen Foster Folk Culture Center State Park canoe launch, Woods Ferry, Spirit of the Suwannee Campground, Fox Trail, Boys Ranch, and Gibson Park.

Northwest of Stephen Foster Folk Culture Center State Park, beneath both US 41 and County Road 25A, is Swift Creek, a tributary of the Suwannee. It also has exposures of the Suwannee Limestone and is a good canoe run. Much agatized coral has been recovered from the banks and channel of Swift Creek, and numerous quarry sites indicate that Native Americans exploited this valuable chert resource.

The contact between the Suwannee Limestone and Statenville Formation is spectacularly exposed less than 1 mile upstream of Fox Trail landing, on the north side of the river. Here, channels and karst depressions in the Suwannee Limestone are filled in with Hawthorn Group rocks. Several additional exposures of the contact can be seen downstream of the US 129 bridge to about 0.5 mile below Fox Trail landing. Just upstream of the US 129 bridge is a bathhouse at Suwannee Springs, a second magnitude spring. The turn-of-the-century bathhouse is constructed of Suwannee Limestone. One can easily access the bathhouse by road. Follow US 129 for 4.3 miles north of I-10, then turn right on old US 129 and proceed 0.5 mile to a graded road that leads to the spring.

The strata of the Suwannee Limestone are undulatory in outcrops, almost as if they were gently folded. But this gentle waviness appears to be a natural depositional surface, perhaps slightly deformed after burial, and not the result

Thinly bedded Hawthorn Group rock of Miocene age along the Suwannee River.

The Suwannee Springs bathhouse, constructed of Suwannee Limestone of Oligocene age.

of tectonic forces. The Suwannee consists of lime mudstone to grainstone. The mudstone can be very finely laminated, and the grainstone is rich in foraminifera, especially miliolids, which resemble a tiny cluster of bananas or rice grains, and cone-shaped foraminifera of the genus *Fallotella.* Fragments of calcareous green algae and coral are also present. All of these fossils indicate an extremely shallow marine depositional environment—from the intertidal zone to the shallow subtidal zone, no more than about 30 feet deep.

Larger invertebrate fossils in the Suwannee include the distinctive echinoid *Rhyncholampas gouldii,* which often occurs in thin, concentrated layers called *lag deposits.* Also present are tubes of a bivalve mollusk of the genus *Kuphus.* This very unusual mollusk lived within a long, tapering tube—not the true shell of the mollusk, which tends to be a smaller, more complex apparatus—buried deeply in the sand. Modern species of *Kuphus,* such as *Kuphus arenarius,* live in the Indo-Pacific and Red Sea and may reach over 5 feet in length and up to 2.5 inches in diameter. *Kuphus* tubes are reported as far back as Eocene time but are common in the Suwannee Limestone and other formations of Oligocene age in the southeastern United States and Caribbean. They have also been found in the Miocene-age Chipola Formation. *Kuphus* species

The Suwannee Limestone of Oligocene age along the Suwannee River.

are found almost exclusively in calcareous sands deposited in shallow subtidal waters and back-reef lagoon areas.

Agatized coral occurs sporadically in the Suwannee River, having been concentrated in the river channel. It is also not uncommon to find projectile points, adzes, scrapers, or flint knives along the riverbanks in places where the Suwannee Limestone has been silicified into chert. Weathering clays in the overlying Hawthorn Group are probably the source of the silica for the chert and agate.

On the river, rock exposures have a quite different appearance than they do in other outcrops, quarries, or drill cores. This difference is due to what might be called "river karst." Acidic soils along the riverbanks and the tannic waters of the river naturally dissolve the limestone's calcium carbonate, radically altering the rock and creating features reminiscent of Swiss cheese or honeycombs. Trees help create the holes, extending large roots through the dissolved cavities and further enlarging them. Giant fibrous root mats also contribute to weathering as they grow over the porous limestone. When on the river, note the beautiful mirror-image optical effect as the calm tannic water of the Suwannee River reflects images of the trees, making them appear upside down on the water. It also reflects the image of the rock outcrops, making them appear twice as thick.

There are numerous limestone and dolostone exposures along the river at Suwannee River State Park. This is the area where geologists first named and described the Suwannee Limestone. From US 90 in Ellaville, take County

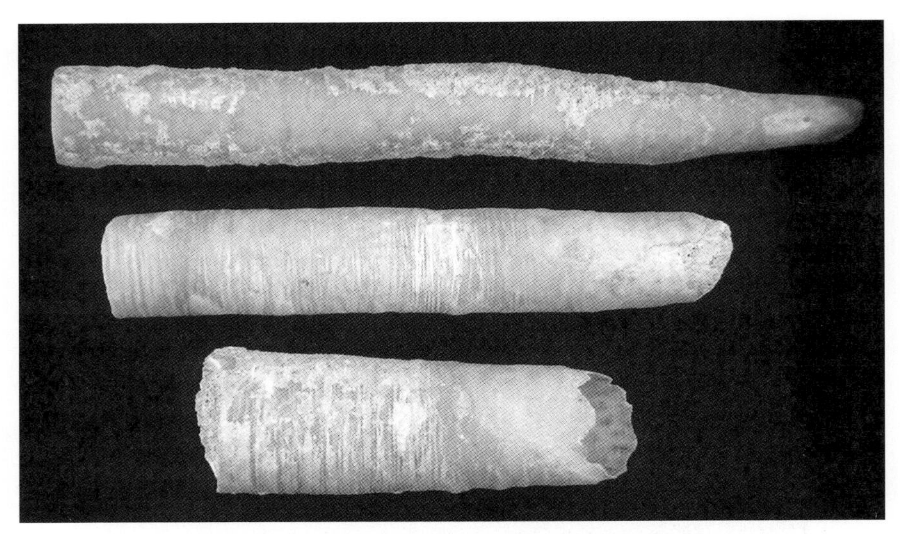

Tubes of unusual bivalve mollusks of the genus Kuphus *from the Suwannee Limestone of Oligocene age.*

Road 132 to the park entrance. Over the years the limestone bluffs along the river have been given a variety of formation names, but today they are considered as they were originally defined—the Suwannee Limestone. Depending on water levels, up to 10 feet or more of rock is exposed at various places from the confluence of the Suwannee and Withlacoochee rivers down to the US 90 bridge. One can easily reach these bluffs by canoe from the park. One can also reach Ellaville Spring by boat, about 0.3 mile downstream of the boat ramp at the state park. On the Withlacoochee River, just upstream of its confluence with the Suwannee, is the old stone bathhouse surrounding Suwannacoochee Spring, a second magnitude spring. An underwater cave system connects Ellaville and Suwannacoochee springs.

The lower part of the Suwannee Limestone is exposed at Suwannee River State Park. It is primarily dolostone but also includes cross-stratified calcareous sands, laminated mudstone, and limestone conglomerate, all of which indicate the limestone was deposited in shallow subtidal and intertidal mudflat conditions. The limestone conglomerate is called an *intraformational conglomerate* and is composed of flat mud pebbles, or chips, in a mud matrix. Layers of mud on a tidal flat were exposed to the air, so they dried, cracked, and hardened into mud chips. The mud chips were then incorporated into sediments that were deposited on top of them, forming the conglomerate. Such mud pebbles are common in the geologic record. They are also called *rip-up clasts*, *intraclasts*, or *inclusions*.

Also common in the lower Suwannee Limestone here is the sand dollar *Clypeaster rogersi* and molds of the gastropod *Turritella martinensis*. These rock types and fossils may be observed in several places: below the Suwannee River bridge on US 90 at Ellaville, about 0.1 mile north of the bridge along the east side of the river, under the I-75 bridge over the river, along the lower part of the Withlacoochee River before it joins the Suwannee River, and in boulders and outcrops in Suwannee River State Park near the boat ramp.

Falmouth Spring is located at Falmouth Spring Recreation Area, off US 90 about 2.9 miles west of the I-10/US 90 intersection. This first magnitude spring is a karst window—a collapsed cave system. It is called a "window" because it reveals the underlying aquifer and former cave. The water flows to the Suwannee River.

Fossil molds of Turritella martinensis *in the Suwannee Limestone of Oligocene age at Suwannee River State Park.*

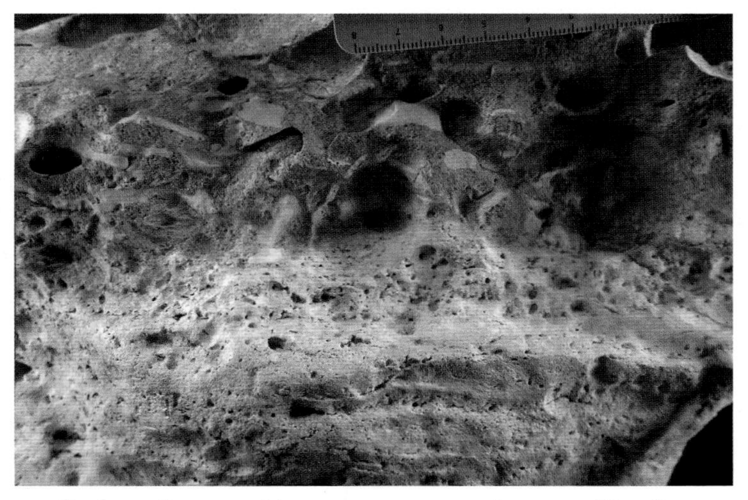

The lower Suwannee Limestone as seen near Suwannee River State Park, with laminated mudstone and rip-up clasts in the overlying dolostone. The finely layered mudstone formed in a tidal mudflat environment and represents a former shoreline.

Living Fossil Fish in Florida Waters

Because of their prehistoric pedigree, there are three very primitive fish in Florida waters that deserve honorable mention. All three have an ancestry reaching back to Cretaceous time, and they have remained essentially unchanged ever since. These living fossils are the sturgeon, bowfin, and gar.

The sturgeon is characterized by its elongated snout and five rows of thick scales, also called *scutes* or *ganoid scales*, which run down its body. There are two sturgeon species in Florida waters. The shortnose sturgeon (*Acipenser brevirostrum*) and the Atlantic sturgeon (*Acipenser oxyrhynchus*) are both found along Florida's east coast. The gulf sturgeon (*Acipenser oxyrhynchus desotoi*), a subspecies of the Atlantic sturgeon, is found exclusively in the Gulf of Mexico, from the Mississippi River to the Suwannee River. It is especially abundant in the Suwannee. Gulf sturgeon spawn in freshwater but spend most of their adult lives in the gulf. Despite their living fossil status, only fragmentary remains have been found in Florida, in sediments of Pleistocene age.

The bowfin (*Amia calva*), characterized by its long dorsal fin, is the only living representative of the order Amiiformes. It lives in freshwater lakes and streams in the eastern United States, from the Great Lakes area to the Gulf of Mexico. It prefers sluggish, slow-moving water filled with abundant vegetation. An air-breather, it must constantly gulp air from the surface. Bowfin remains are fairly common in Pleistocene-age freshwater sediments in Florida.

Living fossils in Florida waters (a) sturgeon; (b) bowfin; (c) gar.
—Artwork by Duane Raver, Jr. and courtesy of the U.S. Fish and Wildlife Service

Like sturgeons and bowfins, gars are perfectly prehistoric in appearance, with their overlapping, bony ganoid scales. Today, gars are found from Costa Rica to eastern North America and in Cuba and the Isle of Pines off the coast of Australia. Like the bowfin, gars are air-breathers and will drown if kept submerged. Four species of gar live in Florida waters. The longnose gar (*Lepisosteus osseus*) is the most widespread species in the state, reaching as far south as Lake Okeechobee, and may grow up to 5 feet in length. The Florida gar (*Lepisosteus platyrhincus*) is found east of the Apalachicola River basin and in peninsula waters. The closely related spotted gar (*Lepisosteus oculatus*) occurs in the Florida Panhandle west of the Ochlockonee River, replacing the Florida gar. It reaches up to 2.5 feet in length. But the largest species, reaching up to 10 feet in length and over 200 pounds, is the wide-snouted alligator gar (*Atractosteus spatula*). Today, this is primarily a Mississippi Basin species, but it is found as far east as Bay County. It is especially common in the Choctawhatchee River. But in Pleistocene time, the alligator gar reached into the peninsula as far as Tampa Bay and to the Atlantic coast. Gar remains are common in the fossil record of Florida, from Late Oligocene to Pleistocene time. They are very common at the Love Bone Bed (Miocene age) and the Leisey Shell Pits (Pleistocene age).

US 90 and Interstate 10

Lake City—Jacksonville

60 miles

North of Lake City, Falling Creek, a tributary of the Suwannee, falls 10 to 12 feet over an exposure of the lower Hawthorn Group. Falling Creek Falls and Falling Waters at Falling Waters State Park in the panhandle are among the few significant waterfalls in the state. Take Exit 301 on I-10 and follow Florida 131 north for about 0.8 mile to the falls.

US 90 and I-10 skirt the southern margin of the Okefenokee Basin—the headwaters of the Suwannee and St. Marys rivers. The St. Marys flows to the Atlantic Ocean, the Suwannee to the Gulf of Mexico. The Okefenokee Swamp is primarily in Georgia, but a portion of it extends into the Osceola National Forest in Baker County. Trail Ridge forms the eastern margin of the swamp. The Okefenokee is a freshwater peat swamp, analogous in many ways to the ancient coal-forming swamps of Pennsylvanian time (318 to 299 million years ago) that have provided us with much of our coal and natural gas. Carbon-14 dating has shown that the oldest peat here accumulated around 7,000 years ago. So the swamp formed at the end of the Holocene Transgression, as rising water tables flooded the low, ridge-bound, and formerly semiarid terrain of Pleistocene time.

The abundant peat is of two types: peat dominated by cypress (genus *Taxodium*), which forms in shallower waters, and peat dominated by water lily (genus *Nymphaea*), which forms in slightly deeper water. Interestingly, the swamp hosts a freshwater sponge (*Spongilla fluvialis*) that contributes silica to the peat deposits in the form of its tiny spicules, which are the skeleton of the sponge. Most sponges are marine organisms. Escaping methane from the decay of vegetation unearths large patches of peat that become floating islands of vegetation. These masses of floating peat, called *batteries*, are firm enough to walk on. They also are the source of the word Okefenokee, a Native American word meaning "land of the trembling earth." Natural fires are common in the swamp during drought cycles, and land managers employ controlled burns to maintain the area's ecological balance.

A curious geobotanical phenomenon that is often directly related to karst topography is the cypress dome, a dome-shaped stand of cypress trees. Shallow solution sinkholes, or dolines, formed by the dissolution of bedrock, commonly impound a pond of water. Cypress trees grow in these depressions because of the standing water. Larger, taller trees grow in the middle and shorter trees surround them, so their combined crowns form a dome shape. Cypress domes are common in karst-related swamps of Florida. Some of the Okefenokee Swamp's cypress domes can be seen along Florida 2 in Baker County near the Florida-Georgia border, west of Baxter.

Just east of Macclenny, I-10 and US 90 cross Trail Ridge, a linear sand body that extends 130 miles from the Altamaha River in Georgia to the northwest corner of Putnam County in Florida. The odd promontory of Georgia that seems to intrude into Florida between Baker and Nassau counties is, in fact, the southern end of the Georgia portion of Trail Ridge, which is skirted by the St. Marys River. Elevations along the ridge can reach almost 240 feet above sea level but typically range between 150 and 160 feet. Trail Ridge is a beach ridge of early Pleistocene (possibly Late Pliocene) age that formed as longshore currents reworked sands eroded from older deposits in Georgia and the Northern Highlands of Florida. Some dune sands are also present. Trail Ridge consists of a lower sequence of organic-rich sands with woody debris and peat lying on the Cypresshead Formation, and an upper layer of quartz sand rich in heavy minerals. A bed of brown coal, or lignitic peat, underlies a large part of the Florida portion of the ridge and is associated with larger plant debris, such as tree trunks, stumps, and limbs. Upright tree trunks can even be found in the peat, evidence that a forested coastal wetland was buried as the rising sea shifted the shoreline inland.

Trail Ridge and other sand deposits in the nearby Green Cove Springs area along the St. Johns River contain economically important

Silurian Sea Scorpions

Two highly unusual fossils of extinct eurypterids (sea scorpions) of Late Silurian age were recovered from deep oil wells in this region. Drillers came across *Acutiramus suwanneensis*, with an estimated length of 18 inches (perhaps a juvenile specimen), at a depth of 3,462 feet in a well in Columbia County. And at a depth of 3,571 feet in a well in Suwannee County, drillers found *Pterygotus floridanus*, a giant among eurypterids. Having reconstructed its body length, paleontologists estimate it was 4 feet 7 inches long!

The genera *Pterygotus* and *Acutiramus* belong to a family of widely distributed eurypterids that lived in the ocean from Ordovician to Devonian time, 488 to 359 million years ago. As a rule, eurypterid fossils are quite rare. So it is rather surprising that they are regularly encountered in deep wells in the

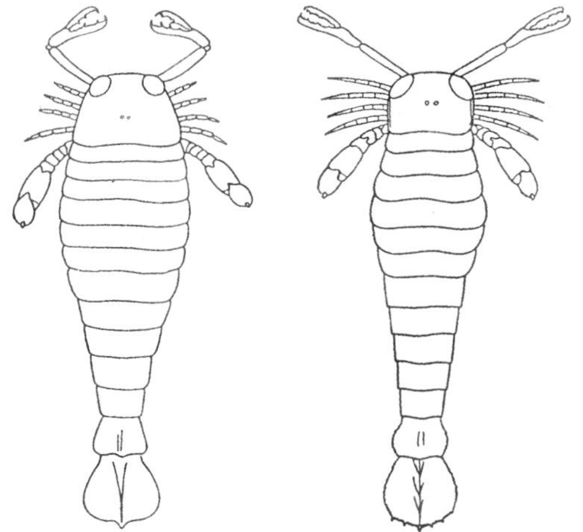

Representatives of the eurypterid genera Pterygotus *and* Acutiramus. *The specimen on the left resembles the giant* Pterygotus floridanus, *which was found in an oil well in Suwannee County. The specimen on the right resembles* Acutiramus suwanneensis, *which was found in a well in Columbia County.* —From *Treatise on Invertebrate Paleontology*, courtesy of and ©1955, the Geological Society of America and the University of Kansas

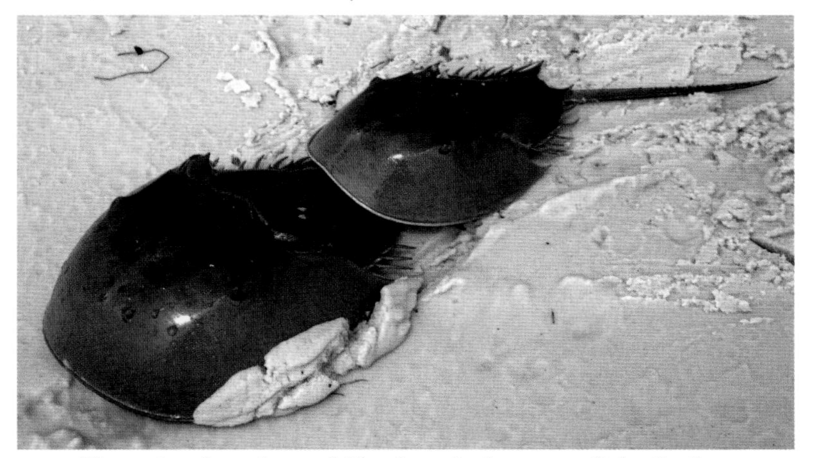

The modern horseshoe crab Limulus polyphemus—*a living fossil species. The smaller individual is a male pursuing a larger female. Adult females are always larger than adult males.*

northern Florida Peninsula (oil wells do not exactly recover large samples of fossils). So far, they have been recovered in Suwannee, Columbia, and Nassau counties. This suggests that during Silurian time Florida was literally crawling with sea scorpions!

Eurypterids belong to the same class of arthropods as modern horseshoe crabs. There are four living species of horseshoe crab, which are called living fossils because they have remained essentially unchanged for long spans of geologic time. The North American species, *Limulus polyphemus*, is found along the Atlantic Coast from Maine to the Yucatan Peninsula and is common in Florida's coastal waters. Eurypterids and horseshoe crabs are nearly the only marine examples of what are called merostome arthropods, and the horshoe crabs are the closest thing to a sea scorpion that lives today.

concentrations of heavy-mineral sands. Heavy minerals are sand-sized grains of minerals with a specific gravity that exceeds quartz (specific gravity is the ratio of a substance's mass to an equal volume of water). Quartz has a specific gravity of 2.7, while heavy minerals have a specific gravity greater than 2.9, meaning they are at least 2.9 times heavier than an equal volume of water. About 45 percent of the heavy minerals of Trail Ridge are the titanium-rich rutile, ilmenite, and leucoxene. Other heavy minerals present in these sands include staurolite, zircon, kyanite, sillimanite, tourmaline, spinel, topaz, corundum, and monazite. The titanium-rich minerals are used in the production of titanium metal alloys, fluxes, welding rods, glass fibers, and a variety of pigments for paint, paper, and plastics. Staurolite is used as a source of iron and alumina in cement production, and as an abrasive. Zircon is used as a foundry sand, refractory, and abrasive, and to manufacture chemicals, ceramics, and zirconium metal. The phosphate mineral monazite contains the rare earth elements cerium, yttrium, lanthanum, praseodymium, neodymium, samarium, europium, gadolinium, and thorium. Rare earth elements are used as a catalyst in petroleum refining, as metallurgical additives (alloys), and in ceramics, polishing agents, electronics, and optical glass. Florida leads the nation in the production of titanium and zirconium concentrates and is third in the nation in the production of rare earth elements. Production will continue for many years, but this is a nonrenewable resource.

Heavy-mineral sands are common in the panhandle and northern peninsula of Florida but occur in economically valuable concentrations (about 4 percent or greater) only in Trail Ridge and vicinity. The heavy minerals were originally derived from the weathering and erosion of igneous and metamorphic rocks of the Appalachian Mountains, and during transport by rivers and the sea they were abraded and rounded. The ocean concentrated them in linear beach ridges during Pleistocene time.

East of Trail Ridge and west of Jacksonville is the Duval Upland, a terrace of Pliocene and Pleistocene age with beach ridges. The upland consists of the cross-stratified sands of the Cypresshead Formation, which were deposited in shallow marine conditions. The Cypresshead Formation extends farther south along the peninsula, forming several terraces and ridges, including the Lake Wales Ridge. East of the Duval Upland is the Eastern Valley, largely a beach ridge plain, which extends along most of the eastern coast. The north-flowing St. Johns River lies in the Eastern Valley and is bordered by uplands and ridges of the Central Lakes District to the west and the Atlantic Coastal Ridge to the east. The coastal ridge is a continuous strip of beach ridges and dunes draped over the coquina rock of the Anastasia Formation. This area of the state is part of the Barrier Island Sequence District, which extends south into the southern peninsula.

US 27

Tallahassee—Perry—High Springs

120 miles

The Cody Scarp is especially evident along Florida 59 about 3 miles south of US 27, where the highway descends to the Woodville Karst Plain at Wacissa. The scarp is steeper (but more vegetated) just to the east, along County Road 259 between US 27 and Wacissa. After descending the Cody Scarp, US 19/27 traverses the flat terrain of the Gulf Coastal Lowlands, passing over dolomitic Suwannee Limestone and, east of Perry, over the Ocala Limestone. Sands of Quaternary age blanket most of these limestones. Some of the best fossil collecting in Florida is had in the Ocala Limestone, and numerous exposures of the Ocala can be found along this segment of US 27. The Ocala is especially rich in larger foraminifera, mollusks, and echinoids.

In Perry, US 27/19 splits. US 27 continues east, and US 19 joins US 98 and heads south. See the **US 98/US 19 St. Marks—Inglis** road log later in this chapter for information about that route. US 27 passes the famous Mayo Quarry, one of the best exposures of the Ocala Limestone in the state. It is located 24.1 miles east of Perry, or about 4.5 miles northwest of Mayo, on the west side of US 27. The quarry has gone by many names in the past, for example, Dell Limerock Mine. Limestone quarries in Florida frequently change ownership, and some owners will not grant permission to access quarries and collect geological samples. The Mayo Quarry exposes over 36 feet of the Ocala Limestone and is densely fossiliferous.

The slope of the Cody Scarp down to the Woodville Karst Plain (and Wicomico Terrace) as seen looking south on County Road 259 just south of Waukeenah.

An exposure of the upper Ocala Limestone of Late Eocene age at Dowling Park.

It is one of the few localities in Florida where remains of some of the earliest and most primitive whales—the archaeocete whales—have been recovered.

Not far from the Mayo Quarry, one can observe and collect the Ocala Limestone along the Suwannee River from the boat ramp beneath the Florida 51 bridge. From US 27 in Mayo, take Florida 51 north for 3.1 miles to the boat ramp on the south side of the river. Several quarries in the Ocala Limestone can be seen along US 27 between Mayo and High Springs. This region also has a high concentration of springs, most of which supply fresh groundwater to the Suwannee and Santa Fe rivers.

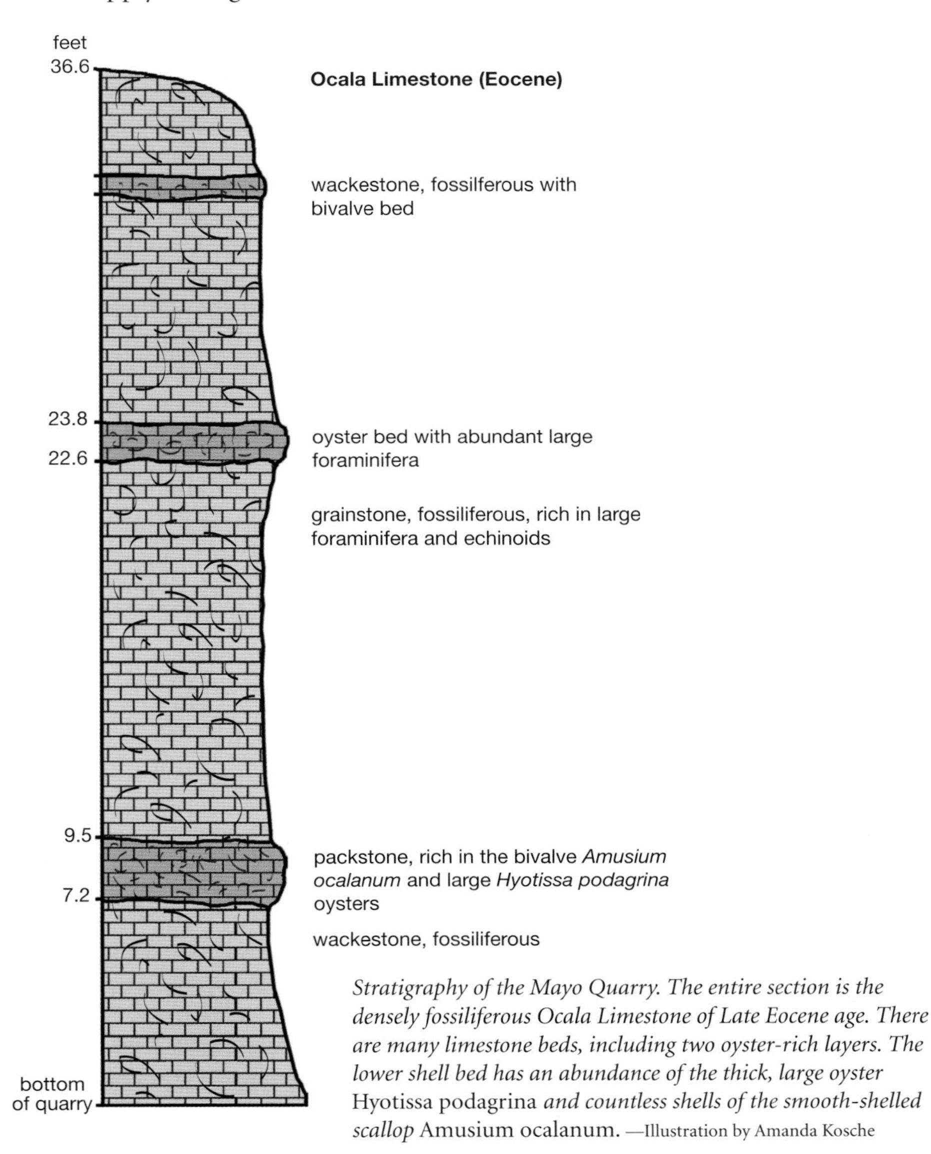

Stratigraphy of the Mayo Quarry. The entire section is the densely fossiliferous Ocala Limestone of Late Eocene age. There are many limestone beds, including two oyster-rich layers. The lower shell bed has an abundance of the thick, large oyster Hyotissa podagrina *and countless shells of the smooth-shelled scallop* Amusium ocalanum. —Illustration by Amanda Kosche

Fossil Echinoids

Popularly known as sand dollars, sea urchins, sea biscuits, and pencil urchins, the beautiful echinoids are among the most abundant and obvious of fossils in the Ocala Limestone, and they are well represented in many other Florida formations as well. The sturdy, calcite skeleton of most species of echinoids fossilizes well, so they are frequently used as index fossils—biological time markers for particular periods of geologic time. About forty species are known from the Ocala Limestone, making it the richest echinoid assemblage in Florida's fossil record.

Like most other echinoderms, echinoids are curiously complex creatures, almost alien in their anatomy. Their skeleton, or test, is composed of intricately interconnected plates and covered with large to small rounded bumps, called *tubercles*, upon which the creature's spines are mounted. The round, radially symmetrical forms are called *regular echinoids* or, more commonly, *urchins*. They are typically epifaunal, meaning they live on top of the seafloor, and they may wedge themselves between rocky ledges. The spines of many regular echinoids are quite large and often appear as fossils. Some very large-spined urchins are called *pencil urchins* because their thick spines were once used as pencils on slate writing tablets. Bilaterally symmetrical echinoids are called *irregular echinoids* or, more commonly, *sea biscuits* and *sand dollars*. They typically are infaunal, meaning they burrow into the substrate, and have very tiny spines that rarely appear as fossils. Irregular echinoids are far more common in the fossil record.

Some Fossil Echinoids of Florida

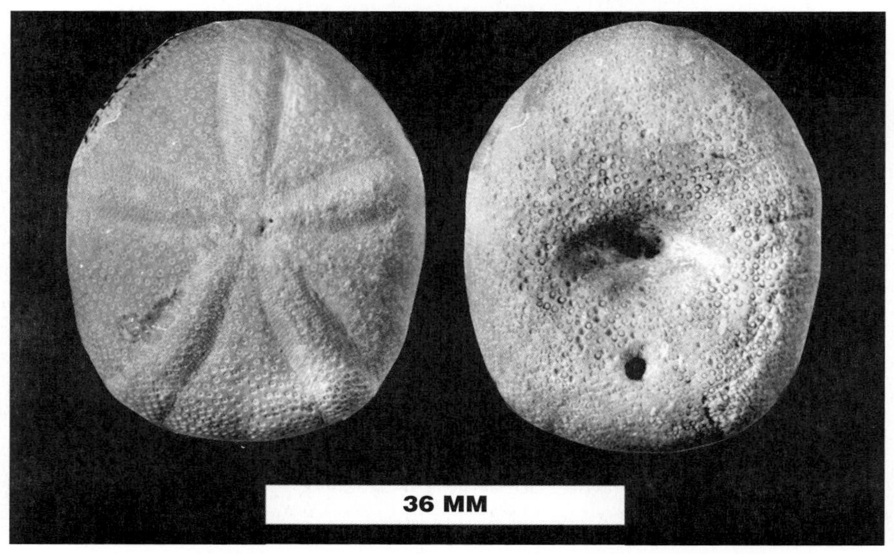

Oligopygus wetherbyi, *Ocala Limestone, Eocene age.* —Courtesy of the Invertebrate Paleontology Division, Florida Museum of Natural History, University of Florida

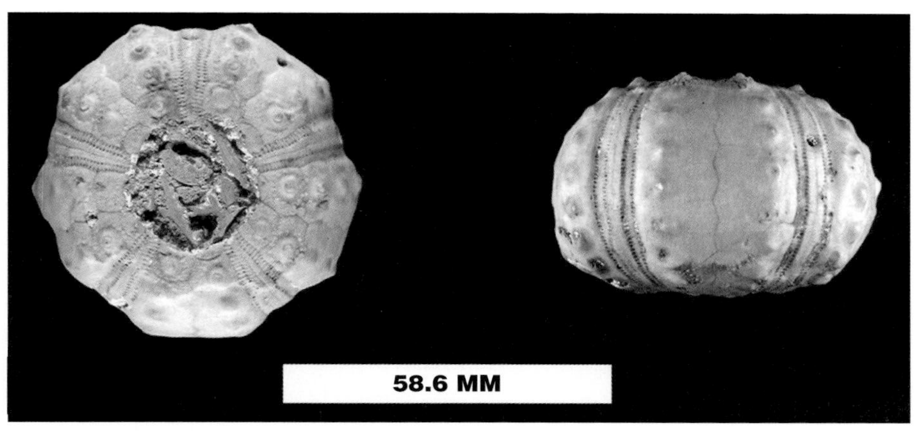

Phyllacanthus mortoni, *a regular echinoid, Ocala Limestone, Eocene age.* —Courtesy of the Invertebrate Paleontology Division, Florida Museum of Natural History, University of Florida

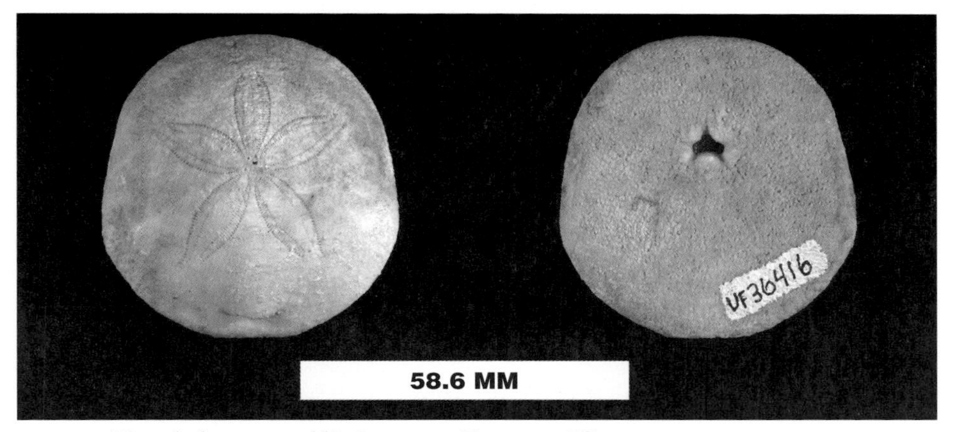

Rhyncholampas gouldii, *Suwannee Limestone, Oligocene age.* —Courtesy of the Invertebrate Paleontology Division, Florida Museum of Natural History, University of Florida

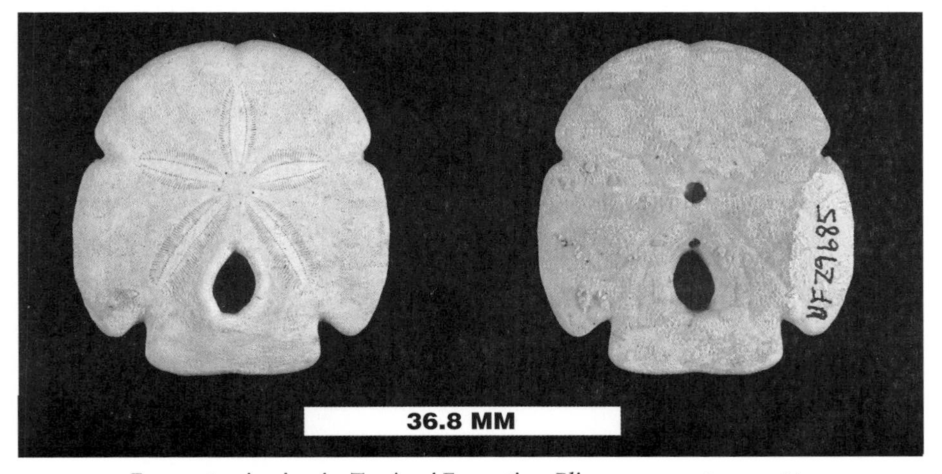

Encope tamiamiensis, *Tamiami Formation, Pliocene age.* —Courtesy of the Invertebrate Paleontology Division, Florida Museum of Natural History, University of Florida

Some Things Never Change— The Remarkable Bryozoan *Nellia tenella*

Known to zoologists as *moss animals*, bryozoans have a fossil record that reaches back to Ordovician time, 488 to 444 million years ago. Because of their small size, many fossil collectors overlook these colonial marine invertebrates, but they can be so abundant that they compose 100 percent of some limestones. Bryozoans are very common in the Ocala Limestone, sometimes in fairly large colonies.

The bryozoan *Nellia tenella* is tiny—an entire sticklike colony is only about 0.25 inch high. It is commonly found in the Ocala Limestone and still lives today in tropical to subtropical waters around the world. The same species has also been identified in Jamaica in strata of Late Cretaceous age, which is more than 65 million years old. If these fossils and the living bryozoan are indeed all the same species, as they appear to be, that would make *Nellia tenella* as much as 70 million years old, one of the longest-living species known on Earth. Of course, scientists cannot determine with certainty whether the fossils are truly the same as the modern species. But it is truly remarkable that such an unchanging form has persisted for so long. *Nellia tenella* is a classic, if unsung, living fossil.

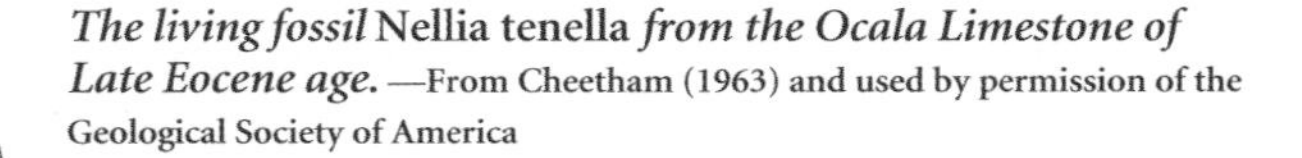

The living fossil Nellia tenella *from the Ocala Limestone of Late Eocene age.* —From Cheetham (1963) and used by permission of the Geological Society of America

Exposures and Springs on the Lower Suwannee, Santa Fe, and Ichetucknee Rivers

Along the Suwannee River, younger strata are exposed upstream and older strata downstream. The upper Suwannee exposes the Miocene-age Hawthorn Group and the Oligocene-age Suwannee Limestone. South of Ellaville, the Suwannee Limestone is sporadically exposed along the riverbank, and the Eocene-age Ocala Limestone makes its first appearance. One can easily observe an exposure of the uppermost Ocala Limestone along the Suwannee River at the town of Dowling Park. From US 27 between Perry and Mayo, take County Road 53 north (from I-10, take County Road 53 south) and turn east onto County Road 250. Drive about 2 miles to the Suwannee River and turn left immediately after crossing the river and entering Dowling Park. Walk north along the riverbank from the boat ramp. The limestone along the bank is the uppermost Ocala Limestone and is characterized by abundant larger

A residual boulder of the Suwannee Limestone of Oligocene age overlying Ocala Limestone near the Suwannee River. —Courtesy of Rick Green and the Florida Geological Survey

foraminifera of the genera *Lepidocyclina* and *Nummulites*, mollusks, bryozoans, beautifully preserved echinoids such as *Schizaster*, and a peculiar worm-tube fossil called *Rotularia vernoni*, which is an index fossil for the uppermost Ocala Limestone in the northern peninsula. This zone of the Ocala contains fossils with both Eocene- and Oligocene-age affinities, so its exact age has been disputed.

Between Branford, which is on US 27 southeast of Mayo, and the Suwannee River's confluence with the Santa Fe River, exposures of the Ocala Limestone are especially common on the Suwannee River, as are residual, silicified boulders of the Suwannee Limestone. These cherty boulders contain abundant specimens of the echinoid *Rhyncholampas gouldii*. The lone boulders testify that the Suwannee Limestone once covered this region. Over time, chemical weathering (dissolution) has almost completely removed all traces of this limestone. South of the confluence of the Santa Fe and Suwannee rivers, the Ocala Limestone is periodically exposed along the Suwannee River as far south as Manatee Springs, and there are few exposures of any kind south of there. However, the Ocala is exposed along the river bottom, and when the river is clear, many fossils can be found. Echinoids from both the Ocala and Suwannee limestones are abundant in potholes along the river bottom, and the remains of vertebrate animals of Pleistocene age are also present. The lower Suwannee River is very broad and lazily meanders to the Gulf of Mexico.

Ichetucknee Springs State Park is about 8 miles east of Branford on US 27. It encompasses several springs that are the source of the Ichetucknee

River, which flows just 6 miles before joining the Santa Fe River. The park is perhaps the most popular tubing run in the state. The Ocala Limestone is exposed in the springs and along the entire river channel, and the river bottom is also a classic locality for concentrations of mammal bones of Pleistocene age. Collecting is no longer permitted, but the river has produced remains of mammoths and mastodons, giant beavers, camels, llamas, glyptodonts, giant armadillos, several species of sloth, dire wolves, and a specimen of the Florida lion (*Panthera atrox*). An abandoned hard-rock phosphate pit, mined in the early 1900s, can be seen along the Trestle Point Trail.

Along much of the Santa Fe River from its confluence with the Ichetucknee to its confluence with the Suwannee, the Ocala Limestone is semicontinuously exposed, and some unusual Quaternary-age strata overlie it. These deposits, informally called the Santa Fe Marl, consist of carbonate sediment, clay, and sand with freshwater molluscan fossils and, rarely, bones of Pleistocene-age mammals. These strata are exposed near a bridge in Gilchrist County. From US 27 about 4 miles east of Branford, take US 129/Florida 49 south 2.8 miles to the bridge over the Santa Fe River. At the southeastern corner of the bridge, 6 to 8 feet of stratified marl, consisting of black clay and mollusk-rich and mollusk-poor layers, is exposed. The fossil mollusks include freshwater snails still found living in the spring runs today. Above the natural levee of the river and the marl, in the adjacent wooded area, more Ocala Limestone is exposed. There are small caves in the limestone, which appear to represent an earlier springhead. The spring was probably active when the water table was higher and the Santa Fe Marl was being deposited.

Freshwater mollusks from the Santa Fe Marl of Pleistocene age: (a) *apple snail* (Pomacea paludosa), (b) *Seminole rams-horn* (Planorbella duryi), (c) *banded mystery snail* (Viviparus georgianus).

Close-up of the Santa Fe Marl.

The Santa Fe Marl, a freshwater deposit of Pleistocene age overlying the Ocala Limestone.

An excellent exposure of the Ocala Limestone—the thickest exposure along the Suwannee River—is visible next to the boat ramp at Rock Bluff Landing, just south of the County Road 340 bridge over the Suwannee River, where 20 feet or more of the limestone is exposed. Additional exposures occur along the river from just north of the bridge up to Rock Bluff Springs. At Branford, take Florida 349 south 10.5 miles to County Road 340. Turn east and drive about 3.3 miles, crossing the river, to the public boat ramp on the right.

One of the highest concentrations of springs in Florida lies along the Suwannee River and its two major tributaries, the Santa Fe and Ichetucknee rivers. There are numerous first and second magnitude springs, and most of them issue from the Ocala Limestone. Some can be reached only by boat and some are on private land, but there are many accessible parks with beautiful springs and good exposures of the Ocala Limestone, especially along the lower Suwannee River.

- **Lafayette Blue Springs State Park**—From its intersection with Florida 51 in Mayo, follow US 27 northwest for 4.9 miles and turn north on County Road 251B. Proceed 2.1 miles and turn right onto a dirt road that leads to the park and this first magnitude spring.

- **Telford Spring**—From its intersection with US 27 in Mayo, follow Florida 51 north for 4.8 miles, crossing the Suwannee River, then turn east (right) at the flashing light in Luraville onto Peacock Springs Road (180th Street). Take the first graded road south (right) and drive almost 1 mile to a fork. Turn right at the fork and proceed 0.1 mile to the spring.

- **Peacock Springs State Park**—From Luraville, 4.8 miles north of Mayo on Florida 51, follow Peacock Springs Road (180th Street) east for 2.5 miles to the park entrance. Peacock Springs contains one of the longest underwater cave systems in the United States. Over 33,000 feet of cave passages have been mapped through the Ocala Limestone. This is a popular destination for highly experienced cave divers. Sinkholes are common in the park, and Bonnett Spring is also located here.

- **Mearson Spring**—From its intersection with Florida 51 in Mayo, follow US 27 about 6.6 miles east and turn east (left) on County Road 251. Follow it for 2.7 miles to a boat ramp. This second magnitude spring is 0.8 mile downstream.

- **Troy Spring State Park**—Follow US 27 4.8 miles northwest of the Suwannee River bridge in Branford and turn north (right) on County Road 425. Follow it for 1.2 miles to the park entrance. An underwater cave system has been mapped at this first magnitude spring, and the Civil War steamship *Madison* lies at the lower end of the spring run, purposely sunk by Confederates.

- **Little River Springs**—From Branford, follow US 129 north for 3.1 miles and turn west (left) on County Road 248. Follow it for 1.7 miles. The spring is on the south side of the road.

•**Branford Springs**—Located in Ivey Memorial Park, in Branford, near the Suwannee River bridge on US 27.

•**Hart Springs**—From US 27 about 3.5 miles east of Branford, turn south on US 129 and proceed about 19 miles to County Road 344. Turn west (right) and proceed about 5.5 miles. Ocala Limestone is exposed in the spring.

•**Gilchrist Blue Spring**—From the junction of US 441/41 and US 27 in High Springs, follow US 41/27 southwest for 0.8 mile and turn west (right) on County Road 340 (Poe Springs Road). Follow it for 4.6 miles and turn north (right) onto Blue Spring Road. Proceed 1.1 miles to the parking area.

•**Ginnie Spring**—The directions are the same as for Gilchrist Blue Spring, but follow County Road 340 for 6.6 miles and turn north onto a graded road. Follow it for 1.2 miles to the spring entrance. Follow the road behind the office and turn north (left) at the bathhouse and continue down to the spring. An extensive cave system has been mapped at Ginnie Spring. Nearby Devils Ear Spring can be accessed from Ginnie Spring. Turn right just before the bathhouse and follow the sand road to the spring.

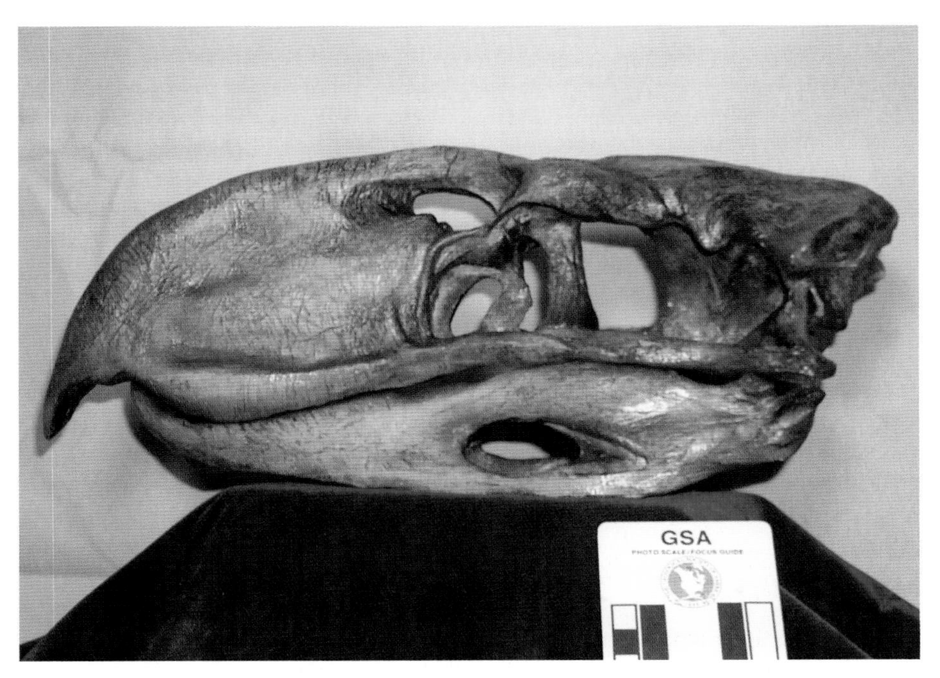

A cast of the skull of Phorusrachus, *a large, flightless predatory bird of Pliocene age from South America that was similar to* Titanis walleri. *The skull is 14.5 inches long.*

This region of the northern peninsula is especially noteworthy for its abundance of vertebrate fossils and rare Paleoindian remains. On the Santa Fe River, mammoth vertebrae with butchering marks have been recovered, indicating that Paleoindians were hunting megafauna here in late Pleistocene time. Projectile points and even the bone hook portions of atlatls (spear-throwing devices) have also been recovered in the Santa Fe. Fragmentary remains of the nightmarish *Titanis walleri* are another spectacular find from this river. This giant, 6-foot-tall, predatory flightless bird was a South American immigrant during the Great American Biotic Interchange. The genus *Titanis* is most closely related to flightless predatory cranes still living in South America. Vertebrate remains of Eocene, Pliocene, and Pleistocene age are concentrated in gravels of the Santa Fe River and can be collected while snorkeling and scuba diving. (Canoe launches are located at the Florida 47 bridge, 6 miles south of Fort White, or the US 27 bridge, 10 miles upstream of the Florida 47 bridge and 5 miles northwest of Fort White.)

Thomas Farm

Southwest of Fort White, near the confluence of the Suwannee and Santa Fe rivers, is one of the most famous vertebrate fossil sites in eastern North America—Thomas Farm. In 1931, Florida Geological Survey geologist J. Clarence Simpson was investigating a reported Native American graveyard on the farm of Raeford Thomas, where some bones were uncovered while the farmer was plowing. As it turned out, the bones were not human but belonged to the extinct three-toed horse *Parahippus.* The site was later worked by crews from the Florida Geological Survey, Harvard University, and most recently the University of Florida, which has owned the land since 1942 and continues to excavate the site.

Based on the recovered land mammals, paleontologists have determined that the site is Early Miocene in age, about 18 million years old. It has yielded over eighty vertebrate species, including freshwater turtles, a variety of lizards, boa constrictors, alligators, numerous birds, three species of horse (*Parahippus leonensis* is the dominant animal in the entire fossil assemblage), two species of rhinoceroses, *Hemicyon* (the first known bear in Florida), two species of the strange and extinct beardogs (amphicyonids), and other large mammals. Many smaller vertebrates, such as frogs, salamanders, rodents, and bats, are also present. Fish are very rare. The fauna indicates that much of north-central Florida during Early Miocene time was forested, with grassland savannahs intermixed. Certain bats and reptiles indicate that the climate was more tropical or subtropical than today.

The geological origin of the Thomas Farm Site is fascinating. Deposition occurred in a rather large sinkhole (115 feet in diameter and 98 feet deep) that developed in the Ocala Limestone. The fossils filtered through the sinkhole—much like sand trickling through an hourglass—into a larger cave system below, accumulating in a large debris cone. Bone orientation and other sedimentological evidence indicate that water periodically flowed through the system, and the abundance of bat remains indicates that it was a dry cave environment at times as well. At some point, the roof and walls of the cave collapsed, evidenced by two distinct limestone boulder layers in the cone, enlarging the sinkhole. Erosion and lowering of the surrounding limestone after Miocene time eventually exposed the sediment cone near the modern surface. Ancient, deeply eroded sinkholes like this are sometimes called *beheaded sinkholes.*

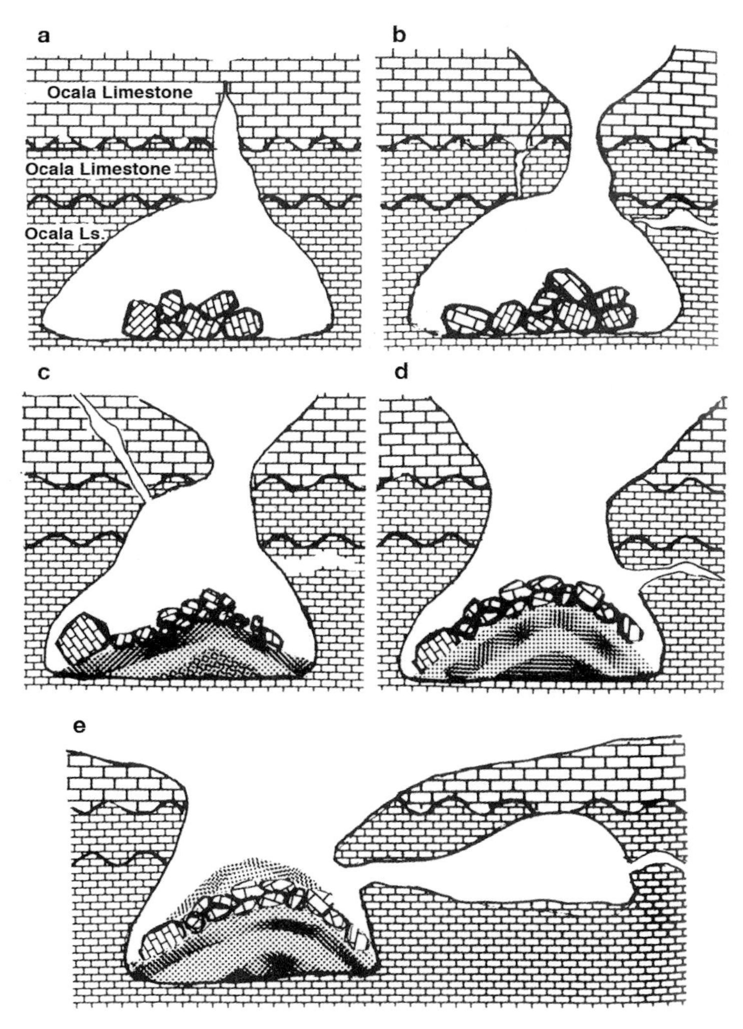

The origin of Thomas Farm as a sinkhole cone, cave, and stream environment: (a) formation of a cave in the middle Ocala Limestone; (b) collapse of part of the cave roof, forming a debris cone of limestone; (c) deposition of sand and clay with the debris cone, and further roof collapse; (d) collapse of the uppermost Ocala Limestone; (e) continued cave formation and sediment deposition. —Modified from Pratt (1990) and courtesy of the Florida Museum of Natural History

Fossil Horses

Horses first appear in the North American geologic record in Early Eocene time with the tiny *Hyracotherium* (formerly *Eohippus*), and the horse's complex evolution is well documented in North America. Their fossils are found in the entire terrestrial geologic record of Florida, from Late Oligocene to Pleistocene time, and they can be the dominant animal at vertebrate fossil sites. Horses compose one of the most diverse vertebrate fossil groups in the state, with over fifty species known. They belong to an order of odd-toed ungulates (hoofed animals) called *perissodactyls*, which includes rhinoceroses, the piglike tapirs, and some extinct groups. During Eocene and Oligocene time, horses generally had teeth with low crowns and were browsers, feeding primarily on soft, leafy vegetation. But during Miocene time, as global climates were cooling and drying, grasslands became widespread. Horses adapted by developing teeth with higher crowns that could withstand a grazing diet of coarse, abrasive grasses. This increase in tooth-crown height, also seen in other grazing mammals, is an evolutionary trend known as *hypsodonty*.

The modern single-toed (hoofed) genus, *Equus*, first appears in Florida in Pliocene-age sediments, and it diversified into several lineages throughout Pliocene and Pleistocene time. Although horses originated and had their greatest success in North America, they became extinct on the continent near the end of Pleistocene time, about 11,000 years ago. They had dispersed into Eurasia during Pliocene and Pleistocene time and were domesticated about 6,000 years ago, but they only made their reappearance in North America in the sixteenth century, when reintroduced by the Spanish. There are at least eight living species of horse, all *Equus*, including the wild Asian horse *Equus caballus*, or *Equus przewalskii*, and its well-known domesticated descendant, also *Equus caballus*; four species of ass; and three species of zebra. A fourth zebra, *Equus quagga*, became extinct in the early 1900s.

US 98/US 19

St. Marks—Inglis

145 miles

There are many names for the coastal terrain between St. Marks and Inglis. The region is within the Ocala Karst District. It is also known as the Gulf Coastal Lowlands and is often referred to as the Big Bend coast or the Nature Coast in tourist brochures. The landscape is a flat, karst surface of carbonates of Eocene, Oligocene, and Miocene age with a thin veneer of Quaternary-age sediment. It has many dunes, coastal ridges, and marine terraces. Unlike most of the Florida coastline, there is a notable absence of barrier islands here. This is due to the paucity of quartz sand, the primary building material of barrier islands. Most of the rivers that discharge into the gulf along the Big Bend are spring fed and flow over limestone, so quartz sand is in short supply. Apalachee Bay, the broad, shallow embayment below St. Marks, is considered a wide estuary even though it's not enclosed by land. However, wave action is very limited here due to the broad, shallow seafloor, which disperses waves and limits the effect they have on the coastline.

The flat terrain was a seafloor during Eocene, Oligocene, and part of Miocene time that was exposed and karstified during Late Oligocene to Holocene time and was also a sea bottom, or terrace, during Pliocene and Pleistocene time. The surface was flooded by the sea for the last time during the Holocene Transgression. The karst surface continues offshore as part of the shallow West Florida Shelf, which is a drowned karst surface with submarine springs and sinkholes. Freshwater issues from some of the nearshore springs. The extensive network of river channels on this submerged surface developed during late Pleistocene and early Holocene time, when sea level was substantially lower. Several prehistoric archaeological sites have also been discovered on the submerged shelf, which was once a coastal lowland just as the modern Big Bend coast is today.

The sea off the Big Bend coast is a modern example of what geologists call an *epicontinental sea*—a shallow sea that floods the margins of a continent and usually extends far inland. Epicontinental seas were common throughout much of Paleozoic, Mesozoic, and early Cenozoic time but are rare during glacial times, which includes the present. Large ice sheets, of which the Antarctic and Greenland ice sheets are the only modern examples, lower sea level substantially. The shallow ocean setting of the West Florida Shelf is one of the few modern examples of what has historically been a common geologic environment.

In western Taylor County, which is east of the Aucilla River, several companies mine the Suwannee Limestone. In some of the mines, the Suwannee is almost entirely a dolostone. The mines produce aggregate and agricultural dolomite, which is a soil additive, as well as other products. Boulders of the Suwannee Limestone lie along several roads north and south of US 98 in this area.

Folk legends abound in the coastal lowlands. There is just something about a marsh or swamp—foreboding, dangerous, inaccessible—that spawns such tales. In addition to the tale of Tates Hell in the Apalachicola area, we have the curious legend of the Wakulla Volcano in the

Wacissa-Aucilla Swamp, through which US 98 passes between Newport and Nutall Rise. To this day, some people still search, so far unsuccessfully, for the source of a mysterious column of smoke that once continuously rose from the swamps near the Wacissa River in southern Jefferson County. The smoke was first reported in the 1830s and was supposedly seen by Native Americans, Spaniards, sailors at St. Marks, and others. What was the smoke? Perhaps it was related to the fires of Native Americans living deep in the swamp, or a band of pirates, or runaway slaves, or Confederate deserters, or possibly muck fires, or maybe . . . a volcano? This was a popular explanation for several years, and the Wakulla Volcano appeared in Maurice Thompson's 1881 novel *A Tallahassee Girl.* But alas, the smoke reportedly stopped rising the day of the Charleston, South Carolina, earthquake of August 31, 1886. Tremors from the quake are thought to have opened a sinkhole, swallowing the volcano!

Although a volcano in this or any other part of Florida is presently a geological impossibility, there are indeed volcanic rocks beneath these swampy coastlands, but they lie some 7,000 feet below the surface. During Triassic time, lava erupted here when Pangaea began its global break-up some 230 million years ago. It is also likely that Florida received volcanic ash from eruptions in the western United States or the Caribbean Islands during the last 65 million years, but it has been many eons since volcanoes actually erupted in what is now Florida. If there is any validity to the tale of smoke rising from this coastal swamp, smoldering peat and muck is a reasonable hypothesis.

The mysterious, dark waters of the Aucilla River gently flow beneath the US 98 bridge at Nutall Rise. The lower reaches of the Aucilla, which define the Taylor-Jefferson county line, pass over the Suwannee Limestone. There are many rocky exposures and white-water shoals along this stretch, and it is a favorite run of experienced canoeists and kayakers. These rapids occur south of US 19/27, and especially south of County Road 257. Beginning several miles upstream from Nutall Rise, the Aucilla disappears at swallets and resurfaces many times—a classic example of a disappearing stream in karst topography. The Suwannee Limestone is often spectacularly exposed in the areas where the river rises to the surface and flows through karst collapse features. The Big Bend portion of the Florida Trail, a system of more than 1,400 miles of hiking trails, passes through this area. From the trail and side hikes you can see numerous sinkholes and collapse features.

Just above the US 98 bridge, the Aucilla rises for the last time from a large sinkhole called Nutall Rise. Above this rise are a number of karst features that have collapsed to reveal an underground portion of the

Aucilla. Several rises are long enough to be named, with Half-Mile Rise and Little River being two of the longer stretches. At two places along the Aucilla, the spring-fed Wacissa River joins in, and there are a number of first magnitude springs in this area. These karst rivers are especially noted for their archaeological and paleontological bounty. Both contain some of the highest concentrations of archaeological sites in the eastern United States, as well as numerous accumulations of vertebrate remains of Pleistocene age.

Both the Aucilla and Wacissa rivers contain evidence of Florida's first human inhabitants—the Paleoindians. These people arrived some 13,000 years ago, as evidenced by chert, bone, and ivory tools. The presence of preserved bone tools is incredibly rare. Most above-water archaeological sites lack bone implements because exposure and aggressive dissolution quickly disintegrate bone; however, the dark waters of the Aucilla and Wacissa protected their treasure of artifacts from these destructive agents.

The University of Florida and the Florida Bureau of Archaeological Research undertook an intensive, fifteen-year study along the Aucilla River called the Aucilla River Prehistory Project. Researchers looked for stratified archaeological sites in the riverbed. The submerged sinkholes that occur along the course of the river contain a sedimentary record of Florida's human prehistory and have produced some of the most exciting evidence of interactions between humans and large animals during late Pleistocene time. The Page-Ladson and Sloth Hole sites preserve a record spanning from late Pleistocene time to the present. The Sloth Hole Site has produced the largest number of ivory tools of any site in North America and is the third-oldest Clovis site discovered thus far in the Americas. Both sites are sinkholes that experienced intermittent water flow, which attracted both animals and humans. During times of drought, these sinkholes may have acted as oases, providing sources of cool springwater.

The Wacissa River also preserved numerous archaeological and paleontological sites from the same time period. The Ryan-Harley Site, a stratified site eroding out of a portion of braided channel in the Wacissa, contains the remains of large animals and Paleoindian points. The site is of great importance because to date no points of this age and associated tools have been found in stratigraphic context with remains of extinct megafauna, such as tapirs and horses. If radiocarbon or other dating methods confirm the estimated time frame, it would indicate that certain Pleistocene-age animals persisted in the southeast region after having gone extinct in the western United States.

Paleoindians and Pleistocene Megafauna

When the first humans set foot in Florida, it was twice its present size. This was at the end of the last glacial maximum of Pleistocene time—the Wisconsinan glacial stage. The coverage of continental ice in North America had never been so extensive, and this resulted in sea levels so low that the Florida shoreline was near the modern edge of the continental shelf. Exactly when the first humans arrived in Florida is not known, but they were certainly here by 13,000 years ago. Their artifacts are found in numerous rivers, springs, and even offshore locations in Florida.

In 1982, a partial skeleton of an extinct species of Pleistocene-age buffalo, *Bison antiquus*, was recovered in the Wacissa River, apparently with the broken tip of a chert projectile point embedded in the posterior of the skull. Using radiocarbon dating, scientists determined that the skull is approximately 11,000 years old, placing it in the Paleoindian time period. Paleontologists have speculated that the extinction of so many species of large mammals, such as mastodons and mammoths, at the end of Pleistocene time may have been linked to the warming climate. But archaeologists have proposed that the extinctions could be related to excessive hunting by Paleoindians. For example, even historic Indian tribes are known to have stampeded herds over cliffs. Of course, one pierced bison skull does not prove this overkill hypothesis, but it does demonstrate that large game were likely an important source of sustenance for the first inhabitants of Florida.

Many Paleoindian sites have been discovered and excavated in Florida. Some of the more spectacular sites include Warm Mineral Springs, Little Salt Spring, Wakulla Springs, and Silver Springs. In fact, most large springs in Florida are prehistoric archaeological sites, as they offered all the necessary natural resources to sustain the early inhabitants: water, food, and abundant chert for stone tools. They were also life-giving environments for countless prehistoric animals. So the next time you take a refreshing swim in a Florida spring, let your geological imagination roam, and consider that Floridians have been enjoying these waters for more than 13,000 years!

Fossil Elephants

Proboscideans, or elephants, are some of the best-represented vertebrate animals in Floridan strata, and their fossils are large and impressive. Worldwide, the oldest elephants uncovered thus far were found in strata of Early Eocene age in northern Africa (Algeria), but proboscideans spread rapidly and diversified. They had arrived in North America by 20 million years ago, in Miocene time. Remains of all three proboscidean families are found in Florida—the mastodons, gomphotheres (shovel-tusked mastodons), and mammoths.

*The Columbian mammoth (*Mammuthus columbi*) of late Pleistocene age, found in the Aucilla River and now exhibited at the Florida Museum of Natural History.*

The oldest mastodons and gomphotheres have been found in Polk County deposits of Middle Miocene age. At least eight species of the strange gomphotheres have been found in Florida, including the genera *Gomphotherium*, *Rhynchotherium*, and the giant *Amebelodon*. Common in Miocene- and Pliocene-age rock, the last of the gomphotheres became extinct in Florida in mid- to late Pleistocene time. Three species of mastodons are found in Florida. The common and widespread American mastodon (*Mammut americanum*) lived from Late Pliocene time to the end of Pleistocene time. Several nearly complete skeletons have been recovered in Florida.

The extinct mammoths (genus *Mammuthus*) belong to the still-living family of true elephants, which includes the African elephant (*Loxodonta africana*) and the Asian (or Indian) elephant (*Elephas maximus*). The oldest specimens of all three of these elephants were found in African strata of Early Pliocene time. The Asian elephant became extinct in Africa by the end of Pleistocene time but lives on in India, Indochina, Sri Lanka, and Indonesia. Mammoths quickly spread through Europe, northern Asia, and North America but were extinct by the end of Pleistocene time.

Mammoths crossed into North America twice—first in early Pleistocene time and again in late Pleistocene time. The famous cold-adapted woolly mammoth (*Mammuthus primigenius*) actually represents the second immigration, and its remains are not found in the southern United States. The two species of mammoth found in Florida are descended from the mammoths of the first

*Artist's rendering of the American mastodon (*Mammut americanum*).*
—Courtesy of the Florida Geological Survey

immigration. The smaller species, *Mammuthus haroldcooki*, is found in early Pleistocene-age deposits. The larger Columbian mammoth (*Mammuthus columbi*) is found in mid- to late Pleistocene sediments, and its remains are widespread in the state. It became extinct by about 11,000 years ago. One of the largest and most complete mammoth skeletons ever found in North America, a mature male Columbian mammoth, came from the Aucilla River in 1968. It is now mounted at the Florida Museum of Natural History in Gainesville. It stands over 14 feet tall and in life would have weighed an estimated 22,000 pounds. Radiocarbon dating indicates it's about 16,000 years old.

Large assemblages of reef-forming corals have been collected from the Suwannee Limestone along the Econfina River, which runs through Econfina on US 98. These are frequently found associated with the index echinoid *Rhyncholampas gouldii.* Longtime collector Jerry Giles of Perry has recovered large coral heads from the Suwannee Limestone in the middle of flat pine forests in this region. These fossils occur just beneath the thin soil that overlies the limestone. Mr. Giles's method for locating them involves pushing a metal rod into the soil, tapping the limestone, and listening for the characteristically hollow sound that indicates the presence of agatized specimens. The corals formed small patch reefs during Oligocene time. It is difficult to determine the exact species of the corals because they are usually thoroughly agatized, but representatives of *Astrocoenia*, *Porites*, *Goniopora*, and other genera are present.

Modern oyster reefs constructed by *Crassostrea virginica* are common in most shallow coastal waters, estuaries, and tidal channels across the state, and they are especially abundant along nearshore areas of the Big Bend coast. Rocky karst pinnacles, reduced salinity, and good tidal pumping make this an ideal setting for oyster growth. The oyster reefs tend to grow perpendicular to tidal currents, nearly parallel to the shore, but eventually they interconnect, forming extensive networks of oyster shoals. Oyster reefs and biostromes, or layers of oyster shells, are commonly fossilized in the geologic record from Triassic time to the present. Knowledge of the ecology and distribution of modern oyster accumulations, such as these along the Big Bend coast, has made it possible for scientists to properly interpret fossil oyster deposits.

In Perry, US 19 joins US 98, and the route turns in a more southerly direction. At Tennille, on the Taylor-Dixie county line, US 98/19 crosses the Steinhatchee River, which one can plainly see flows over karst limestone. River-drained karst is a characteristic terrain of Florida. The Steinhatchee River is a well-known and rich source of Eocene- and Pleistocene-age vertebrate fossils, which become concentrated in the river bottom. Snorkelers are able to collect them when the river is low. The river is accessible from Steinhatchee Falls Recreation Area. Go 1.8 miles south of US 98/19 on Florida 51 and turn left onto an unpaved road at the recreation area's sign. Drive 0.3 mile, go right at the fork, and proceed almost 1 mile. When the river is low, the small but beautiful falls of the Steinhatchee flow over the Steinhatchee Dolomite Member of the Ocala Limestone.

There is a natural bridge over the Steinhatchee River, also near Tennille. On US 98/19 just south of its intersection with Florida 51, the road crosses the Steinhatchee, which is often a dry riverbed. Immediately after crossing the river, turn east (left) on a dirt road. Drive approximately 0.5 mile to a small pond on the south (right) side of the road. Here you will see the river emerging from a cave. There are a number of small karst windows, openings where you can see into the cave from above.

Between Cross City and Chiefland, US 98/19 crosses the Suwannee River, and Fanning Springs State Park is conveniently located at this point. The spring in this park is a beautiful first magnitude spring and popular swimming hole with good exposures of the Ocala Limestone in and around it.

Manatee is another easily accessible first magnitude spring issuing out of the Ocala Limestone. Just south of the town of Fanning Springs, near Chiefland, take Florida 320 west for 6 miles to Manatee Springs State Park. From the main springhead, a long boardwalk follows the spring

Oyster reefs in a tidal marsh, Little Talbot Island State Park.

Cluster of modern Crassostrea virginica.

The Salt Marsh Environment

The salt marsh is a crucial geobiological environment that is especially well developed in this region. Along the Big Bend coast from Apalachicola to Cedar Keys is the most extensive salt marsh coverage in the state. Salt marshes are nontropical, muddy, vegetated coastal wetlands usually composed of salt-tolerant grasses such as common smooth cordgrass (*Spartina alterniflora*), saltmeadow cordgrass (*Spartina patens*), black needle rush (*Juncus roemerianus*), and Jamaica swamp sawgrass (*Cladium jamaicense*). Along tropical coasts, salt marsh vegetation is largely replaced by mangroves, although it may exist in a narrow band between the mangroves and freshwater swamps.

Salt marshes serve as nurseries for many marine species and as feeding grounds for various birds and terrestrial animals. They also filter surface water before it's discharged into estuaries and other low-energy coastal waters. Geologically, salt marshes are very effective sediment traps, slowly building up the coastal surface. This can be evident in salt marsh vegetation, which may display a unique succession of species as marsh sediments thicken. For example, *Spartina* grasses can thrive even with regular tidal inundation, but *Juncus* species will replace *Spartina* as the volume of trapped sediments increases and the ground rises above the average tidal range. Decaying salt marsh vegetation also produces peat—a precursor to coal. One can view salt marsh environments up close at Econfina River State Park, south of US 98 on County Road 14, and at Cedar Key Scrub State Reserve and Waccasassa Bay Preserve State Park, on Florida 24 at Cedar Key.

A coastal salt marsh.

run through a cypress swamp and floodplain to the Suwannee River. An extensive underwater cave system extends 2,000 feet from the main spring. The park's Sink Trail, which winds 0.6 mile though a very shady hardwood hammock, offers views of several dry sinkholes. William Bartram visited "Manate Spring" in 1774 and was elated with the natural beauty and wildlife of the area. Manatees are commonly encountered at both Fanning and Manatee springs during the winter months.

Paleodunes are present along much of this drowned karst coastline. These relict dunes are found near the coast in Taylor County and extend sporadically to the Tampa Bay area (they are especially visible along US 19 and Florida 50 in the Weeki Wachee area). The islands of Cedar Keys, southwest of Chiefland, are actually part of a dune field that was flooded during the Holocene Transgression. The highest dune is some 50 feet in elevation. The dunes have been in place for at least 6,000 years and were once undoubtedly larger since they have been under the influence of erosion since that time .

The route crosses the Waccasassa River at Gulf Hammock. Like the Steinhatchee, the Waccasassa River is another karst stream in which Pleistocene-age vertebrate remains are preserved. The Waccasassa also contains the rare remains of vertebrates of Pliocene age. Geologists think the Waccasassa has followed collapse features (sinkholes) along its

Fossil fragments of calcareous green algae, which indicate that the lower Ocala Limestone was deposited in very shallow water in a tropical climate during Late Eocene time. Each specimen is about 1.5 millimeters in length or diameter.

length, eroding fossiliferous sediments that formerly filled the sinkholes and concentrating them in its bed. To canoe the Waccasassa and collect fossils, put in at the Florida 24 bridge north of the town Otter Creek (12 miles south of Chiefland) and float downstream to US 98/19, or put in at US 98/19 and take out at the landing at County Road 326.

Near Gulf Hammock, mining operations along US 98/19 are extracting the oldest rocks exposed at the surface in Florida—the Avon Park Formation of Middle Eocene age. Here the upper Avon Park is a dolostone and contains numerous organic-rich beds. Some of the organic-rich zones contain the carbonized remains (carbon residue or film) of marine grasses. Occasionally, leaves from terrestrial trees have been recovered.

Early geologic exploration of Florida occurred in and around the town of Inglis, on the Withlacoochee River. Some of the best exposures of the lowermost Ocala Limestone, formerly called the Inglis Limestone, are found in this area. The lower Ocala is quite distinct from the upper. It consists of grainstone that is rich in nearshore-dwelling miliolid foraminifera and the remains of calcareous green algae, both of which indicate the limestone was deposited in water depths of about 30 feet or less. The mud-free grainstone, composed of coarse, rounded grains, also indicates that water conditions were energetic during deposition, removing fine sediments and rounding the coarser sediments by tumbling them. One can see the lower Ocala Limestone along many banks and boat ramps on the Withlacoochee River, but the most accessible exposures are found south of Inglis, near Crystal River, along the Cross Florida Barge Canal (see the **US 98/US 19** road log in the **Central Peninsula** chapter).

Rainbow Springs State Park

Numerous springs feed the Rainbow River, which is east of Inglis. From its headspring area at Rainbow Springs State Park, north of Dunnellon off Florida 40, the river flows approximately 5.7 miles to its confluence with the Withlacoochee River. Crystal clear springwater flows over bare Ocala Limestone for most of the river's length. There are some spectacular underwater outcrops to be inspected by either scuba diving or snorkeling, especially near spring vents. The river is best seen via canoe and can be accessed in many places along its course.

Between the inbound and outbound roads in the park there is an abandoned hard-rock phosphate pit. Hard-rock phosphate was mined in this area from the late 1800s until the mid-1960s. Rainbow Springs was an enormously popular tourist destination in the 1930s and postwar years.

Interstate 75 and US 441

Lake City—Gainesville—Ocala

80 miles

There is no faster way to traverse northern and central Florida than the I-75 corridor, but be sure to stop and see the geology! There are numerous karst streams and lakes in this region.

I-75 and US 441 roughly parallel each other between Lake City and Ocala, crossing each other several times. Several parks and preserves along the banks of the Santa Fe River are easily accessed from US 441. O'Leno State Park, first developed by the Civilian Conservation Corps in the 1930s, features many exposures of the Eocene-age Ocala Limestone in wet and dry sinkholes and an abandoned quarry. At O'Leno, the Santa Fe River disappears through a large swallet. Along the 5-mile-long Parener's Loop hiking trail, there are several karst windows—areas where the limestone bedrock has collapsed, revealing the river flowing underground. The park is 5 miles south of where US 441 and I-75 first cross each other.

About 4.7 miles south of O'Leno State Park is the entrance to River Rise Preserve State Park, where the Santa Fe River reemerges from underground. Hiking and horse trails connect O'Leno and River Rise. About 0.25 mile south of River Rise is Camp Kulaqua, the location of Hornsby Spring, a first magnitude spring that discharges into the Santa Fe River. Divers have mapped an underwater cave system at this spring.

Off County Road 232 (NW 53rd Avenue), northwest of Gainesville and sandwiched between the routes, is Devil's Millhopper Geological State Park, home of a 120-foot-deep collapse sinkhole. Access to the sinkhole is limited to the designated trails and a boardwalk to the bottom. The Devil's Millhopper primarily exposes the Hawthorn Group and a few feet of the Ocala Limestone at its base. The sinkhole has long been considered an important exposure of the Hawthorn Group of Early to Middle Miocene time (23 to 10 million years ago) due to the thickness of the exposed section. Approximately 105 feet of phosphatic sands, clays, and dolostones, with a few feet of undifferentiated sands at the surface, compose the section. Though covered with lush vegetation, Devil's Millhopper remains one of the best exposures of the Hawthorn Group in Florida.

Throughout much of the state, Hawthorn Group rocks function as water-bearing aquifers or, more often, as low-permeability aquitards, layers of rock or sediment that inhibit the flow of groundwater. They are also mined for phosphate in northern and central Florida. The sinkhole at Devil's Millhopper formed when sediments overlying the Ocala Limestone collapsed into cavities that groundwater had dissolved in the limestone. The elevation at the lip of the sinkhole is approximately 175 feet above sea level.

In Gainesville, it will be absolutely necessary for every geologist, paleontologist, archaeologist, or naturalist of any stripe to visit the Florida Museum of Natural History. Chartered in 1917, the museum is the largest natural history museum in the Southeast, with more than 25 million specimens of recent and fossil plants, invertebrates, vertebrates, and archaeological artifacts. Strong research programs focus on the natural history of Florida, the southeastern United States, and the Caribbean. From I-75, take Exit 384 and travel east on Florida 24 (Archer Road). Turn north (left) on Florida 121 (SW 34th Street). At the third traffic signal, turn east (right) on Hull Road and travel 0.25 mile. The entrance to the University of Florida Cultural Plaza is on the south (right) side of Hull Road. From US 441, travel west on Florida 24 to Florida 121.

Several creeks within Gainesville have yielded impressive Miocene-age vertebrates from the Hawthorn Group, including an abundance of shark teeth and dugong ribs. US 441 crosses over Hogtown Creek, probably the most productive of these fossil sites.

Numerous limestone quarries dot the landscape in Alachua County and surrounding areas. Located in the Newberry area, about 14 miles west of Gainesville on Florida 26 and then 3 miles north on County Road 235, is the famous complex of limestone pits known to geologists

Ocala Limestone of Late Eocene age exposed at the Haile Quarry complex.

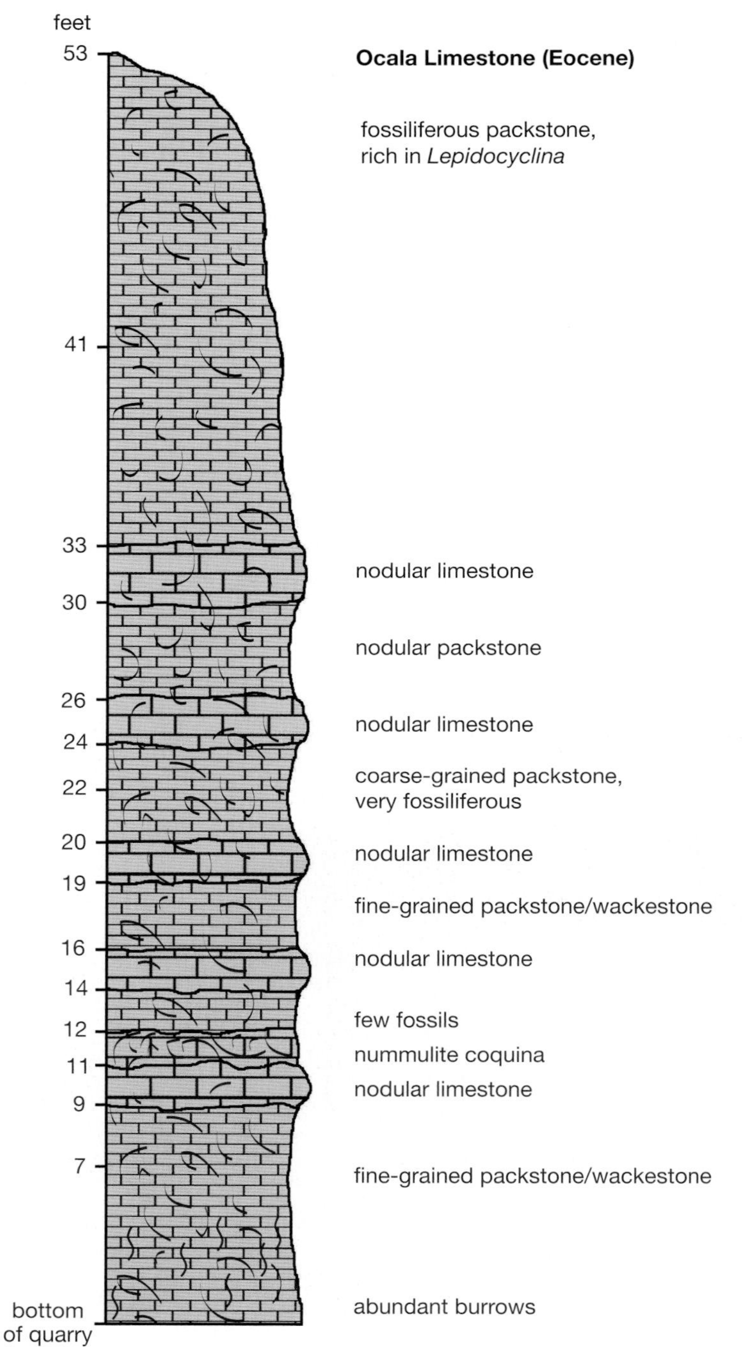

Stratigraphy of the Ocala Limestone at the Haile Quarry complex.
—Illustration by Amanda Kosche

as the Haile Quarry complex. These active mines have extensive and continuous exposures of the Ocala Limestone and have produced an extremely diverse fauna of larger foraminifera and other marine invertebrates. Through careful study of the strata and fossils in this complex, geologists deduce that the Ocala Limestone in these pits preserves several deposition cycles of sandy (carbonate), fossiliferous limestone followed by nodular limestone. There are also layers of nummulitic coquinas of the large foraminifera *Nummulites*, which represent shoal or bar deposits. The rocks also have other larger foraminifera in abundance, including the genera *Lepidocyclina*, *Pseudophragmina*, and *Asterocyclina*. These fossils indicate that sea level was rising throughout the time the strata were accumulating. Such "deepening-up" of strata is called a *transgressive sequence.*

The Haile Quarry complex has also been a rich source of vertebrate fossils of Miocene to Pleistocene age. These fossils are preserved in chimneylike sinkholes and other karst features that developed in the Ocala Limestone. One Pliocene-age site, discovered in 1986 and known as the "sloth hole," produced ten skeletons of a new species of giant ground sloth, *Eremotherium eomigrans.* The sloth hole was a large sinkhole lake

Fossiliferous Ocala Limestone of Late Eocene age, rich in the larger foraminiferan Lepidocyclina ocalana, *from the Haile Quarry complex.*

The piglike peccary Platygonus cumberlandensis *being pursued by the giant jaguar (*Panthera onca*). These fossils of middle Pleistocene age were recovered at the Haile Quarry complex and are on display at the Florida Museum of Natural History, Gainesville.*

or pond that also preserved many other vertebrates, including amphibians, alligators, turtles, birds, rabbits, horses, camels, and giant tortoises. Another site at the Haile Quarry complex, called "hog heaven," preserved numerous Pleistocene-age peccaries, piglike creatures that were extinct in Florida by the end of Pleistocene time. These creatures survive today, primarily in the Central and South American tropics.

After leaving Gainesville, the routes travel through a region that has some well-known fossil localities. These archived a great deal of North America's animal life that lived from Late Oligocene to Pleistocene time. The first is located on the south side of Gainesville, where Florida 121 crosses I-75. Although the site is no longer visible, it was at this intersection that, in 1965, construction workers uncovered a sediment-filled fissure while building the interstate. The deposit, about 16 feet in diameter and 6 feet deep, preserved the oldest land animals ever found in Florida. The I-75 Fauna of Late Oligocene age contained a variety of vertebrates, some up to 30 million years old. The ancient sinkhole held the fragmentary remains of the horse *Merychippus*, two carnivores, the deerlike ungulates *Leptomeryx* and *Hypisodus*, a variety of other mammals (including two oreodonts, *Eporeodon occidentalis* and *Oreodontoides oregonensis*), and several smaller species of reptiles and mammals,

Fossil Ground Sloths

Perhaps the strangest animals living today are the small tree sloths of the Central and South American rain forests. There are two genera, distinguished by having two claws versus three claws, and both are arboreal, nocturnal, slow-moving vegetarians with low internal body temperatures and special adaptations for digesting plant cellulose. They are so adapted to their arboreal habitat that they are unable to walk. When on the ground, they must drag themselves by their long, clawed arms. But their ancient cousins were primarily adapted for life on the ground.

Ground sloths first appear in the geologic record in South American rocks of Oligocene age. They reached North America by Late Miocene time, and they appeared in Florida during this time. The arrival of sloths in Florida from South America was the beginning of the Great American Biotic Interchange, which continued through Pleistocene time. Twelve species of ground sloths,

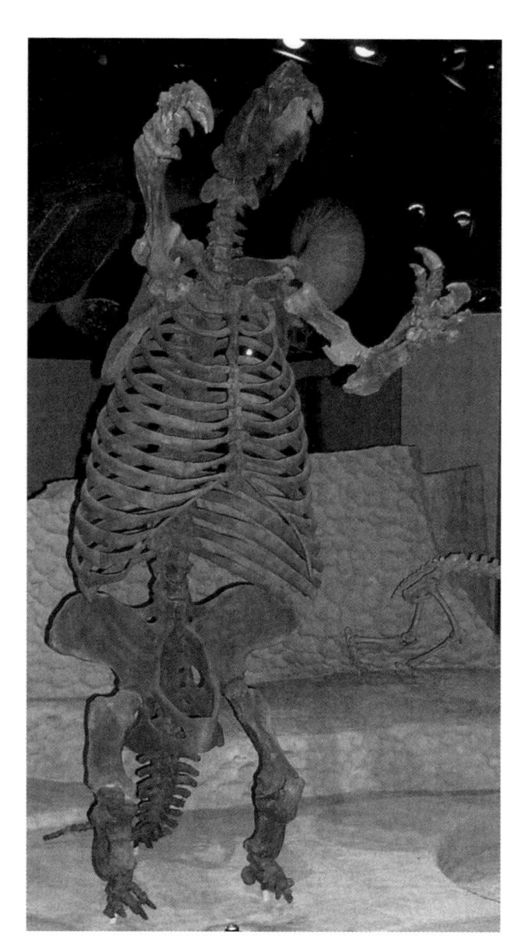

Thinobadistes segnis *from the famous Mixson's Bone Bed of Late Miocene age in Williston.*

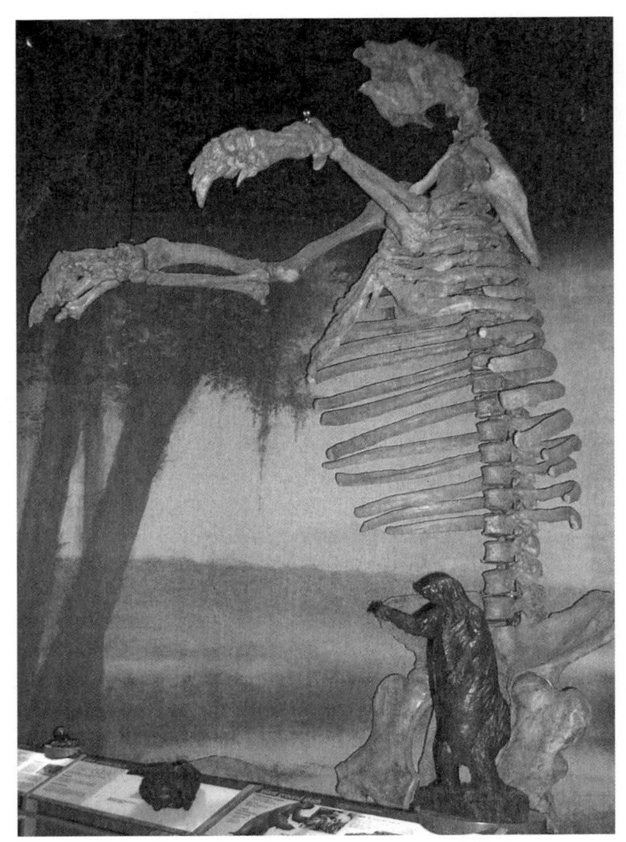

The "Sloth from Haile," the giant Eremotherium eomigrans *found in sediments of Late Pliocene age in the Haile Quarry complex.*

Artist's rendering of ground sloths and the armadillo-like glyptodonts.
—Courtesy of the Florida Geological Survey

representing three families, have been found in Florida in rocks of Late Miocene, Pliocene, and Pleistocene age. From their distribution, it appears that they were primarily coastal dwellers. They became extinct in North America at the end of Pleistocene time.

Because of their large claws, most sloths walked on the sides of their feet. They used their tails for balance in a tripod stance when standing erect so they could reach and tear down the tree branches on which they fed. One species found in Florida, *Megalonyx jeffersoni*, was first described by President Thomas Jefferson in 1797. One of the most remarkable sloths ever found in Florida was the gigantic *Eremotherium eomigrans*, discovered in a sinkhole deposit in limestone at the Haile Quarry complex. This species was the largest sloth ever to have lived. Adult males of this species weighed an estimated 6,600 pounds.

Fossil Rhinoceroses and Tapirs

It is sometimes hard to imagine just how strange prehistoric life in Florida truly was. The mighty rhinoceroses of Africa, India, and Indonesia, of which there are five species, are the last surviving members of an ancient lineage that included the Pleistocene-age woolly rhinoceros from Eurasia (depicted in cave paintings) and the largest land mammal of all time, the 16-foot-tall *Baluchitherium* from central Asia. This rhino of Oligocene and Miocene time had a skull that was nearly 4 feet long. Rhinoceroses had a successful tenure in

Teleoceras proterum, *a short-limbed, hippopotamus-like rhino of Miocene age from McGehee Farm, Alachua County. The specimen is at the Florida Museum of Natural History.*

North America from Late Eocene to Early Pliocene time, and ten or more species, including the genera *Menoceras, Floridaceras, Aphelops,* and *Teleoceras,* have been recovered from Floridan strata. They are especially common in several Miocene-age fossil localities in the northern peninsula.

Like horses and rhinos, the peculiar tapirs are also part of the perissodactyl order of odd-toed ungulates (hoofed animals). There are four living species of the genus *Tapirus,* three in Central and South America, and one in Southeast Asia. They are easy to recognize because of their very short, elephant-like proboscis, which the animal uses to gather food. Several species of tapir have been found in Florida formations of Early Miocene to late Pleistocene age, when they became extinct in North America.

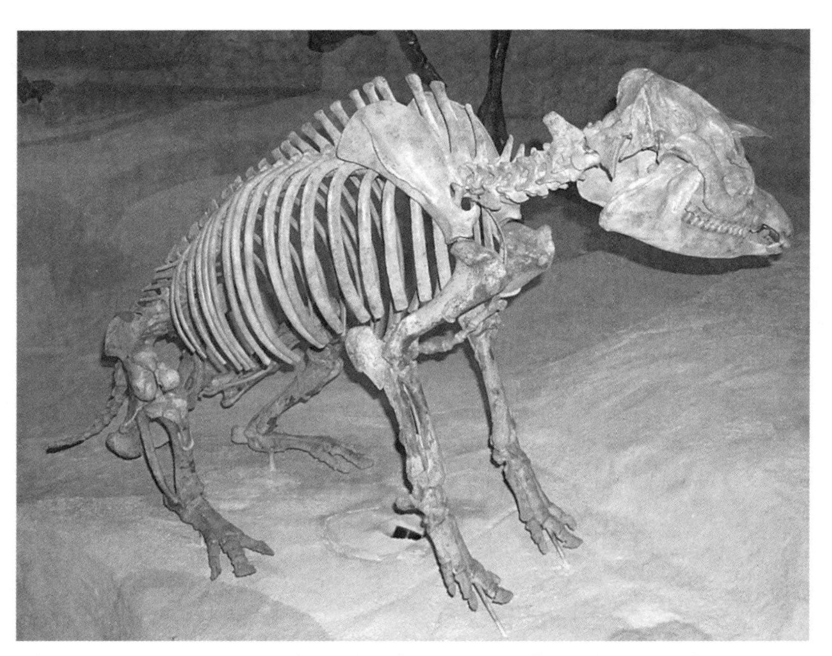

The tapir* Tapirus veroensis *from Florida sediments of middle to late Pleistocene age. The specimen is at the Florida Museum of Natural History.

including an insectivore, an opossum, and bats. Interestingly, the site also preserved about ten species of marine fish. This association of terrestrial and marine species indicates that the site must have been very close to a shoreline in Late Oligocene time. Since the sinkhole was filled with Oligocene-age fossils, geologists know that karst processes were in effect in Florida as early as Oligocene time. Terrestrial fossil localities of this age have also been found near Tampa, Brooksville, and Live Oak.

Southwest of Gainesville, and just north of the town of Archer (on Florida 24), is one of the richest vertebrate fossil sites of Late Miocene age (about 9 million years old) in the United States. Mr. Ron Love discovered the bone bed in 1974 while tilling his okra crop. He turned up a tibia of the extinct hippopotamus-like rhinoceros *Teleoceras*. Extensive excavations in the 1970s revealed that the Love Bone Bed, as it came to be called, represented the channel of an ancient stream that flowed over karst.

The bottom bed of the site is a type of bone deposit called a *bone breccia*, which represents deposition that occurred in the bottom of a river channel, where bones were concentrated by currents. The bed is dominated by abraded and worn bones and is especially rich in freshwater turtle and gar remains. The middle bed consists of trough cross-stratified sands, also representing river channel deposits, and is the most fossiliferous bed. The top bed is composed of clay-rich sands, representing floodplain deposits. It contains well-preserved, unabraded material, including several nearly complete turtle shells, indicating that this bed was laid down in quiet water. Large blocks of Ocala Limestone also appear in the old stream channel. The site could not have been far from coastal or estuarine environments because remains of marine vertebrates occur at the base of the deposit, including the extinct great white shark *Carcharodon megalodon*, other sharks, the dugong *Metaxytherium*, and even fragmentary whale remains.

Researchers have uncovered nearly one hundred species of vertebrates at the Love Bone Bed, including giant tortoises; the extinct crocodile *Gavialosuchus*; a variety of birds; peccaries; tapirs; the rhinos *Teleoceras* and *Aphelops*; numerous species of horses, including *Cormohipparion*; the gomphothere *Amebelodon*; and early camels, including *Procamelus* and the giant giraffe camel *Aepycamelus major*, which stood over 16 feet tall and browsed the tops of trees. Two of the most spectacular finds at the Love Site are specimens of extinct catlike carnivores (nimravids): *Nimravides galiani* and *Barbourofelis loveorum*, the so-called false saber-toothed cat.

About 3 miles south of Gainesville the routes enter Paynes Prairie Preserve State Park, a 21,000-acre preserve over a karst lake. In 1774, the naturalist and explorer William Bartram visited the area and called it the "great Alachua Savanna." He described it this way:

> *The extensive Alachua savanna is a level green plain, above fifteen miles over, fifty miles in circumference, and scarcely a tree or bush of any kind to be seen on it. It is encircled with high, sloping hills, covered with waving forests and fragrant Orange groves, rising from an exuberantly fertile soil. The towering magnolia grandiflora and transcendent Palm, stand conspicuous amongst them.*
>
> —*Travels of William Bartram*, 1791

The ridge on the north end of the prairie is the easternmost expression of the Cody Scarp, and it marks the southern terminus of the older Northern Highlands physiographic feature. Paynes Prairie is essentially an erosional gap where a portion of the Central Highlands has been removed, connecting three lower areas (sometimes called the Western, Central, and Eastern valleys) in an east-west corridor running from Newberry to Palatka, and just south of Gainesville.

During the 1800s, the area was mostly a dry savannah or marsh, depending on rainfall. During times of heavy rainfall an ephemeral lake formed, but it drained rather quickly. The main drain for the area is known as Alachua Sink. The sink has drained water from Paynes Prairie since before the area was settled. Locals enjoyed visiting the sink—and throwing logs into the swirling waters and watching them disappear. Geologists think this activity eventually plugged the sink, causing a lake to form. A small steamboat operated on the lake from 1871 to 1891. Eventually, the logjam disintegrated and the lake drained, creating the large marshy area that remains today. Like many intermittent karst lakes, water level is a function of precipitation, groundwater level, and sedimentation. Both the northern and the southern margin of the karst plain are prominent on I-75 and US 441, rising up to 40 feet above the prairie floor. The prairie is about 2 miles wide from north to south. The I-75 rest areas on the northern side of the basin provide excellent views of the prairie.

Paynes Prairie, a karst basin south of Gainesville.

On the southern edge of Paynes Prairie, the land rises onto Bolen Bluff, an important Native American site. Around 9,000 years ago, prehistoric Floridians flourished here and made tools that indicate they were of the early Archaic Period. They manufactured a type of projectile point called a *bolen*, named after this site. They lived during a period of rapid climate change and shortly after the extinction of many large animals at the end of Pleistocene time. Their tool kits are much different from those of the Paleoindian cultures that preceded them and were designed for hunting smaller game. The shift from large, lanceolate projectile points to smaller projectile points with notched sides or corners was a dramatic change in hunting technologies.

About 11 miles south of Paynes Prairie on US 441, and about 1 mile south of the quaint, historic town of McIntosh, is a scenic overlook from a sandy karst bluff. To the east is Orange Lake, a large alligator-infested karst lake. Essentially, the lake represents the water table—the top of the saturated zone—of the Floridan aquifer system in this region, and water levels in this lake may fluctuate dramatically during times of diminished rainfall.

About 14 miles west-southwest of Orange Lake, near the towns of Williston and Morriston, was a vertebrate fossil locality of Late Miocene age known as the Moss Acres Racetrack Site. As with many fossil sites, Moss Acres was serendipitously discovered. In 1984, workers uncovered fossil rhino bones of the hippopotamus-like *Teleoceras* while grading a horse racetrack on the site of a clay-filled sinkhole. The sinkhole also contained peat, indicating marsh and swamp conditions, and partially articulated skeletal material, which is extremely rare in Florida. The site entombed many turtles, alligators, and several mammals, including specimens of *Amebelodon britti*, a giant shovel-tusked mastodon (a gomphothere); the ground sloth *Pliometanastes*; about eight species of horse; a second rhinoceros called *Aphelops*; peccaries; camels; llamas; antelope-like ruminants, such as the strange, three-horned *Pediomeryx*; the carnivorous beardog *Ischyrocyon*; and the giant otter *Enhydritherium*.

Just north of Williston is another famous fossil locality—the popular diver's haunt Devil's Den, a sinkhole with vertical walls that made it a natural trap. It preserved numerous terrestrial vertebrates of late Pleistocene age in water about 70 feet deep. Many small mammals are present, including opossums, shrews, moles, bats, squirrels, gophers, mice, rats, voles, raccoons, weasels, skunks, and muskrats. Large mammal remains include dire and red wolves, ground sloths, black bears, spectacled bears, bobcats, jaguars, saber-toothed cats, mastodons, horses, peccaries, pigs, deer, buffalo, and even humans. It is clear from numerous sites in Florida that entrapment in sinkholes or spring-fed water bodies was a common way that terrestrial animals were fossilized from Oligocene time to the present day.

The region around Ocala rests on the Ocala Limestone, which was named for exposures near the town. There are many active and inactive mines in the area, as well as caves, but access to both is extremely limited. Many springs issue out of the Ocala Limestone here. There was once a privately owned attraction called Ocala Caverns, but it is now closed.

No visit to Ocala would be complete without a stop at Silver River State Park (Silver Springs), one of Florida's oldest tourist attractions. It is 3.5 miles east of Ocala off Florida 40. The Silver River area has been inhabited and utilized by humans since the time of the Paleoindians, some 11,000 years ago. Many archaeological and paleontological sites exist along its banks and in the riverbed. The Silver River Museum contains a wonderful exhibit on the history of Silver Springs and the Silver River.

Commerce along the river began in the 1850s as steamboats paddled upriver, exchanging goods along the way. During the late 1800s the first glass-bottomed boats, for which Silver Springs is so famous, were put into use. Visitors were awed by the beauty and clarity of the massive quantity of water that discharges from more than fourteen underwater vents along the upper mile of the Silver River. The section of the Silver

Silver River, the spring run of Silver Springs, feeds the Oklawaha River.

River from Silver Springs to the Oklawaha River is one of the most scenic canoe runs in the state.

Silver Springs was used for many TV programs because of the great clarity of its water. Over twenty movies have been filmed at or near the springs, including several Tarzan features of the 1930s and 1940s, *The Yearling* with Gregory Peck (1946), and some James Bond films. Today, Silver Springs remains a popular tourist destination. Having one of Florida's largest springs, it is another geological must-see.

Florida 100

Lake City—Palatka

76 miles

Between Lake City and Starke, Florida 100 passes through the Okefenokee Basin District, lying over the Coosawhatchie Formation, which is the uppermost unit of the Miocene-age Hawthorn Group in much of northern Florida. Although it is not well exposed in the area, the formation consists of sandy, phosphatic dolostones and clays. The same unit is exposed at Devil's Millhopper, near Gainesville. The Coosawhatchie Formation gently dips to the northeast and thickens toward Jacksonville.

About 6 miles northeast of Keystone Heights on Florida 21 is Mike Roess Gold Head Branch State Park. Developed by the Civilian Conservation Corps in 1939, it was one of the first state parks in Florida. Here, a deep ravine, up to 75 feet deep, cuts through undifferentiated sediments of Pleistocene age and the Cypresshead Formation of Pliocene age. Rumor has it that minute quantities of gold were once panned from the Gold Head Branch, but gold has never been documented here. Others claim that a Civil War gold shipment was hidden nearby. While either story may account for the original name, neither seems likely to be factually true. However, the topography of this part of the Central Highlands (it is the southern end of Trail Ridge) as seen in this park is impressive!

Sheelar Lake, also located in the park, has an extremely important sediment and pollen record that reaches back 14,000 years. Much windblown sand fell into the lake during the last glacial maximum, the Wisconsinan glacial stage, and the pollen record shows the transition from drier vegetation to modern forest conditions of early Holocene time.

Farther southeast on Florida 100, along the west bank of the St. Johns River in Palatka, is Ravine Gardens State Park. Located on Twigg Street, 0.5 mile south of Florida 20 in the historical district, the park features several beautiful ravines that cut deeply into Pleistocene-age sands—up to 100 feet or more. A paved road winds around the ravines, and there are numerous hiking trails. Located on the other side of the St. Johns River, off US 17, is Dunns Creek State Park, which also has spectacular, deep erosional ravines in Pleistocene-age sands. Archaeological sites are common in this area.

South of Palatka, Florida 19 passes through the Ocala National Forest. This forested region—the Big Scrub—lies within the broad Central Lakes District. The Cypresshead Formation of Pliocene age and undifferentiated sands of Quaternary age are exposed in various hillsides and sand pits. This area was the setting for the Pulitzer Prize–winning 1938 novel *The Yearling*, by Marjorie Kinnan Rawlings. There are several exceptional springs in the Ocala National Forest, most flowing through Hawthorn Group sediments, with the ultimate source of the water being the Ocala Limestone of the Floridan aquifer system.

Interesting sites in and around the Ocala National Forest

•**Salt Springs Recreational Area**—On Florida 19 about 22 miles south of Palatka. Saline springwater emerges from Hawthorn Group carbonates.

•**Silver Glen Springs**—About 6 miles north of Florida 40 on Florida 19. An underwater cave system has been mapped here.

•**Juniper Springs Recreation Area and Fern Hammock Springs**—About 4.5 miles west of Florida 19 on Florida 40. Beautiful Juniper Springs is a very popular camping area and canoe run. Fern Hammock, accessible by hiking trail, is located downstream of Juniper Springs along Juniper Springs Run. The spring emerges from Hawthorn Group carbonates.

•**Alexander Springs Recreation Area**—From its intersection with Florida 40, take Florida 19 south 9.4 miles into Lake County. Turn east (left) on County Road 445 and drive about 5 miles. This first magnitude spring has a very large vent and spring pool and well-exposed Hawthorn Group limestone. The spring run flows about 8 miles to the St. Johns River.

•**DeLeon Springs State Park**—East of the Ocala National Forest and in Volusia County, about 8 miles northeast of DeLand. From US 92 in DeLand, take US 17 north for 5.9 miles to the park entrance. The Miocene-age Coosawhatchie Formation is exposed in this second magnitude spring, which runs to Lake Woodruff and Lake Dexter, and then to the St. Johns River. While in DeLand, check out the Gillespie Museum of Minerals on the campus of Stetson University.

•**Blue Spring State Park**—From Florida 44 in DeLand, take US 17/92 south for 5.3 miles. Turn west (right) on County Road 4142 (French Avenue) at Orange City and drive about 2.5 miles. Volusia Blue Spring is a first magnitude spring and a favorite winter home of Florida manatees, which make their way from the Atlantic Ocean and up the St. Johns River to this point. It is not uncommon to see fifty to one hundred manatees leisurely swimming in the river. The clear springwater makes the gray manatees appear as giant blue submarines. It is truly an amazing and even prehistoric sight. The naturalist John Bartram explored the St. Johns River and Volusia Blue Spring in 1766.

•**Hontoon Island State Park**—From its intersection with US 17/92 in DeLand, follow Florida 44 west about 6 miles. A ferry shuttles visitors to this island between two channels of the St. Johns River. The island has a rich archaeological history, including Timucuan Indian mounds that can be observed from the nature trail.

Florida A1A (I-95 and US 1)

Fernandina Beach—New Smyrna Beach

140 miles

The Atlantic Coastal Ridge, that long, narrow, rather stable complex of raised dunes, berms, beaches, coquina rock, and oolitic limestone of Pleistocene and Holocene age, stretches between Amelia Island and Biscayne Bay—almost the entire length of the Florida Peninsula. The Atlantic Ocean continues to modify this ridge, and it is a fascinating study in coastal geological processes. The backbone of the ridge is the unique coquina of the Anastasia Formation of Pleistocene age, which grades into the Miami Limestone (also of Pleistocene age) in southern Palm Beach County. The Miami Limestone composes the ridge along its southern portion. The Anastasia is exposed in many places along the coast, but it is often covered by a thin drape of Holocene-age sand. US 1 and Florida A1A were largely constructed atop this ridge.

Just inland of the ridge is the Eastern Valley, through which flows the St. Johns River. Ancient dune fields and beach ridges are common in this valley. I-95 and US 1 follow the Atlantic Coastal Ridge and its boundary with the Eastern Valley from near Jacksonville almost to Palm Beach. From Palm Beach to Miami, I-95 and US 1 are on top of the Atlantic Coastal Ridge. The modern coast is armored by a thickly vegetated foredune ridge. The Florida segment of the Atlantic Coast is perfectly situated for coastal dune development because it extends into latitudes that are influenced by the strong easterly trade winds. This eastern spine of Florida also made a great roadbed for Henry Flagler's Florida East Coast Railway. Beginning in 1886, this railway was created through purchases of existing rail lines and construction of new rail sections, finally reaching Key West in 1912.

The north-flowing St. Johns River, which some refer to as a "river of lakes," has its headwaters in St. Johns Marsh in Brevard and Indian River counties and becomes an estuary in Jacksonville. It flows over 300 miles in the Eastern Valley, making it the longest river in Florida.

The origin of this river and its interesting course have been the subjects of much geologic speculation. It appears the river formed over an ancient coastal lagoon or beach ridge plain, both topographical features in which river valleys readily form because they are relatively low compared to surrounding areas. The river is impounded on the east by the Atlantic Coastal Ridge and is otherwise surrounded by Quaternary-age dune and beach ridge plains or ridges of the Cypresshead Formation. Regional faulting or jointing (fracturing) patterns may also be responsible for the flow pattern.

Offshore of the Atlantic Coastal Ridge from Jacksonville to the St. Lucie Inlet near Port Salerno, and in depths of about 230 to 330 feet, are the deepwater coral reefs of the Oculina Banks. Ivory tree coral (*Oculina varicosa*) constructs these reefs, forming dense thickets of colonies up to nearly 5 feet tall. They can cover several acres or occur as small pinnacles, mounds, or ridges. In many cases, the corals grow over oolitic and algal limestones, probably of Pleistocene age. In shallower water closer to shore in the Jacksonville and Daytona Beach areas, many rocky ridges or ledges of the Anastasia Formation or its equivalent rise up to 10 feet or more off the seafloor and host a variety of marine life. These are not true reefs, but they are popular spots for scuba diving.

Oculina, *a common deepwater reef constructor.*

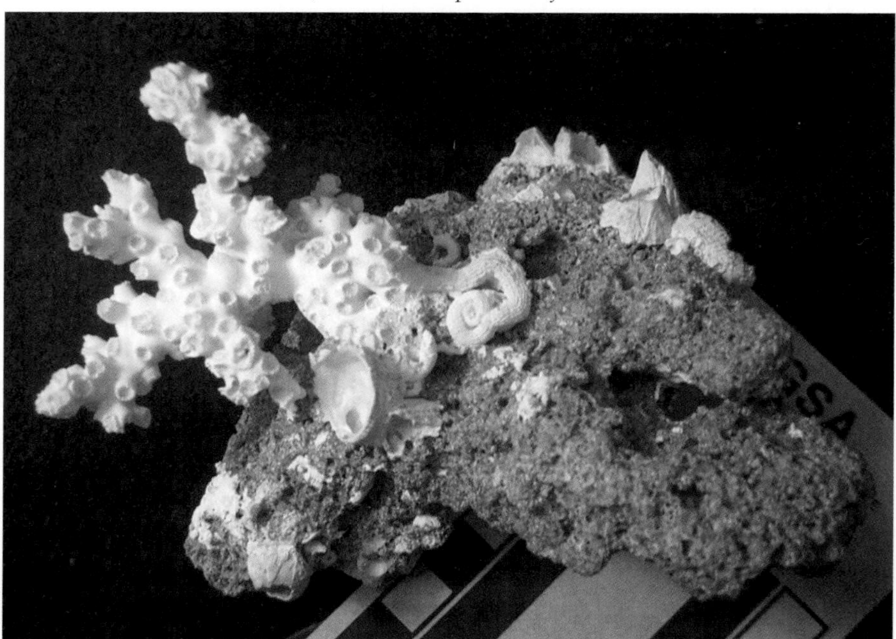

The coastal region of the northeast corner of Florida is a wonderfully complex geological and ecological environment. The town of Fernandina Beach is on Amelia Island, which, along with Big Talbot and Little Talbot islands, are the southernmost Sea Islands. These barrier islands, which are numerous along the Georgia and South Carolina coasts, are composites of both Pleistocene- and Holocene-age sediments. They formed due to abundant and direct input of river sediment and strong reworking of that sediment by waves, longshore currents, and mesotidal tides (exceeding 6 feet). Onshore Atlantic winds form dunes on these islands, and at Fernandina Beach paleodunes may exceed 50 feet in elevation. Extensive salt marshes reach far inland due to the influence of relatively high tidal ranges over this coastal lowland. They are visible along A1A just east of Fernandina Beach, as the road passes over the Amelia River. Oyster reefs or shoals are common in the tidal creeks.

Farther inland and along the St. Marys River, which forms the Florida-Georgia border in this region, a Quaternary-age barrier island and dune sands are preserved along Reids Bluff, Roses Bluff, and Bells Bluff, upriver from St. Marys, Georgia. Fossils in these ancient barriers include the oyster *Crassostrea virginica*, quahog clam (*Mercenaria mercenaria*), and the shrimp burrow trace fossil known as *Ophiomorpha nodosa*. Also, remarkably preserved at the base of Reids Bluff are large, upright cypress tree stumps buried in their original growth position with preserved roots and soil. Radiocarbon dates indicate they range from 25,000 to more than 38,000 years old.

Amelia Island consists of beach and dune deposits. After crossing the Nassau River, A1A passes onto Talbot Island. Big Talbot Island State Park, at the southern tip of this island, is a good place to see humate sand, a fine sand that is weakly cemented by a black or brown organic material called *humate*. It is visible along the park's Black Rock Trail and is extensively exposed along the beach. Humate is derived from decaying plant debris and transported by groundwater. As groundwater percolates through the porous coastal sand, the carbon-rich humate comes out of solution and may cement the sand into a crumbly black sandstone. Many fallen trees and driftwood can be seen on this eroded coast at "boneyard beach." There are Timucuan Indian mounds on Talbot Island and in surrounding areas.

Florida A1A next crosses Simpson Creek and travels onto the pristine Little Talbot Island, the southernmost of the Sea Island chain. At the historical Fort George Island Cultural State Park, 3 miles south of Little Talbot Island State Park, people of the late Archaic Period built a mound, now called Mount Cornelia, that is over 60 feet high. It is said to be one of the highest points along the Atlantic Coast south of North Carolina.

Two miles south of Jacksonville Beach is the town of Ponte Vedra Beach, where it is not uncommon to find fossil shark teeth and other vertebrate remains that have washed ashore. Be sure to check the surf zone and any shell piles on the beach. (A good collecting spot is at Mickler Landing, or anywhere just south of there along the shoulders of A1A.) Ponte Vedra was formerly known as Mineral City because of heavy-mineral mining here. South of Mickler Landing and just north of St. Augustine is Guana River State Park, a national estuarine preserve and possibly the site of Ponce de Leon's first landing in Florida.

Though they can be found in many areas across the state, the peculiar fulgurites are common along the Atlantic Coast. Fulgurites are sandy and glassy cylindrical tubes that form when lightning strikes the sand. The superheated quartz sand melts and resolidifies as silica glass, forming the granular tubes. Unmelted sand grains are fused into the glass as well. In some cases, fulgurites look like fossilized root systems, and this may indeed be the case. Geologists think roots may provide a template of

A fulgurite.

sorts for the formation of fulgurites because they contain water, which could conduct the electrical charge.

Any visit to northeast Florida must include a tour of the truly historic and charming St. Augustine. It has been said that when the English Pilgrims landed at Plymouth Rock in 1620, the town of St. Augustine (founded in 1565) was already fifty-five years old and up for urban renewal! St. Augustine was indeed the first city Europeans permanently settled in the United States. Residents of the western Florida Panhandle boast that Pensacola was, in fact, founded earlier—in 1559. This is true, but Pensacola was not *permanently* settled until 1698.

The geologic foundation of St. Augustine is the coquina of the Anastasia Formation, first named in 1912 for exposures on nearby Anastasia Island. The Anastasia is a molluscan coquina with interbedded, calcite-cemented quartz sand. It extends along most of the eastern seaboard of Florida from St. Augustine to Boca Raton. The best exposures lie between St. Augustine and Ormond Beach, Cocoa and Eau Gallie Beach, and Stuart and Boca Raton. In some areas, it is an uncemented sand and shell mix. The shell fragments of the coquina are primarily bivalve mollusks with some gastropods, nearly all of which are nearshore species still living today. The shell grains were broken, tumbled, and waterworn in an ancient surf before being cemented together to form the coquina. Cross-stratification is very evident in the formation, further indicating that deposition occurred in high-energy coastal waters. Besides mollusks, other fossil invertebrates are found in the Anastasia, as are marine and terrestrial vertebrate remains, including those of whales, horses, giant armadillos, tapirs, turtles, and sharks. Geologists have dated the formation as being 110,000 years old, or Pleistocene in age, but parts of it may have actually been cemented as late as Holocene time. Some geologists believe that the deeper portions of the formation are significantly older than 110,000 years.

Anastasia Formation coquina was the building stone for many historic structures in St. Augustine and surrounding areas. Most famous is the impressive Spanish fort Castillo de San Marcos, started in 1672 and essentially completed in 1695. Twice, in 1702 and 1740, the coquina fort successfully withstood English attacks. The geological secret to this military success is the coquina itself. Rather than crumbling, it absorbs cannonballs "as though you would stick a knife into cheese," as one early observer noted. Coquina rock of the Anastasia Formation was used as a building stone for numerous other structures in the area, including houses and walls. Unconsolidated coquina gravel has been used as road base and concrete aggregate, and as an additive in chicken feed (for calcium). William Bartram wrote the following description of the coquina and its use in construction in his *Travels*:

Castillo de San Marcos, constructed of coquina from the Anastasia Formation of Pleistocene age.

> *The works are constructed with hewn stone, cemented with lime. The stone was cut out of quarries on St. Anastatius Island, opposite St. Augustine: it is of a pale reddish brick colour, and a testaceous composition, consisting of small fragments of sea-shells and fine sand. It is well adapted to the construction of fortifications. It lies in horizontal masses in the quarry, and constitutes the foundation of that island. The castle at St. Augustine, and most of the buildings of the town, are of this stone.*
>
> —*Travels of William Bartram*, 1791

One can observe the Anastasia Formation at an old, abandoned coquina quarry at the entrance to Anastasia State Park off Florida A1A. The park entrance is 1.3 miles east of the east end of the Bridge of Lions, which crosses the Matanzas River. The quarry, which is marked, is on the west side of the park's entrance road. It reveals about 10 feet of the Anastasia Formation, consisting entirely of coquina. There are other coquina quarries near St. Augustine. Anastasia State Park also features many Holocene-age dunes along the seaward side of Salt Run lagoon, and older Quaternary dunes can be explored along the Ancient Dunes

Nature Trail, near the camp store. Anastasia Island was isolated from the mainland during the Holocene Transgression and is separated from the peninsula by the Matanzas River lagoon. Anastasia State Park is also the entry point to Conch Island—a sand spit on the northeastern end of Anastasia Island. Inland from Anastasia Island, along the mainland, is an extensive beach ridge and dune field.

It was along the shores of Anastasia Island that, in late 1896, the St. Augustine Monster washed ashore and became lodged in the sand. The "monster" was a gigantic blob, reportedly 21 feet long, 7 feet wide, and 4 feet high. First found by two boys and then reported to a local physician, the blob aroused the interest of some leading zoologists of the day, including A. E. Verrill at Yale and W. H. Dall at the National Museum of Natural History, and the strange specimen is widely discussed in popular literature (and some textbooks) to this day. The carcass was first suspected to be the remains of a giant squid (*Architeuthis*), then a giant octopus (the name *Octopus giganteus* was even given to it), then a whale. As one scientist described it in 1897, "The substance looks like blubber, and smells like blubber, and it is blubber, nothing more or less." Recent

Early quarrying of the Anastasia Formation. —Reproduced by permission of the St. Augustine Historical Society

biochemical work on preserved samples indicates that the St. Augustine blob was the remains of a whale, a mass of collagen.

This curious bit of cryptozoological history is an important lesson in the taphonomy (process of fossilization) of large marine animals. Fossil remains of marine mammals are common in many parts of the state and range from Eocene to Pleistocene age. Subfossil (nonmineralized) remains have been found in sediments of Holocene age. When a large marine animal dies, if it is not immediately scavenged it may float due to the accumulation of decay-related gases before sinking to the seafloor, or it may wash ashore or into a bay. During this period of decay, the carcass may decompose and fragment so badly that the original creature becomes unrecognizable. This is one reason why fossil whales, manatees, and other large marine animals may consist of only isolated parts, such as skulls, limbs, or partial skeletons. If the carcass sinks quickly, however, the whole skeleton may be fossilized. The modern ocean floor is littered with such ghostly skeletal remains. Large carcasses also account for many "sea monsters" of uncertain identity that people have seen floating in the ocean, washed ashore, or caught in fishing nets. As with the St. Augustine monster, this may be the explanation for some other "sea serpent" sightings and encounters in Florida waters over the years.

Fort Matanzas National Monument is at the southern tip of Anastasia Island. Like Castillo de San Marcos, Fort Matanzas is constructed of coquina of the Anastasia Formation. Built between 1740 and 1742 to protect St. Augustine, the fort was constructed at the site where Pedro Menendez de Aviles slaughtered hundreds of French soldiers who had been shipwrecked by a hurricane. The Anastasia Formation can be seen at a prominent outcrop at Matanzas Inlet and also along the beach at Washington Oaks Gardens State Park, 2 miles south of Marineland. Exposures of coquina of the Anastasia Formation alternate with sandy beaches along this part of the Florida coast, down to the vicinity of New Smyrna Beach. Geologists have suggested that this pattern may be caused by large-scale circulation cells. As rock is exposed by converging, erosive longshore currents, beach sediments accumulate between the rock outcrops.

Living and Fossil Sea Turtles

There are eight species of sea turtle living in the world. All are threatened or endangered, largely due to the loss of nesting grounds. Five species nest on Florida shores: the hawksbill (*Eretmochelys imbricata*), Kemp's ridley, (*Lepidochelys kempii*), loggerhead (*Caretta caretta*), green (*Chelonia mydas*), and leatherback (*Dermochelys coriacea*). The black (*Chelonia agassizi*), olive ridley (*Lepidochelys olivacea*), and Australian flatback (*Natator depressus*, found only in the eastern hemisphere) aren't found in Florida waters.

Sea turtles nest on subtropical to tropical sandy coasts. Scientists and volunteers record between forty thousand and seventy thousand nests annually along the Florida coast, making it the most important nesting coast in the United States. A female may deposit several nests during a season but will nest only every two to three years. After about two months of incubation in the warm sand, hatchlings scurry to the sea. Survivors may live fifty years or more. Most sea turtles feed on small arthropods, fish, sponges, or jellyfish. Sadly, many turtles mistake plastic bottles or bags for food, with pitiful and tragic results.

The largest of the living sea turtles, the amazing leatherback, can approach 8 feet in length and weigh as much as 2,000 pounds. It can dive to over 4,000 feet and remain submerged for well over 30 minutes. The distribution of the leatherback is nearly global. They have been observed swimming near icebergs, and individuals are known to migrate across the entire Atlantic Ocean, from South America to Africa. Each year, one hundred to two hundred leatherback nests are recorded in Florida.

Sea turtles are truly ancient mariners, with a distinguished history reaching back to Jurassic time, more than 150 million years ago. Some were gigantic, such as the *Archelon* of Cretaceous time, which grew up to 12 feet in length and almost certainly swam in Florida waters. Unfortunately and surprisingly, the fossil record of sea turtles in Florida is rather poor, but remains have been recovered in strata of Eocene to Pleistocene age. (There was also one fortuitous discovery of a Cretaceous-age specimen in a deep oil well in Collier County, near Lake Okeechobee.) Only fragmentary remains have been found in deposits of Eocene and Oligocene age. Five genera have been recovered in sediments of Early Pliocene age in the Bone Valley Member of the Peace River Formation, including an extinct species of leatherback (genus *Psephopherus*), and undescribed species of loggerhead (genus *Caretta*), green (genus *Chelonia*), and ridley (genus *Lepidochelys*). Fossils of today's green and loggerhead turtles have been found in Pleistocene-age sediments.

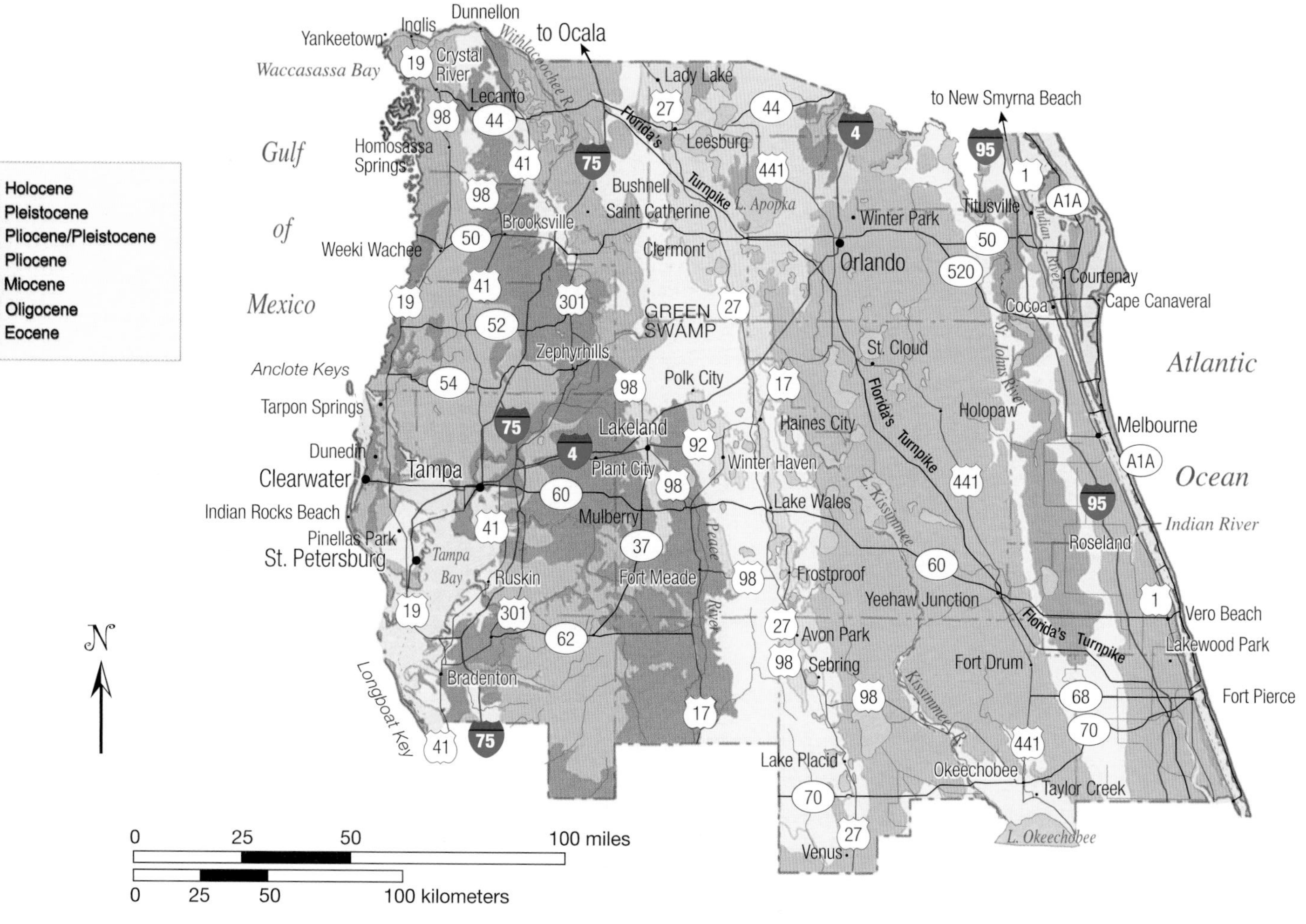

Road log map of Central Peninsula. —Map by Mountain Press

Central Peninsula

We Dig Phosphate

The central peninsula has a surprising variety of geological features. First on the list is the most important economic mineral resource of the state—phosphate! Since discovery of phosphate along the Peace River in the 1880s, Florida has led the nation in its production. Presently, Florida provides about 75 percent of U.S. and 25 percent of global needs. About 90 percent of Florida's phosphate is used in agricultural fertilizers. The remaining 10 percent is used in many products, including food preservatives, dyes, steel alloys, oil and gasoline additives, toothpaste, plastics, optical glass, photographic film, insecticides, soft drinks, fire-extinguishing compounds, and more. A useful by-product of phosphate processing is fluorine, which is used to fluoridate public water. No wonder so many bumper stickers in central Florida boast, "We Dig Phosphate!" See **Central Florida Phosphate District** later in this chapter for more information.

The central peninsula is also famous for its paleontological bounty. There are the unique mollusks of Eocene age found in the Inglis and Crystal River area, the invertebrates of Oligocene and Pleistocene age found in the Tampa Bay region, and the incomparable vertebrates of Miocene and Pliocene age found east of Tampa, in the Central Florida Phosphate District. The most visible geologic aspect of the central peninsula is its long ridges, which parallel the trend of the peninsula. Often referred to as the Central Highlands, these ridges actually extend from Georgia to Lake Okeechobee. To avoid any confusion between the central peninsula and the Central Highlands, we want to emphasize that the Central Highlands is a north-south-trending geomorphic feature extending from the Georgia border down roughly two-thirds of the Florida Peninsula. "Central peninsula," on the other hand, is a fairly arbitrary division of the Florida Peninsula made for the purposes of this road guide, although the southern end of the central peninsula does coincide with the southern extent of the Central Highlands.

The ridges of the Central Highlands are especially evident in aerial and satellite photos and extend over several geomorphic areas, including the Ocala Karst District, the Central Lakes District, the Barrier Island Sequence District, and the Sarasota River District. The most prominent ridges are Trail Ridge, Brooksville Ridge, Mount Dora Ridge, and Lake Wales Ridge. The question of the origin of the ridges is an intriguing one. It seems almost intuitive that it must have something to do with former shorelines and higher sea levels, and that is indeed the case.

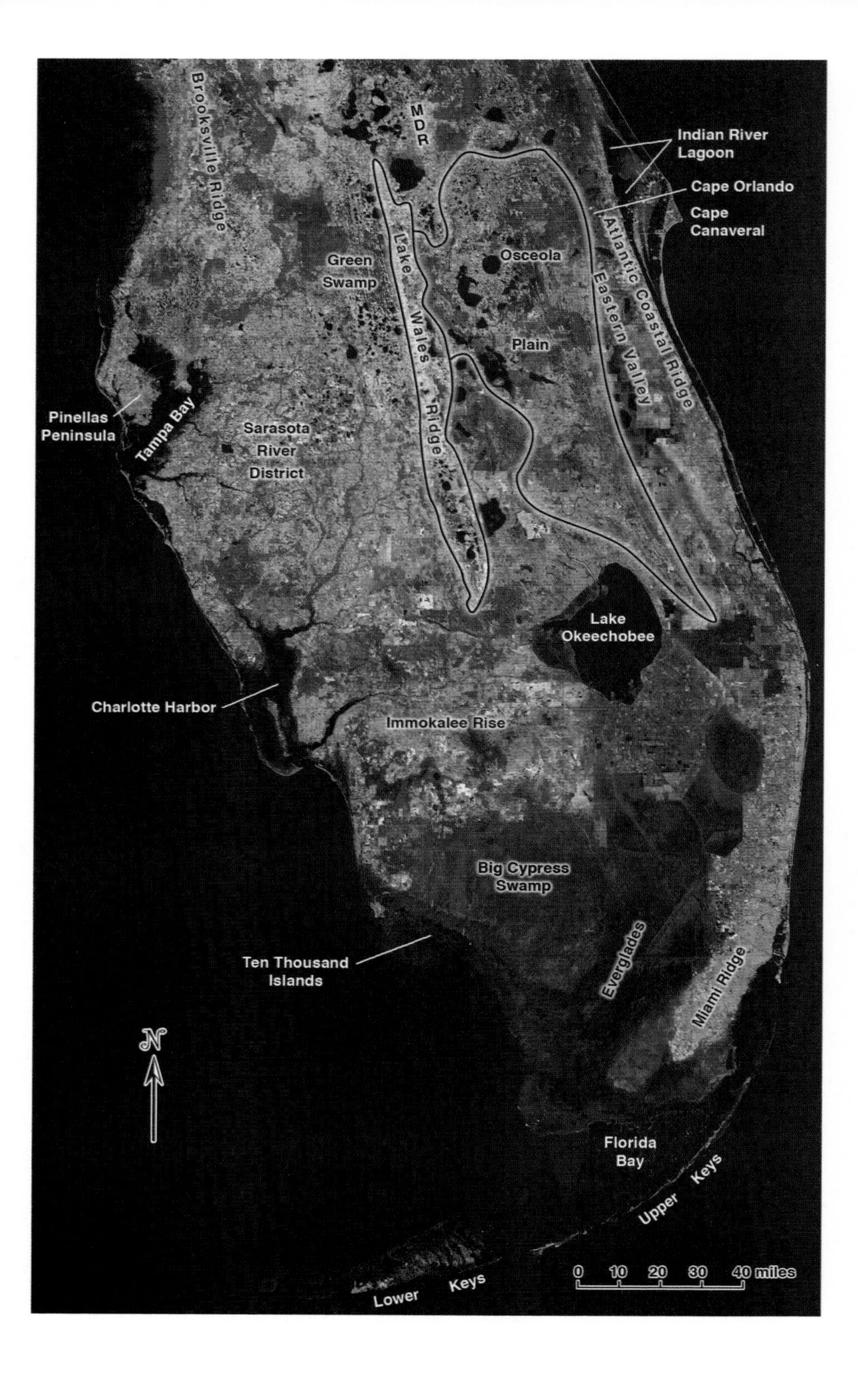

Brooksville Ridge
M D R
Indian River
Lagoon
Cape Orlando
Cape
Canaveral
Atlantic Coastal Ridge
Eastern Valley
Osceola
Plain
Lake
Wales
Ridge
Green
Swamp
Pinellas
Peninsula
Tampa Bay
Sarasota
River
District
Lake
Okeechobee
Charlotte Harbor
Immokalee Rise
Big Cypress
Swamp
Ten Thousand
Islands
Everglades
Miami Ridge
N
Florida
Bay
Upper Keys
Lower Keys
0 10 20 30 40 miles

Geologists have interpreted most of the ridges as being stranded coastal landforms, such as barrier islands, spits, dunes, erosional escarpments, or some combination of these. Most of the ridges consist of sediments and rock that are much older than the actual geomorphic features, the ridges themselves. So the ridges appear, at least in part, to be erosional in origin, such as escarpments cut into bedrock by the ocean, or they may have formed as the ocean reworked and modified older material. According to the coastal ridge hypothesis, the ridges formed at times of higher sea levels. The sea flooded the peninsula from both the Gulf of Mexico and the Atlantic, reaching far inland. At a variety of levels, the sea formed terraces, cut escarpments, and reworked sediments into beach ridges and barrier islands. As the sea retreated, the ridges were left as escarpments or other coastal landforms. Sand dunes were later

Satellite image of the southern two-thirds of the Florida Peninsula, showing many distinctive geomorphological features that testify to Florida's coastal and submarine past. Modern barrier islands are evident along both the eastern and western coasts. Modern coral reefs are just offshore and south of the Keys. A majority of lakes in the peninsula are karst lakes. —Courtesy of the U.S. Geological Survey (labeled by authors)

- Atlantic Coastal Ridge, *composed of coquina rock of the Anastasia Formation of Pleistocene age and draped by modern beach sands.*
- Big Cypress, *a freshwater peat swamp and part of the Pamlico Terrace of the last interglacial period of high sea level (the Sangamonian).*
- Brooksville Ridge, *a terrace.*
- Cape Canaveral, *a cuspate foreland and beach ridge plain.*
- Charlotte Harbor, *a drowned river valley estuary flooded by the Holocene Transgression.*
- Cape Orlando, *a former cuspate foreland.*
- Everglades, *a Holocene-age freshwater peat marsh, a Pleistocene-age lagoon, and part of the Pamlico Terrace of the Sangamonian.*
- Eastern Valley, *a former lagoon and the location of the St. Johns River.*
- Florida Bay, *a modern carbonate lagoon and former freshwater peat marsh during an early stage of the Holocene Transgression (carbonate mud mounds are visible).*
- Green Swamp, *a beach ridge plain of Pleistocene age.*
- Immokalee Rise, *a slightly elevated plain and former island or shoal during the Sangamonian.*
- Indian River Lagoon, *a beach ridge plain flooded during the Holocene Transgression.*
- Lower Keys, *composed of the Miami Limestone, an ooid shoal during late Pleistocene time.*
- Lake Okeechobee, *part of a Pleistocene-age lagoon behind the Miami Ridge, and part of the Pamlico Terrace during the Sangamonian.*
- Lake Wales Ridge, *a former narrow peninsula or barrier island.*
- Mount Dora Ridge (MDR), *similar to the Lake Wales Ridge.*
- Miami Ridge, *the southern end of the Atlantic Coastal Ridge and composed of the Miami Limestone, a giant, Pleistocene-age ooid shoal.*
- Osceola Plain, *an extensive beach ridge plain of Pleistocene age;*
- Pinellas Peninsula, *the central part of this peninsula, a cuspate foreland, was surrounded by the Pamlico Sea during the last interglacial period.*
- Sarasota River District, *a broad plain of former marine terraces that developed over sediments of Miocene, Pliocene, and Pleistocene age.*
- Tampa Bay, *a karstic drowned river valley estuary flooded by the Holocene Transgression.*
- Ten Thousand Islands, *mangrove islands built upon ancient oyster shoals and mounds of tubular shells of wormlike vermetid gastropods.*
- Upper Keys, *composed of the Key Largo Limestone, a coral reef during late Pleistocene time.*

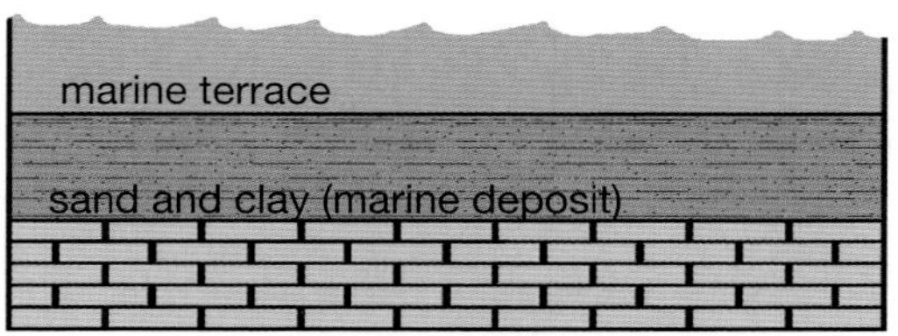

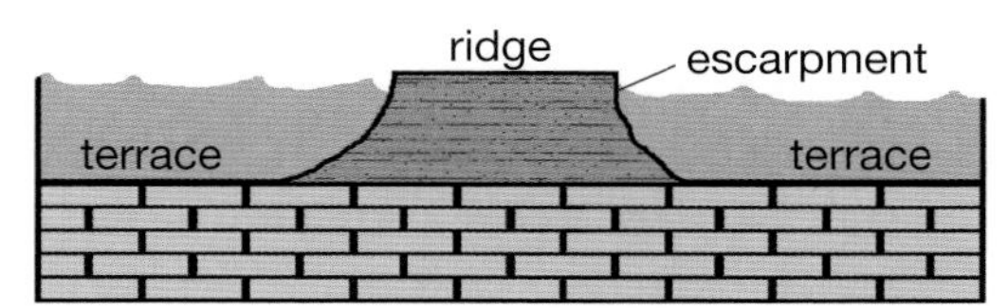

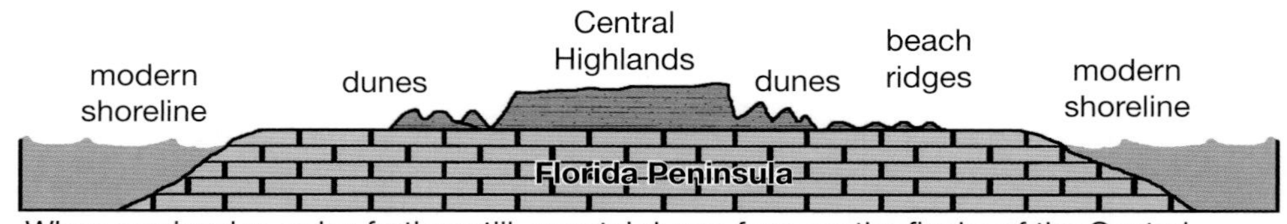

Coastal ridge model for the origin of the Central Highlands. —Illustration by Amanda Kosche

added to the ridges. The intervening valleys between the ridges may represent former lagoons, tidal passes, or shallow subtidal seafloors. Along the eastern edge of the peninsula, beach ridge plains are adjacent to the ridges. The ridges and valleys have been modified by erosion over the last 10,000 years or more.

Geologists have proposed another plausible hypothesis, called *topographic inversion*, for the origin of the Brooksville Ridge. In this model, the ridge began as a karst surface of Eocene- and Oligocene-age limestone deposited between 56 and 23 million years ago. The karst had many solution valleys, across which streams flowed. These valleys were most likely related to fracturing (jointing) or faulting of the limestone, which accounts for their linear nature (north-south-trending valleys). As the sea rose over the peninsula in Miocene and Pliocene time, it deposited sediments of the Hawthorn Group and the Cypresshead Formation over

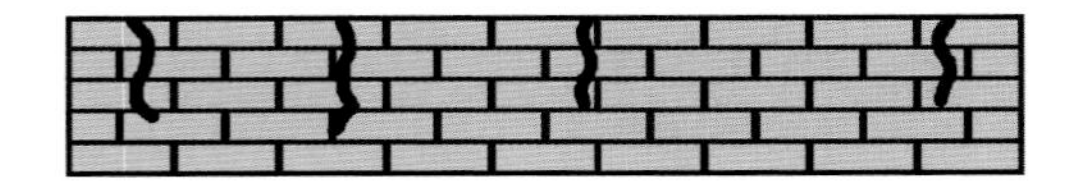

Exposed limestone surface is heavily fractured and faulted.

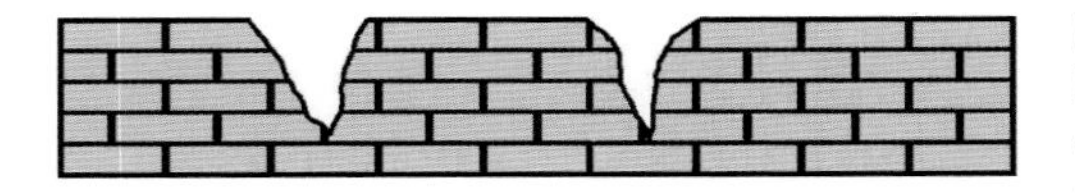

Long, linear fractures or faults become enlarged by karst dissolution since water flows more readily through fractures.

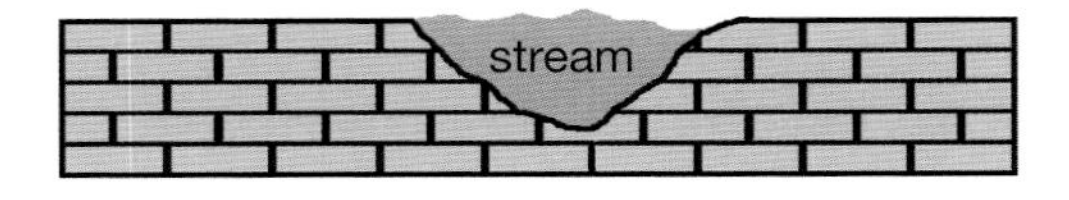

An enlarged fracture becomes a river channel for a karst stream.

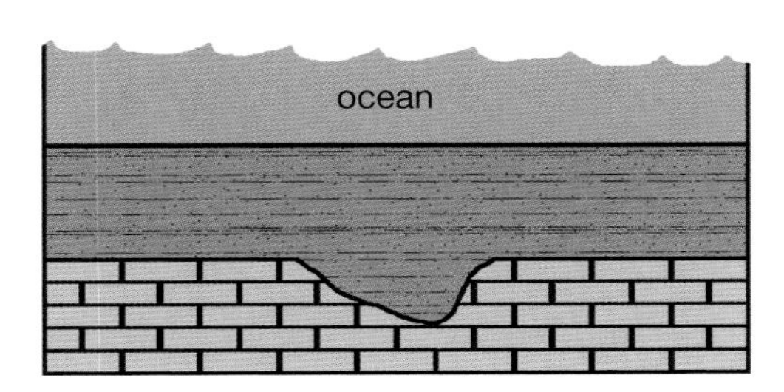

During a subsequent sea level rise, the entire surface is buried in marine clay and sand, with thicker strata in the former stream channel.

After sea level falls the thinner clay strata is eroded and karst dissolution resumes. But thicker strata in the old stream channel resist erosion, protecting the old stream channel from further karst dissolution and forming a long ridge.

Topographic inversion model for the origin of the Central Highlands. —Illustration by Amanda Kosche

the karst. Thick, clay-rich sediments filled the old karst stream valleys. As sea level fell, erosion and karst dissolution resumed. But paradoxically, the karst valleys, with their thicker sediment cover, were protected from extensive groundwater dissolution. The limestone surrounding the karst valleys was lowered by dissolution, forming lower valleys while the older, sediment-filled karst valleys were left standing high.

Windblown dunes are associated with the ridges, but usually they occur along the flanks of the ridges and not on top of them. The ridges acted as natural barriers that trapped windblown sand after sea level dropped in latest Pleistocene time. It seems likely that both the coastal ridge and topographic inversion theories may be part of the explanation for the origin of the peninsula's ridges and valleys.

US 98/US 19

Inglis—St. Petersburg—Bradenton

125 miles

According to tourist brochures, the area between Inglis and Bradenton is Florida's "Gold Coast," a reference to the fact that, historically, it has been a retirement destination for many Americans (the golden years). The geologic attractions of this area are many, although they are not immediately evident when driving down congested US 19!

Between Inglis and Crystal River are exposures of the oldest surface rocks in the state—the lower member of the Ocala Limestone and the Avon Park Formation, both of which are Eocene in age. The lower Ocala is noted for its unusual mollusks, which are similar to Tethyan mollusks of European fossil assemblages. It also contains an abundance of Florida's (unofficial) state fossil—the sea biscuit *Eupatagus antillarum.* Up to 14 feet of the lower Ocala Limestone and 1 to 2 feet of the Avon Park Formation lie exposed along the banks of the Cross Florida Barge Canal, about 1 mile south of Inglis.

The purpose of the barge canal was to link Yankeetown on the Gulf Coast to the St. Johns River on the Atlantic Coast, creating a quicker shipping route between the gulf and the Atlantic Ocean. King Philip of Spain originally proposed the idea of a cross-peninsula canal in 1567. However, actual construction of the canal began much later, in 1935, east of Dunnellon, then ceased in 1936 as funds ran out. Construction resumed in 1964 and continued until early 1971, when President Nixon suspended work. The economic appeal of a cross-state canal was undeniable, but the project was deauthorized by congress in 1979 due to public outcry over its negative environmental impact.

Both the north and south sides of the barge canal are accessible from US 98/19, and outcrops of the lower Ocala can be reached in several areas. These outcrops are conveniently located along the Withlacoochee Bay Trail, which is the westernmost extension of the Marjorie Harris Carr Cross Florida Greenway. This trail, accessible on the south side of the US 98/19 bridge over the barge canal, extends for 5 miles from US 98/19 to the Gulf of Mexico. A vehicle access road follows the bike trail for 4.3 miles, almost to the gulf. There are several picnic pavilions along the trail. Nearly continuous, easily accessible, and fossiliferous spoil piles of the lower Ocala Limestone, dredged from the canal, occur along this road. (There are also islands off Yankeetown composed of debris dredged from the barge canal.) Large fossiliferous boulders of the lower Ocala ornament the access area beneath the US 98/19 bridge as well. Along the banks of the canal is one of the few places where the lower Ocala Limestone can be seen in stratigraphic position and undisturbed, and also where Florida's state fossil can be found. This easily accessible location should not be missed!

The Avon Park Formation is exposed in dolostone quarries near the barge canal. In the now inactive Dolime Quarry, located on the southeastern side of US 98/19 and the barge canal, the Avon Park preserves rare and exceptional sea grass fossils. Overlying the sea grass bed is a cross-stratified dolomitic sand with *Ophiomorpha nodosa* burrows. The cross-stratification indicates the limestone was deposited in high-energy, coastal waters. Sea grasses can grow only in very shallow water, usually 30 feet or less, so their presence indicates extremely shallow depositional conditions. Geologists have identified six species of fossil sea grasses in the Avon Park Formation, the most common of which are *Thalassodendron auricula-leporis* and *Cymodocea floridana.* Today, these genera are only found in the Indian and southwestern Pacific oceans. The sea grass bed also contains an assortment of small marine organisms that lived on the grasses, such as algae, foraminifera, worms, barnacles, mollusks, bryozoans, echinoids, star fish, and brittle stars. Dugongs and sea turtles were common inhabitants of areas with sea grass today and have also been recovered in the Avon Park. Geologically, sea grasses are effective sediment trappers. Modern turtle grass (*Thalassia testudinum*), for example, can extend 2 feet or more below the surface of the seafloor, stabilizing the substrate and capturing suspended sediment. Invertebrate fossils also accumulate and are concentrated in sea grass beds. Sea grass

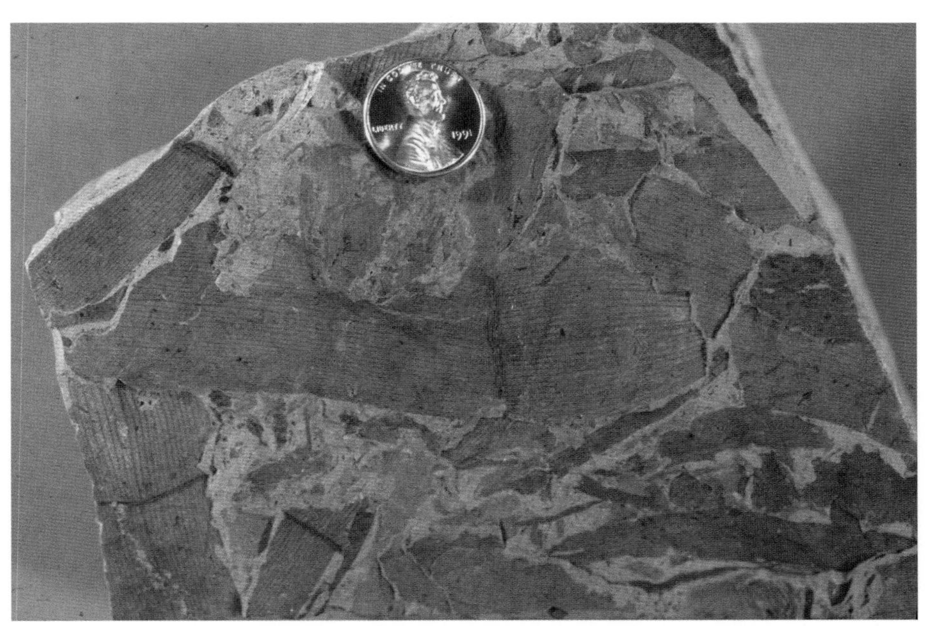

Fossil sea grasses of the Avon Park Formation of Middle Eocene age.

Eocene Mollusks—the Tethyan Connection

Geologists have long recognized that the molluscan fauna of the Ocala Limestone of Eocene age contains many species with so-called Tethyan affinities. The term Tethyan refers to an equatorial seaway—the Tethys Sea—that flowed between the ancient megacontinents of Laurasia and Gondwana. The Mediterranean Sea is essentially what is left of this ancient seaway. Tethyan mollusks of Florida are those species with some relation to species found in the rocks of the ancient Mediterranean region, and many of these fossils are similar to gastropod and bivalve species living in the Indian and western Pacific oceans today. Tethyan mollusks in the Ocala Limestone include the gastropod genera *Ampullinopsis, Athleta, Batillaria, Bellatara, Caricella, Eovasum, Gisortia, Lapparia, Lyria, Pseudoaluca, Seraphs, Tectariopsis, Velates,* and *Voluticella,* and the bivalve genera *Clavagella, Fimbria, Lithophaga, Meiocardia, Nayadina,* and *Pseudomiltha.*

Tethyan mollusks are common in the lower part of the Ocala but become increasingly rare in the upper portion of the limestone. The causes of this large-scale biogeographic change are not fully understood. Some paleontologists suspect the Tethyan species died out in Florida because of gradual cooling from Middle to Late Eocene time, which preceded the onset of even cooler glacial conditions in Oligocene time. This may have restricted a once widespread tropical fauna of mollusks to the tropical refuge of the modern Indo-Pacific seas. In any case, it is clear that with the extinction of many Eocene-age species in Florida, the fauna began to take on a more regional, western Atlantic character.

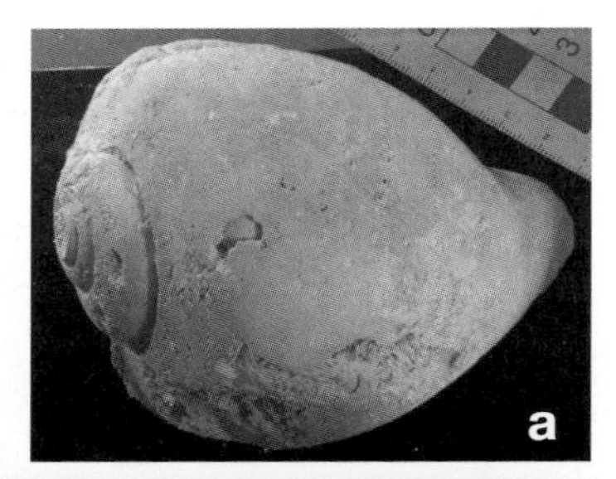

Ocala Limestone mollusks with Tethyan affinities. (a) Sycospira eocenica, *5 inches long; (b)* Agaronia inglisia, *3.3 inches long; (c)* Velates floridanus, *2 inches in diameter; (d)* Gisortia harrisi, *7 inches long (left) and 5.25 inches wide (right).* —Images c and d courtesy of the Invertebrate Paleontology Division, Florida Museum of Natural History, University of Florida

remains are generally very rare in the fossil record, but in Florida they have also been discovered at Marianna in the Marianna Limestone and in the subsurface of Pinellas County in the Suwannee Limestone, both of Oligocene age.

Sea grass beds, or meadows, are found in nearly all coastal waters of Florida, but they are most extensively developed from Apalachee Bay to the Clearwater area, and offshore of the southwest corner of the state from Naples to the Keys. Sea grasses are aquatic flowering plants, with about fifty-two species worldwide. Only seven species grow in Florida waters, including widgeon grass (*Ruppia maritima*), found in both fresh- and saltwater and common in estuaries; shoal grass (*Halodule wrightii*), which usually grows in water that is too shallow for most other species; turtle grass (*Thalassia testudinum*), which is the most widespread species, and also the largest, with leaves up to 1 inch wide and 1 foot long; and manatee grass (*Syringodium filiforme*), which has cylindrical leaves. Less common in distribution are star grass (*Halophila engelmanni*), paddle grass (*Halophila decipiens*), and Johnson's sea grass (*Halophila johnsonii*). Sea grass meadows are some of the most diverse ecosystems in the marine environment. They provide attachment surfaces for countless algae and marine invertebrates and habitat for thousands of other small invertebrates.

The Crystal River Archaeological State Park, a national historic landmark, is just north of Crystal River. This Archaic Period site consists of shell middens, shell mounds, temple platforms, and two burial mounds. The six-mound complex was a major cultural and trade center in the region and was continuously occupied for an estimated 1,600 years. From US 98/19 about 2 miles north of Crystal River, take State Park Street west for 1 mile to Museum Pointe.

A geological tour of Florida would not be complete without a tour of the Crystal River National Wildlife Refuge, one of the winter homes of the Florida manatee (*Trichechus manatus latirostris*). This refuge, accessible by boat only, is composed of small parcels of land and about twenty islands west of the town of Crystal River. The best snorkeling tours leave in the early morning hours of winter (there are several local outfitters). As your boat quietly approaches the palm hammock limestone islands just before sunrise, a rising mist condenses in the cold air over the warmer water that has emerged from local springs, creating a primeval atmosphere. Before long, you will witness a pair of large nostrils breaking the surface of the coastal springwater for air—"There's one!" You have just encountered a herd of the bizarre sea cows. Manatees, along with their cousins the dugongs, have had a prolific history in Florida. It takes little imagination to see oneself back in Miocene or Pliocene time. The scene in the refuge can be perfectly prehistoric!

The amazingly gentle manatees are often just as curious about humans as we are about them. Some will swim up and stare at you eye to eye. Others, like a family pet, will approach to be scratched and may even roll over for a rubdown. This is truly an experience not to be missed. Feeding is not allowed, you may not swim after them or over them, and you may only touch the animals with one hand. You can also observe manatees just south of Crystal River at Homosassa Springs Wildlife State Park, and they are commonly encountered in many of Florida's coastal waters and spring runs.

About 2.5 miles west of Homosassa Springs on County Road 490 (Yulee Drive) is the Yulee Sugar Mill Ruins Historic State Park, the former location of a 5,100-acre sugar plantation and steam-driven mill. This mill operated from 1851 to 1864 and supplied sugar products to Confederate troops during the Civil War. The 40-foot masonry chimney is constructed of local Ocala Limestone. The Ocala continues to be mined extensively at nearby Lecanto. South of Homosassa Springs, US 98/19 splits. US 98 heads southeast, while US 19, the route of the rest of this road log, continues south.

According to legend, manatees were mistaken for mermaids. The name of the order that includes manatees, Sirenia, refers to this confusion and the sirens of Greek myth. While their resemblance to mermaids may be

Remains of the Yulee Sugar Mill, built with local Ocala Limestone.

Fossil Manatees and Dugongs

Popularly known as sea cows, manatees belong to the order of aquatic mammals known as sirenians. There are three manatee species living today: the west African manatee (*Trichechus senegalensis*), Amazonian manatee (*Trichechus inunguis*), and West Indian manatee. The latter consists of two subspecies: the Antillean manatee (*Trichechus manatus manatus*) and the Florida manatee (*Trichechus manatus latirostris*). The dugong, or sea pig, is similar to the manatee. There is only one living dugong species, *Dugong dugon*, which lives in the Indo-Pacific, Indian Ocean, and Red Sea. A gigantic dugong called Steller's sea cow (*Hydrodamalis gigas*), which could reach 25 feet in length, lived in the Bering Sea until it was hunted to extinction in the 1700s.

Manatees made a relatively late appearance in Florida. They evolved in the Amazon Basin and later moved into Caribbean and Floridan waters sometime

*The Florida manatee (*Trichechus manatus latirostris*).*

in the last 5 million years, following the extinction of dugongs. The first well-documented record is from sediments of early Pleistocene age in the Tampa Bay area. Dugongs, however, were extremely abundant in Florida from Middle Eocene to Early Pliocene time, and their dense, heavy rib bones are, along with shark teeth, among the most common fossils of vertebrate animals in the state. Florida is one of the few places in the world where sirenians have continuously resided from their first appearance in Middle Eocene time to the present day.

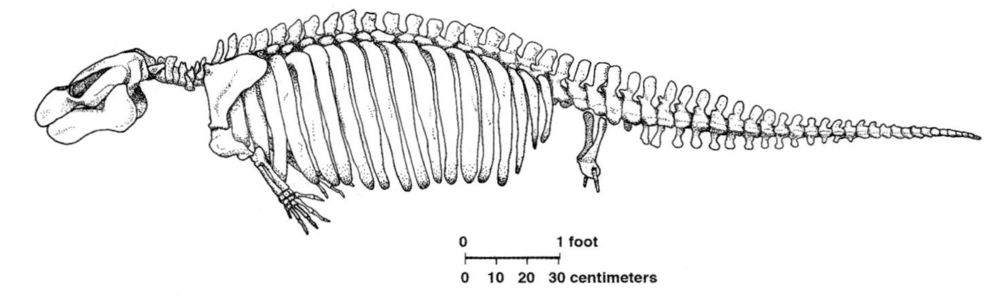

The dugong Metaxytherium crataegense *from the Torreya Formation of Early Miocene age near Quincy.* —Courtesy of the Florida Geological Survey

Fossil Camels

It is a strange biogeographical fact, but the camelid order, including llamas and camels, originated in North America. In fact, camelids remained entirely in North America from Eocene to Pliocene time. Only then did the genus *Camelops* move to Eurasia (via the Bering Land Bridge) and llamas to South America (via the Isthmus of Panama). Geologists recognize more than fifty camelid species in the fossil record, but only six are alive today. There are two species of camel in Asia: the one-humped dromedary and the two-humped Bactrian camel. Most of these camels are domesticated, with very few wild populations remaining. There are four living species of llama: the llama, guanaco, alpaca, and vicuna. The oldest llama fossils found in Florida are about 9 million years old, from Late Miocene time. Llamas first appeared in South America during Pleistocene time in the Argentine Pampas.

The Miocene-age giraffe-camels, so named because of their greatly elongated legs and necks, constituted an unusual group of camels in Florida. *Aepycamelus major* was as tall as a modern giraffe, standing over 13 feet at the shoulder. Fossils of this giant are common at several localities of Late Miocene age in north-central Florida, including the Love Bone Bed. Another unusual giraffe-camel was *Floridatragulus*, found at the Thomas Farm site near Fort White. It had a greatly elongated skull. Other Miocene-age genera found in

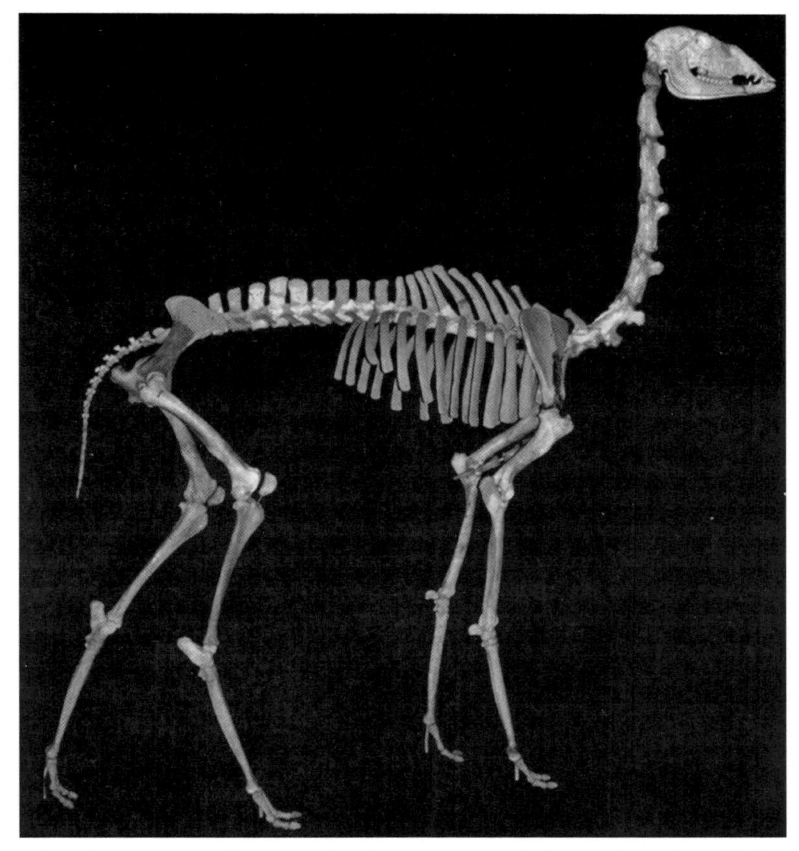

This Pleistocene-age llama Hemiauchenia macrocephala *was found in a filled-in karst fissure in Citrus County. This specimen, located at the Florida Museum of Natural History, is about 9 feet high at the head.*

Florida include *Nothokemas*, *Oxydactylus*, *Procamelus*, and another giant of Miocene and Pliocene time, *Megatylopus.*

By Early Pliocene time, the only camelids that remained in Florida were llamas. They are represented in the fossil record by several species of the genus *Hemiauchenia*, which occur in Miocene- to Pleistocene-age formations. One species, *Hemiauchenia macrocephala*, was found at the Pleistocene-age Leisey Shell Pits, near Tampa, in what appears to be, in part, a death assemblage of a herd of mostly subadult individuals. The Pleistocene-age *Paleolama mirifica*, commonly found with *Hemiauchenia macrocephala*, is the species most closely related to modern South American llamas. About fifteen camelid species are found in Florida in strata of Miocene to Pleistocene age. As with so many other large mammals, all were extinct in North America by the end of Pleistocene time.

debatable, about 25 miles south of Crystal River you can see "real" mermaids at Weeki Wachee Springs. This popular attraction has been entertaining tourists with mermaid shows in its underwater theater since 1947. Weeki Wachee is derived from a Creek Indian term—*wekiwa chee*—meaning "little spring." Despite that diminutive attribution, it is a large, first magnitude spring. It emerges from the Suwannee Limestone—the "mountain underwater" of early promotional brochures—and the exposure is indeed impressive as seen from the underwater theater.

While driving through Weeki Wachee and vicinity, large sand hills are very evident. These are part of the Pleistocene-age Weeki Wachee dune field that lies next to the Brooksville Ridge. (The sandy dune terrain is also evident while driving the Florida 589 Suncoast Parkway/Veteran's Expressway toll road, which parallels US 19 from south of Homosassa Springs to Tampa.)

If you drive about 5 miles east from Weeki Wachee on Florida 50, you will approach the southern flank of the Brooksville Ridge. The ridge is up to 200 feet above sea level here, rising over 100 feet above the lowlands to the west. The town of Brooksville sits on the ridge near its southern end. Many dunes and karst lakes are visible along Florida 50, and in places the topography is very pronounced. The Brooksville Ridge is the largest of the ridges in the Central Highlands. It is roughly the same length as the Lake Wales Ridge (about 110 miles) but is wider and consists of several formations. Most surface exposures on the ridge are Hawthorn Group and Quaternary-age sediments, but these overlie a core of Ocala and Suwannee limestones, which are mined along the ridge's southern end. Sand dunes surround much of the ridge.

Along the eastern edge of the Brooksville Ridge is an area known as the Hard-Rock Phosphate Belt. This belt extends from southern Columbia County southward to Hernando County. The hard-rock phosphate occurs as boulders and pebbles of high-grade phosphorus-bearing sediments in ancient sinkholes. These deposits formed as an older, pebbly phosphate deposit (similar to the deposits mined in the Central Florida Phosphate District) was eroded and then deposited in sinkholes. Phosphorus-laden water moved downward through the sinkholes, depositing the hard-rock phosphate, and occasionally the phosphate minerals replaced the limestone in the sinks. Hard-rock phosphate was mined in the Hard-Rock Phosphate Belt from the late 1800s until the mid-1960s, when the deposits were no longer commercially viable.

About 100 miles offshore of Weeki Wachee, in water depths between 80 and 130 feet, is the Florida Middle Ground, a diverse community of coralline algae, sponges, corals, mollusks, and other invertebrates at the northern limit of coral reef growth in the gulf. Fire coral (genus *Millepora*) dominates the rugged top of the reef. Although it is often classified as the northernmost reef in the Gulf of Mexico, some would prefer to classify the Middle Ground as a hardbottom community, in which bottom-dwelling marine animals grow over older, rocky limestone exposures on the seafloor rather than actually constructing the limestone. Hardbottom communities are common in Florida, particularly across the West Florida Shelf. The Middle Ground occurs as two north-south-trending, somewhat parallel ridges. Each is over 18.5 miles in length, and the two are separated by a flat, 4.5-mile-wide valley. The ridges probably developed as a result of river or karst erosion in the past, or they were reef ridges that developed in Pleistocene time. The current reef growth may have started as early as late Pleistocene time. Several other shore-parallel limestone ridges occur offshore, some with as much as 13 feet of relief above the sea bottom. They are Miocene to Pleistocene in age.

Who says there are no dinosaurs in Florida? Be sure to see one of the great highway landmarks of the Florida Gulf Coast on US 19 at Spring Hill—the 110-foot-long, 47-foot-high Sinclair brontosaurus. Built in 1964 and formerly a Sinclair Refining Company gas station, this giant is now home to Harold's Auto Station.

Of special historic interest is the colorful city of Tarpon Springs, the Sponge Capital of the World, located 25 miles south of Weeki Wachee. For over 100 years, Greek Americans have harvested sponges from the northeastern Gulf of Mexico. Sponging began in earnest here between 1905 and 1907, when Greek immigrants brought ancient Mediterranean sponge harvesting methods to the Tarpon Springs area. This move was motivated by depletion of the historic sponge grounds of the Aegean Sea, and also declines in the Florida Keys, where a prominent sponge industry was centered from 1849 into the 1890s. Harvesting began to move north when the prolific sponge beds of the northeastern Gulf were discovered in 1873. In the Keys, sponges were collected from small boats using a glass box to view the bottom and long, pitchfork-like hooks. But the deeper northeastern gulf was especially suited to the use of the classic Greek diving suits of hard helmet, hose, and air pump. It was dangerous business, and many divers perished. The spongers of Tarpon Springs were romanticized in the entertaining 1953 film *Beneath the 12-Mile Reef*, shot on location at Tarpon Springs and in the Florida Keys. The market for gulf sponges declined after a 1946 red tide devastated the sponge grounds, and after the development of synthetic sponges. Today the sponge industry in Tarpon Springs is modest, but tourism is booming.

Sponges are found worldwide in the ocean, but they are especially common in three settings: the deep sea; the deep fore-reef zone, where coral growth is minimal; and in nearshore waters with a high nutrient content, hard substrates for attachment, and minimal suspended sediment.

The northeastern gulf provides the latter environment. The shallow continental shelf around Florida's Big Bend coast is a rugged, karstic limestone. Sport divers report many prominent north-south-trending rock ledges and drowned sinkholes in this area. And because the rivers that empty into this part of the gulf flow over limestone, very little suspended sediment, such as clay or quartz sand, is delivered to the area.

South of the drowned karst shoreline of the Big Bend area, barrier islands of quartz sand resume along Florida's west-central coast at Anclote Key, west of Anclote, and continue south to Cape Romano in Collier County. Anclote Key is a young barrier, having formed about 1,200 years ago when a sandbar shoaled and emerged from the water, and it is probably the largest barrier island to have formed in this fashion.

The Pinellas County peninsula, which stretches between Tarpon Springs and St. Petersburg, and the Apalachicola delta at Apalachicola are pointed coastline features called *cuspate forelands*. They bracket the drowned karst shoreline of the Big Bend. On Florida's east coast, another cuspate foreland is found at Cape Canaveral. If you drew a line between Cape Canaveral and the Pinellas County peninsula, across the widest part of the Florida Peninsula (from Indian Rocks Beach to Cape Canaveral), you would have marked what geologists call the *cross-peninsular divide*. This line separates most of the Florida rivers that flow north from those that flow south, and subsurface strata south of the divide are considerably thicker than strata to the north. The line also approximates a tectonic suture zone in the basement rocks, where bedrock from other continents was attached during the formation of Pangaea. It may well represent a zone of structural flexure, or bending, in the deep subsurface, which could have long influenced sedimentation and drainage patterns at the surface.

Much of Honeymoon Island, west of Dunedin, as well as the material for Dunedin Causeway, which connects the island to the mainland, is composed of rounded cobbles and small boulders that were dredged locally. The cobbles, derived from the Tampa Member of the Arcadia Formation, are richly fossiliferous. Many of the cobbles are, in fact, fossilized colonies of several species of coral, and some are silicified. More boulders and cobbles are visible in the shallow water along the shore. Mollusks, worm tubes, and other fossils are also common. In Dunedin, take Florida 586 west to Honeymoon Island State Park. Accessible by private boat or ferry from Honeymoon Island, nearby Caladesi Island State Park is one of the few remaining undeveloped barrier islands along this segment of the Florida Gulf Coast.

Like the beach at Honeymoon Island, much of coastal Pinellas County has been enhanced by the addition of fragments of the fossiliferous Tampa Member of the Arcadia Formation dredged from the intracoastal

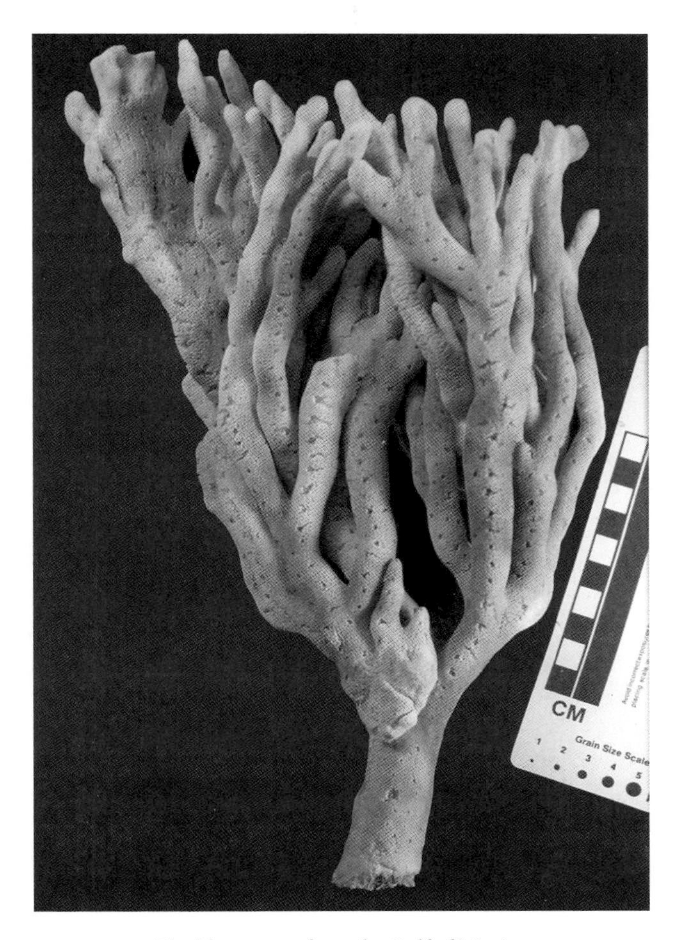

Florida sponges from the Gulf of Mexico.

Fossiliferous cobbles of the Tampa Member of the Arcadia Formation on Honeymoon Island.

waterways that separate the mainland from the barrier island chain. It is not at all uncommon to see large boulders of the Tampa Member used as ornamental rocks around schools and condominiums and on the lawns of private homes. So even though it is not exposed in outcrops, the Tampa Member does make a regular appearance in this area.

The Pamlico Terrace and its escarpment, which formed during the last interglacial period in late Pleistocene time, are well developed in Pinellas County. A barrier island once extended between Tarpon Springs and Seminole, but now it is high and dry. Near Seminole, the southern edge of this island (the Pamlico shoreline) is evident along County Road 694 (Oakhurst Road/78th Avenue/Park Boulevard), where this and nearby roads quickly descend the escarpment, from elevations around 60 feet or more to about 10 feet near the coast. The higher elevations here may preserve even older Pleistocene-age shorelines.

Several important fossil localities have been discovered in Pinellas County, although most of them are now covered by roads or other structures. Probably the most famous site is Seminole Field on St. Joes Creek, a coastal marsh of late Pleistocene age located about 3 miles southeast of Seminole and 2.25 miles south-southwest of Pinellas Park. Paleontologists have documented over forty vertebrate species from this site, including species of saber-toothed cats, mastodons, llamas, horses,

Large boulders of the Tampa Member of the Arcadia Formation are found in many residential areas in Pinellas County.

and sloths, along with many other species typical of late Pleistocene time. The bone-bearing layer overlies a shell bed with shallow-water mollusks. Such shell beds underlie much of Pinellas County, and they are frequently associated with vertebrate fossils, indicating former shoreline environments. Another site, also near Seminole, was exposed in a streambank near the Intracoastal Waterway, south of Indian Rocks Beach near the Imperial Point community. Late Pleistocene remains of various species have been recovered, including a giant tortoise (*Hespero-testudo crassiscutata*), glyptodont (*Glyptotherium floridanum*), alligator (*Alligator mississippiensis*), dire wolf (*Canis dirus*), sloth (*Glossotherium harlani*), mammoth (*Mammuthus* species), horse (*Equus* species), and bison (*Bison antiquus*). The site has since been covered by condominiums. A similar assemblage of Pleistocene-age vertebrates was recently discovered at nearby Boca Ciega Millennium Park on 74th Avenue North, just south of Park Boulevard.

Three other important fossil sites lie in St. Petersburg. Fossil shells are clearly visible in the ponds and drainage ditches at Fossil Park, a city park at 9th Avenue North and 71st Street. Toy Town Dump is a fantastic fossil site that was exposed during city excavations. It is located at the southwest corner of I-275 and Roosevelt Avenue. And at 13th Avenue

Dermal scutes (body armor) of the glyptodont Glyptotherium floridanum *of late Pleistocene time, collected south of Indian Rocks Beach near the Imperial Point community.*

North and 31st Street North, thirty vertebrate species of Late Pliocene age were collected at the St. Petersburg Times Site. Although most of the Pinellas County sites are no longer accessible, they all testify to the county's unusually rich fossil record. Any future excavation or construction will likely uncover more fossil sites, but they will have to be worked quickly to beat the construction!

East of Pinellas Park is the site of one of the first major archaeological investigations in the state, Weedon Island Preserve. First excavated in 1923 and 1924 by Smithsonian Institution archaeologists, this large burial mound and shell midden complex represents the standard that archaeologists use to compare and correlate other living sites of this culture. The Weeden Island culture existed from about AD 300 to 900, with sites along the Gulf Coast from Sarasota to Alabama and into the interior of northern and northwestern Florida. This cultural complex is recognized by distinctive ceramics and burial sites. Head east on Gandy Boulevard (US 92) toward Tampa. Turn south on San Martin Boulevard (if you reach the Gandy Bridge, you've gone too far) and drive about 3 miles to the preserve. The Old Tampa Bay area also harbors numerous

With Love, from South America

Geologists have long recognized that for most of its post-Pangaean history, which began roughly 230 million years ago, South America has been isolated from the rest of the world. Such circumstances usually result in the development of an extremely unusual assemblage of native plants and animals. Such unique and geographically restricted species are referred to as *endemic.* Australia has an abundance of marsupial mammals found nowhere else. The island of Madagascar is the only natural home of lemurs. And modern South America has its New World monkeys, jaguars, tree sloths, and a host of other bizarre creatures.

But when such isolated landmasses do come into geographic contact with the rest of the world, whether due to falling sea level or tectonic movement, interesting biotic exchanges can occur. The Great American Biotic Interchange, which occurred when the Panamanian Land Bridge connected South America and North America (see **Bush Gardens** in **Geological and Paleontological History**), had a profound zoological effect on the southern United States, especially Florida.

In addition to numerous species of ground sloth, several other South American immigrants reached Florida during Pliocene and Pleistocene time. The tanklike glyptodonts are represented in Florida's fossil record by two species: *Glyptotherium arizonae* and *Glyptotherium floridanum.* Both were very large, but *Glyptotherium arizonae* was a giant, reaching nearly 10 feet in length and weighing an estimated 2,000 pounds. Armadillos first appeared in Florida in Pliocene time. *Dasypus bellus* was roughly twice the size of the modern armadillo, but there were also giant armadillos called pampatheres that reached 6.5 feet or more in length. Two species lived in Florida, *Holmesina*

Artist's rendering of Glyptotherium arizonae. —From Gillette and Ray (1981) and courtesy of the Smithsonian Institution

floridanus and what appears to be its descendent, *Holmesina septentrionalis*, which was over twice as large as its predecessor. All armadillos became extinct in Florida in late Pleistocene time. The modern nine-banded armadillo (*Dasypus novemcinctus*), also called the long-nosed armadillo, represents an entirely new species and a second wave of immigration from its native South America during Holocene time.

Three species of porcupine lived in Florida from Late Pliocene through Pleistocene time. The modern species *Erethizon dorsatum*, found in Pleistocene deposits in Florida, lives today in temperate forests of the northern United States. Three species of capybaras, the largest living rodents, are today found only in South America, but their fossils commonly occur in sediments of Late Pliocene through Pleistocene age in Florida. Rare fossils of extinct species of opossum are found in rocks of Late Oligocene and Miocene age in Florida, but remains of the surviving species of opossum (*Didelphis virginiana*) can be found in deposits of middle Pleistocene age to the present day. The opossum is the only marsupial in North America.

The pampathere Holmesina septentrionalis. *Pampatheres could grow up to 6.5 feet long. This specimen is at the Florida Museum of Natural History.*

other archaeological sites of various ages and cultures, including Yat Kitischee and Safety Harbor. Unfortunately, in the early years before they were protected, many shell middens were mined for use as road base.

The long chain of heavily developed barrier islands of Pinellas County represents a very dynamic coastal system. The islands have diverse origins, including spit modification (when breaching of a spit creates a barrier island) and the shoaling of bars. Several of these islands have emerged from the sea only in the past several decades, as sandbars shoaled. Many tidal inlets were opened by hurricane activity, such as Johns Pass north of Treasure Island in 1848, and some former tidal inlets have been closed by barrier island migration. All of these barriers are Holocene in age, most having developed in the past 3,000 years, but deeper sands date at more than 4,000 years of age. The Pinellas coast has been "hardened," or "armored," by seawalls and other structures. Although communities build such features to protect and preserve their beaches, they tend to deprive beaches of sand and accentuate erosion in the long run. Several beaches here have been nourished with offshore sands (dredged sand has been placed on eroding beaches).

There are some beautiful barrier islands in this area, many of them accessible only by private boat or ferry. Be sure to explore the beaches of Fort De Soto Park, located on the V-shaped Mullet Key, west of the Sunshine Skyway Bridge. Fort De Soto, constructed and garrisoned in and around the year 1900, was armed with state-of-the-art cannons and rapid-fire guns. It never saw battle. To reach the fort and park, follow Florida 682 west and then Florida 679 south from the tip of the Pinellas County peninsula. About 3.5 miles north of the park on Florida 679 is a historic marker for Tierra Verde Mound. This AD 1500 burial mound of a nearby village of the Safety Harbor culture is now covered by condominiums.

Beneath a relatively thin veneer of sand, the shallow seafloor off west-central Florida is not unlike that of the Big Bend and other areas of the Florida coast. However, unlike the offshore area to the north, where Eocene- and Oligocene-age limestones lie beneath the sand veneer, younger, Miocene-age limestone and dolostone occur here. Offshore of the barrier island chain are many rocky outcrops, most of them linear ridges with a north-south trend. Some rise up to 12 feet off the seafloor. After storms, numerous shark teeth may be found along the coast, having been eroded from the Miocene-age rocks offshore. There are also many drowned sinkholes, including one 32 miles offshore at a depth of 110 feet. These sinkholes testify to the fact that much of the Florida Platform was exposed during the glacial stages of Pleistocene time, developing into a very broad karst terrain before being submerged by the rising sea.

The Sunshine Skyway Bridge traverses the mouth of Tampa Bay from the Pinellas Peninsula to Terra Ceia Island. Madira Bickel Mound State Archaeological Site lies southwest of US 19 on this island. This is primarily a Safety Harbor culture ceremonial mound, which was used into the time of European contact. But there is evidence that the site had been occupied since the Woodland Period, starting some 2,000 years ago. Madira Bickel was the first designated state archaeological site. About 1 mile south of where I-275 diverges from US 19, turn west onto Bayshore Drive and proceed about 1.5 miles to the site. South of the mouth of Tampa Bay and the Sunshine Skyway Bridge, the barrier island chain continues with Anna Maria Island and Longboat Key.

Interstate 75

Ocala—Tampa—Bradenton

134 miles

South of Ocala, I-75 passes over the Ocala Limestone, which lies just beneath a thin layer of soil and extends past the southern end of the Brooksville Ridge, reaching as far south as Zephyrhills. The Ocala Limestone has been extensively mined in Marion County. South of Bushnell, large quarries expose the southernmost exposures of the Ocala Limestone. At Exit 309, take County Road 673 east 1.5 to 2 miles toward St. Catherine to see them.

About 4 miles north of the US 98 interchange, I-75 crosses over the northwest-flowing Withlacoochee River (this is the southern Withlacoochee, not to be confused with the northern Withlacoochee at the Florida-Georgia border). The headwaters of the Withlacoochee are in the Green Swamp, located approximately between Clermont and Polk City. The Green Swamp is a karst plain erosionally leveled to the water table, with a low beach ridge plain, probably of Pliocene age. The swamp is essentially surrounded by the Brooksville, Lakeland, Lake Wales, and Winter Haven ridges, and beneath it lies the thickest sequence of freshwater-bearing limestones in the state. Here, the freshwater portion of the Floridan aquifer system is more than 2,200 feet thick. Pleistocene-age fossils are commonly found in the Withlacoochee River, and one can collect them while snorkeling. (West of I-75, a good collecting area is the canoe run between the Florida 44 bridge in Sumter County and the Florida 200 bridge in Marion County.)

East of Pasco is one of the more unusual structures constructed of native stone in Florida. Drive about 4.4 miles east of I-75 on Florida 52 to a beautiful trail and garden called the Grotto on the south side of the road and across from St. Leo Abbey, home of the Benedictine monks

The Grotto at St. Leo Abbey, constructed entirely of Oligocene-age reef corals.

of Florida. The public is welcome at this garden, which has two large shrines, or grottoes, constructed of natural, uncut stone. What is truly surprising about the grottoes is that every single stone is a colony of silicified coral. Some of them are quite large, reaching up to 3 feet in diameter. Evidently, the fossil corals were quarried from nearby exposures of the Tampa Member of the Arcadia Formation of Late Oligocene to Early Miocene age.

Abundant sources of silicified coral have been found in Pasco County, particularly near the town of Wesley Chapel, on Florida 54 just east of I-75. It is probably not coincidental that many archaeological sites in this area are associated with stone tools made from this agatized coral. Archaeologists believe that this region may have served as an important source of raw material for peoples of the Archaic Period, and they interpret some of the coral tool sites as quarries or workshops. Many such sites have been found in Pinellas and Hillsborough counties, where the silicified Tampa Member also occurs at the surface. The town of Thonotosassa lies just northeast of Tampa. Its name is a Muskogean Indian word meaning "flint place," another reference to the silicified corals of the region.

An important fossil site called Cowhouse Slough lies in northern Hillsborough County. A small sediment deposit within a fissure in

Close-up of coral colonies at the Grotto, St. Leo Abbey.

the Tampa Member, it contains land mammals of Late Oligocene age, including rodents and other small forms, but also a horse (*Miohippus*), an oreodont, an even-toed hoofed animal (the artiodactyl *Leptomeryx*), a false saber-toothed cat (*Nimravus*), and the remains of several reptiles (boas, tortoises, lizards). This site, along with only a few others of this age, indicates that Florida had emerged from the sea by Late Oligocene time, at least 24 million years ago, and was supporting a robust terrestrial ecosystem.

Hillsborough River State Park is east of I-75 before the route enters Tampa. Of special interest here is a beautiful spot called The Rapids, which features good exposures of the Tampa Member. Most of the rocks are coated with iron oxide. The Tampa Member is also exposed in places along the Alafia River, forming mild rapids, south of Tampa and Plant City.

The estuary of Tampa Bay, with its inland divisions of Old Tampa Bay and Hillsborough Bay separated by the Interbay Peninsula, is an example of a drowned-karst river valley. Tampa Bay is fed by four rivers—the Alafia, Hillsborough, Little Manatee, and Manatee. The scenic Bayshore Boulevard winds along the eastern shore of the Interbay Peninsula (the western shore of Hillsborough Bay), and halfway down the peninsula, it provides access to a city park and fishing pier that today would never be suspected for its paleontological significance. Ballast Point has been discussed in the geological literature since 1846 and was well known before that time. What is of interest here is the Tampa Member of the Arcadia Formation, which was exposed at and below water level before much of the area was built up with seawalls. Today, large boulders of the Tampa ornament the shoreline of the park and fragments can be collected in shallow water.

Ballast Point is famously known as the primary site where Florida's state stone, agatized coral, was preserved. The fossil-rich Tampa Member was long referred to as the Silex Beds because of the silicified marine and terrestrial mollusks and corals found here. Similar localities are nearby, such as Sixmile Creek, at the head of Hillsborough Bay.

Just south of Ruskin, which is on US 41 (Tamiami Trail) west of I-75, and 1.5 miles south of the Little Manatee River, is Cockroach Bay Road, which leads to Tampa Bay and the richest vertebrate fossil locality of early Pleistocene age in North America, the Leisey Shell Pits. At this remarkable site, which you must have permission to enter, scientists have recovered more than two hundred vertebrate species, including fourteen sharks, nine stingrays, fifty bony fish, three amphibians, twenty-six reptiles, fifty-two birds, and forty-nine mammals. Among the larger mammal species recovered were two llamas, two horses, an early mammoth, two ground sloths, a peccary, a giant tapir, a saber-toothed cat, a short-faced bear, and a wolf. One nightmarish find was the bear-sized giant beaver *Castoroides leiseyorum*, one of two giant beavers found in North American and Floridan strata of Pleistocene age. The modern beaver, *Castor canadensis*, has been found in Pliocene- and Pleistocene-age formations and survives today in Florida's wetlands and streams.

The Leisey site was first uncovered in 1983, and a second bone-bearing bed was discovered in 1986. Many avocational and professional crews have since worked the site. Several formations are present: part of the Arcadia Formation of Oligocene to Miocene age is at the base of the site; next is the primary bone-bearing formation, the shelly Bermont Beds of early Pleistocene age; above this is another shell bed, the Fort Thompson Formation of late Pleistocene age; and on top are undifferentiated Quaternary-age sands and soil. Using several methods, geologists have precisely dated the Bermont Formation as being between 1.6 and 1.4 million years old. There are two distinct bone beds in the 23-foot-thick Bermont Formation, each from 2 to 12 inches thick and rich in organic material. Mollusks dominate the invertebrate fossils of these beds, with at least 98 species of bivalves, 113 species of gastropods, and a variety of other forms. Several freshwater mollusks are present, especially in association with the bone beds. There are also pieces of wood and cones or seeds from pine, live oak, cabbage palm, and saw palmetto, and pollen from several other plant species.

The Leisey site was a nearshore setting, at times inundated by the sea, that experienced periods of significant freshwater input, as evidenced by its mix of terrestrial, freshwater, and nearshore-marine vertebrates; shallow-marine, estuarine, and freshwater mollusks; and coastal vegetation. It almost certainly was an estuarine setting, not unlike modern Tampa Bay. Thousands of the Leisey fossils reside in the Florida Museum of Natural History in Gainesville. Many other Leisey fossils are in private collections.

Fossil Saber-Tooths

The term *saber-tooth* is commonly used in reference to a wide variety of extinct cats with elongated upper canines. Most belong to the family Felidae, or true cats. Other saber-tooths belonged to various catlike families of carnivores. Probably no other extinct animal is as popular an icon of the Pleistocene Ice Age as the famous saber-toothed "tigers." Several species of saber-tooths once roamed over Florida from Oligocene to Pleistocene time. One remarkable saber-tooth is found in Miocene-age rocks in Florida. Often called a false saber-toothed cat because it wasn't a true cat, instead belonging to an extinct group of catlike carnivores called the *nimravids*, *Barbourofelis loveorum* nonetheless had retractable claws and was in other respects very catlike.

Among the true cats, there were two groups of saber-tooths in Florida. The scimitar-toothed cats had shorter, broad canines that were normally serrated. An example was the long-legged, jaguarlike genus *Nimravides*, often found with *Barbourofelis* in Miocene-age formations. The dirk-toothed cats had longer, narrower canines, with or without fine serrations. This group included the genus *Megantereon*, found in Pliocene-age sediments in Bone Valley, and the more famous genus *Smilodon*, popularly associated with the Pleistocene-age Rancho La Brea Tar Pits of Los Angeles, California. Two species have been found in Florida. *Smilodon gracilis* occurs in formations of Late Pliocene to middle Pleistocene age. The much larger, lion-sized *Smilodon fatalis*, which is the species found in the California tar pits, lived in Florida from middle to late Pleistocene time. The saber teeth of this large species reached up to 7 inches in length. Worldwide, saber-tooths went extinct at the end of Pleistocene time, about 10,000 years ago.

Several cats with normal-sized canines also lived in Florida during Pleistocene time, including lynxes, lions, jaguars, cheetahs, and the ocelot. Fossils of the lion *Panthera atrox*, a very close relative of the modern African lion but significantly larger, were found in the Ichetucknee River, east of Branford off US 27. The only surviving large cat from this time is the panther (*Puma concolor*), also known as the cougar. Sadly, the Florida population of this magnificent cat is nearing extinction.

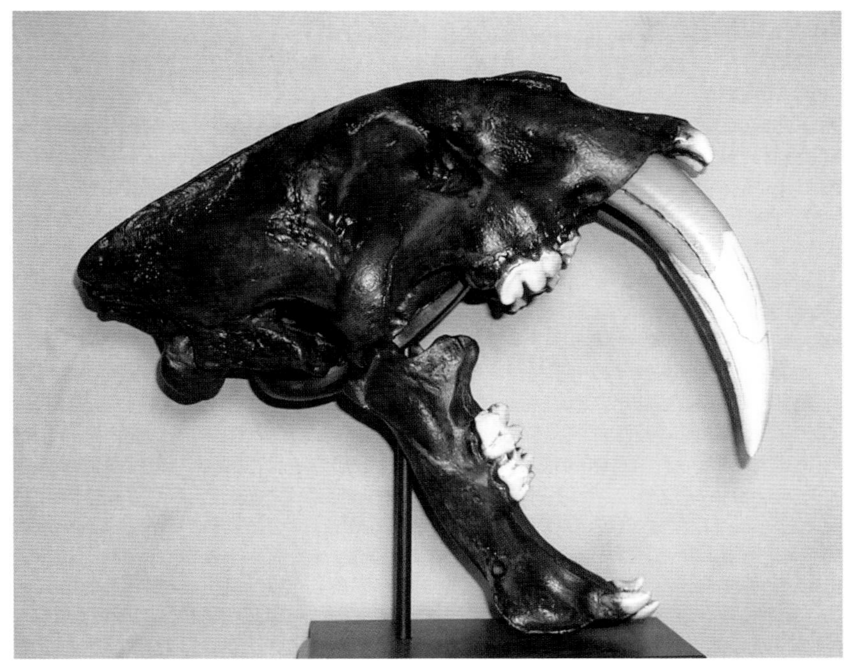

Smilodon fatalis, *a true saber-toothed cat. This model was cast from a specimen recovered from the La Brea Tar Pits.*

Barbourofelis loveorum, *a catlike nimravid from the Miocene-age Love Bone Bed in Alachua County, attacking the short-limbed, hippopotamus-like rhino* Teleoceras proterum.

Central Florida Phosphate District

The Central Florida Phosphate District lies in portions of Polk, Hillsborough, Manatee, Hardee and DeSoto counties in central Florida, and phosphate mining operations are visible along several routes, including Florida 37, Florida 60, US 98, and US 17. Phosphate deposits were initially discovered in the Peace River near Fort Meade in Polk County during surveying for canal construction in 1881, and mining began in the late 1880s. The area is also especially rich in Miocene- and Pliocene-age vertebrate fossils and is better known as Bone Valley among fossil collectors.

Miners found two types of phosphate deposits in Fort Meade: river-pebble phosphate and land-pebble phosphate. As the names imply, river-pebble deposits occurred beneath rivers and streams, while land-pebble deposits (mostly gravel but some rock) were found beneath land. River-pebble gravel was mined with large dredges that pulled the phosphate from the rivers. Land-pebble mining replaced dredge mining in the mid-1890s, since these deposits were more economical to mine. The highest-grade deposits occur in the Bone Valley Formation, now called the Bone Valley Member of the Peace River Formation of the Hawthorn Group, in Polk and Hillsborough counties. As better deposits were depleted, mining migrated south to lower-quality deposits.

The Central Florida Phosphate District has provided a significant portion of the phosphate mined in the United States since the early 1900s. The industry supported the growth of the cities in this region, employing many locals. However, some small towns disappeared as mining companies purchased the land on which the towns had developed and mined through them.

Geologists once proposed that the origin of phosphate in Bone Valley was related to bird guano deposits from large rookeries, but this hypothesis has been discounted. Most feel that marine phosphogenesis explains the origins of Florida's phosphate deposits. Scientists generally agree that phosphogenesis proceeds as follows: Cold, oceanic bottom water contains an abundance of the element phosphorus. Sometimes bottom water currents well up near a continental margin, making the phosphorus available to marine algae and animals. The result is a proliferation of life and the production of huge quantities of organic matter. As the algae and animals die, their remains sink to the ocean floor and accumulate. If conditions are very low in oxygen, the organic material may be degraded by sulfate-reducing bacteria, which release the phosphorus into water within the sediments. Chemical interactions cause small grains of phosphate minerals to form. In Florida's case, sediments containing phosphate deposits were reworked by ocean waves and currents as sea level fluctuated. This concentrated the phosphate grains into regionally extensive, economically important deposits of phosphorite, a sedimentary rock.

Florida phosphate occurs in sand- to gravel-sized particles in a matrix of quartz sand and clay. A minimally economical deposit contains approximately 33 percent of each component. Large draglines working twenty-four hours a day, with booms that stick above the landscape in the district, strip off the overlying sediments to access the phosphate ore zone. The overburden is stockpiled for future use in the reclamation of the mine. The draglines then dig out the ore zone, called the *matrix*, using huge buckets that remove 40 to 70 cubic yards with each scoop. They pile the matrix next to the pit, where workers use high-pressure water guns to slurry the matrix. The slurry is then pumped to a processing plant, which can be several miles from the pit. The phosphate grains are separated from the other sediments through a complex process. The sand tailings are used in the reclamation process, and the clays are pumped into large settling ponds, called *slime ponds*, where the clay separates from the water over time. You can see numerous old slime ponds, surrounded by large dikes, as you drive through the district.

Also dotting the landscape are huge stacks—some say mountains—of phosphogypsum, the main by-product of phosphate processing. As phosphate ore is dissolved in sulfuric acid, phosphogypsum is released. No good use has been found for phosphogypsum because it contains many trace elements, including uranium, that are hazardous to humans. However, mining companies did recover uranium from the phosphate ore in the 1970s, when uranium prices were high.

As the draglines mine, they "walk." Imagine, if you can, a 7-million-pound machine with a boom as long as a football field. How does a machine of this size move? It has no treads or wheels. It uses large walking pads that rotate off the ground, lifting the behemoth up while moving it forward. It is a slow

Phosphate mining with a dragline.

process—draglines move about 10 feet per step. In some cases, draglines have walked for miles from one mine to another.

For many years, companies mined phosphate with no regard for the environment. They pumped huge quantities of water from the Floridan aquifer system, reducing water levels by more than 50 feet. In addition, mines and their spoil piles were simply left behind as mining activities moved on to new areas, leaving a moonscape where flat to rolling land once existed. As Floridians and the rest of the United States became environmentally aware in the 1960s and 1970s, mining companies began reclaiming mined areas and reusing water when processing ore. The state now mandates the reclamation of mines. Prior to mandatory reclamation, nearly 150,000 acres were mined. Since then, 166,722 acres have been mined, with 104,386 acres reclaimed through 2002. Reclaimed areas are typically rolling hills interspersed with lakes. Examples include Alafia River State Park, a 6,000-acre reclaimed area in Lithia, and Lake Louisa State Park, home to one of thirteen lakes connected by the Palatlakaha River.

While visiting the central Florida area and seeing its phosphate mining operations, you can stop by the Mulberry Phosphate Museum in the town of Mulberry. The museum has an extensive collection of fossils found in the phosphate mines and interpretive displays on the history of phosphate mining operations. It is located on Florida 37 one block south of its intersection with Florida 60. The museum is housed in an authentic 1899 train station.

Where the Buffalo Roamed, and the Deer and the Antelope Played

A least three species of buffalo once roamed across Florida. *Bison latifrons*, found in deposits of middle Pleistocene age, was the largest, with horns that spanned as much as 6.5 feet. *Bison antiquus*, perhaps only a subspecies of *Bison bison*, is found in deposits of late Pleistocene age. *Bison bison* is found in recent deposits, those that formed in the last 10,000 years. Like elephants, bison migrated to North America from Asia in two waves—first in Pliocene time, and then in the middle Pleistocene time. The remains of *Bison antiquus* are especially common in river sediments.

Buffalo disappeared from Florida in late Pleistocene time, but modern *Bison bison* reentered the state from the Great Plains during Holocene time. European explorers noted the presence of bison in Florida, but settlers had completely exterminated them by 1875. Small herds of bison have been reintroduced in a few areas,

FLORIDA PALEONTOLOGICAL SOCIETY, INC.
FOUNDED 1978

Hexobelomeryx simpsoni *from Bone Valley sediments of Early Pliocene age, the official mascot of the Florida Paleontological Society.*

such as Paynes Prairie south of Gainesville. Besides bison, Florida has long been home to a great variety of other ruminants, animals that "chew the cud," including several deer and related forms. Three or more species of pronghorn antelope are found in the Pliocene-age rocks of Bone Valley, including the six-horned *Hexobelomeryx simpsoni,* which is the official mascot of the Florida Paleontological Society.

Fossil Whales

Whale fossils are fairly common in Florida, and at least thirty species have been recovered. Eocene-age whales belong to a very primitive group called the archaeocetes, whose fossils are especially abundant in India, Pakistan, northern Africa, and the southeastern United States. Although not abundant in Florida, archaeocete whales can be found in strata of Middle to Late Eocene age. At least three species have been found, mostly in the Ocala Limestone. The most abundant species, *Zygorhiza kochii,* reached lengths of 20 feet. *Pontogeneus brachyspondylus* was a very large whale, but in Florida only isolated vertebrae have been unearthed. Also rare in Florida, but common in Alabama and Mississippi, is the giant *Basilosaurus cetoides,* which grew up to 65 feet long and had a skull nearly 5 feet long. Whales of Oligocene age have not been found in Florida.

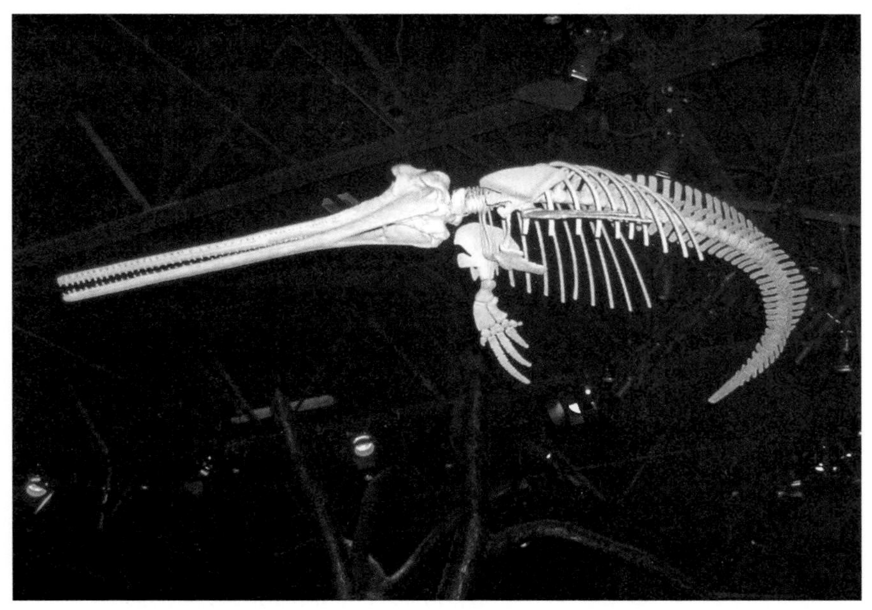

Pomatodelphis inaequalis, *the long-beaked dolphin, from Bone Valley deposits of Middle Miocene age. This specimen is at the Florida Museum of Natural History, Gainesville.*

In Florida, whale fossils are far more common in rocks of Miocene, Pliocene, and Pleistocene age. These fossils represent the two modern whale groups, the odontocetes (toothed whales), and the mysticetes (baleen or whale-bone whales). The most common type of whale remains are isolated vertebrae and teeth, and their unusual solid inner ear bones, called *auditory bullae*, which fossilize extremely well. Toothed and baleen whales are especially common in the Hawthorn Group in the Bone Valley phosphate region, along the southwestern coast from Tampa Bay to Lee County, and in north-central Florida. Toothed whale fossils include the unusual long-beaked dolphin (*Pomatodelphis inaequalis*), which was extremely abundant in Florida waters and distantly related to the modern Asian river dolphin; several other dolphins; and the common sperm whale, usually represented by its teeth. The Pliocene-age whales include killer whales, sperm whales, and *Goniodelphis*, which was related to the modern Amazon River dolphin.

Like toothed whales, baleen whales first appeared in Florida in Miocene time, with rare fossils of several species of a poorly understood, extinct family called cetotheres. *Balaenoptera* species were the dominant toothed whales of Pliocene time. This genus includes the living fin and blue whales. *Balaenoptera floridana* grew up to 36 feet long. By Pleistocene time, most modern whale species were present, and many are found in coastal deposits, particularly along the Atlantic Coast, where Florida's most complete fossil whale skeletons have been recovered.

Whales of the Eocene Seas, *by Charles Knight. The archaeocete whale* Basilosaurus cetoides *grew over 65 feet long and has been collected from the Ocala Limestone along with two other species of this extinct group of whales.* —©The Field Museum, #CK26T

US 441

Ocala—Orlando—Okeechobee

225 miles

US 441/27 passes through Lady Lake, which has some pronounced topography with exposures of the Cypresshead Formation of Pliocene age and dunes of Quaternary age. Karst lakes are also visible. Lake County is appropriately named. It has the highest concentration of lakes in the state—1,345! US 27 heads south in Leesburg, while US 441 heads east, passing several large lakes, including Lakes Griffin, Harris, Eustis, and Dora. US 441 continues through the hilly terrain of Mount Dora, namesake of the Mount Dora Ridge, one of the central peninsular ridges. Although nowhere does it exceed 190 feet, bumper stickers and T-shirts proudly boast, "I climbed Mount Dora." More exposures of the Cypresshead Formation, Quaternary-age dunes, and numerous lakes are common in this sand hill karst terrain, and these extend down to Zellwood.

What is geologically curious about these lakes is that they are associated with the relatively high and sandy Mount Dora Ridge. Just beneath the ridge, though, is karstic limestone. The association of sand ridges and lakes over karstic limestone, referred to as *mantled karst*, *deep-cover karst*, or, more commonly, *sand hill karst*, is common in the state. The lakes form as sand falls into shallow sinkholes, similar to sand falling through an hourglass, and rising groundwater fills the depressions. Sand hill karst is well developed in the Central Lakes District, comprising Osceola, Orange, Lake, Polk, and Highlands counties, and in several other areas across the state.

On the northwest edge of the greater Orlando area are a couple of popular swimming holes and beautiful second magnitude springs, both north of Apopka: Wekiwa Springs State Park and the enormously popular Dr. Howard A. Kelly County Park (commonly known as Rock Springs). Both springs issue out of the Miocene-age Coosawhatchie Formation of the Hawthorn Group, although the water comes from the Ocala Limestone. Better outcrops of the Hawthorn Group carbonates are visible at Dr. Howard A. Kelly County Park. About 20 feet of phosphatic dolostone is beautifully exposed here, and one can snorkel and collect fossil shark and ray teeth from coarse gravels along the rocky, shallow spring run. From Apopka, take County Road 435 north for 5.8 miles and turn east (right) on Kelly Park Road. Proceed about 0.3 mile to the park entrance.

With towns like Winter Beach, Winter Garden, Winter Haven, Winter Park, Winter Springs, and Frostproof, is it any wonder that so many northerners move to the Sunshine State? But be careful where you build your dream home. In May of 1981, the now famous Winter Park sinkhole

Rock Springs issuing out of Hawthorn Group carbonates.

The Winter Park sinkhole.

collapsed in an urban area. Although the 2-acre sinkhole is not exceptionally large compared to other sinks in the region, its collapse initiated an enormous research effort into the geology of sinkholes. The Winter Park sinkhole has become a classic example of a geological hazard in karst terrain, and pictures of it appear in many earth science textbooks.

Beyond Orlando, the segment of US 441 from St. Cloud to Yeehaw Junction passes over the Osceola Plain, part of the Barrier Island Sequence District. It is a Pleistocene-age beach ridge plain composed of quartz sand. The beach ridges are especially well developed east of St. Cloud, and they clearly control the linear and parallel distribution of streams and lakes in this region. The ridges are very noticeable on topographic maps and aerial photos. East of the Osceola Plain but west of the Atlantic Coastal Ridge (between Florida's Turnpike and I-95) is the southern extension of the long Eastern Valley, which extends down nearly the entire east coast of the peninsula. In the northern half of the peninsula, the Eastern Valley is mostly covered with beach ridge plains of Pleistocene age. But in the southern end it is very marshy, and geologists speculate that in this region the valley was a lagoon floor during Pleistocene time. St. Johns Marsh, the headwaters of the St. Johns River, is located here.

About 2.6 miles southeast of Ashton, US 441 passes a dune field, and about 1.7 miles south of Holopaw, linear cypress swamps are evident on both sides of the road. The linear orientation is controlled by the Pleistocene-age beach ridges and intervening low areas, or swales. About 13.5 miles south of the Yeehaw Junction and Florida 60, at Fort Drum, County Road 15C heads east from US 441. This road leads to Rucks Pit, where fossil shells of the clam genus *Mercenaria* can be collected. These fossils are filled with large, yellow calcite crystals. Densely fossiliferous sediments of Pliocene and Pleistocene age are exposed here and at nearby mines. The mine entrance is 2.8 miles east of US 441 on the north side of the road.

Flat, dry prairie habitat surrounds US 441 from Fort Drum to the County Road 68 intersection. What is now prairie grassland was a flat ocean bottom in late Pleistocene time. Kissimmee Prairie Preserve State Park, 14 miles west of US 441 on County Road 724, is a pleasant place to view this ocean-bottom-turned-prairie. About 11.5 miles south of County Road 724, US 441 crosses Taylor Creek. The Fort Thompson Formation of Pleistocene age, consisting of shell-rich sands and sandy limestones, is exposed along the west side of this canal in the spoil bank.

Shell beds of Pleistocene age in the Dickerson Pit near Lakewood Park, St. Lucie County.

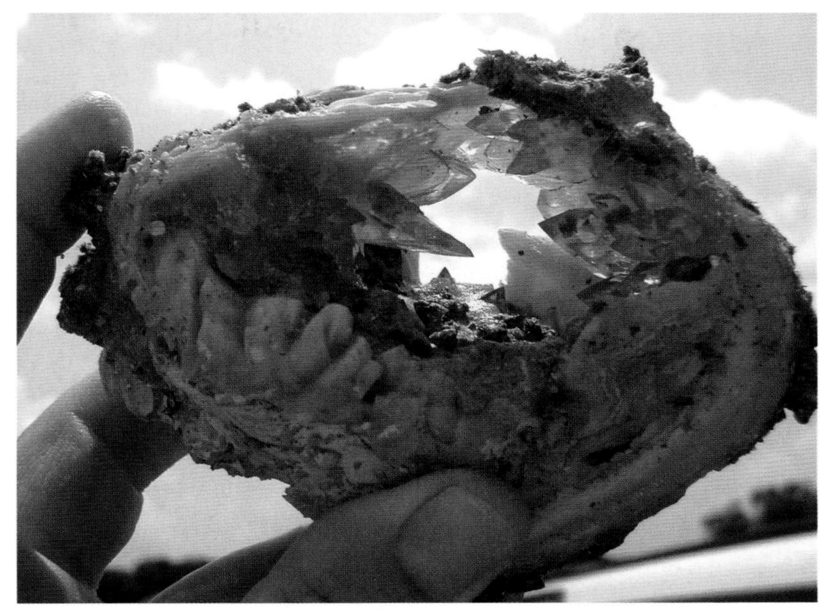

*Calcite-filled fossil of the quahog clam (*Mercenaria*) of Pleistocene age.*

US 27

Clermont—Venus

114 miles

US 27 between Clermont and Lake Placid follows the entire length of what is perhaps the most prominent surface feature of the entire state—the Lake Wales Ridge. The ridge rises up to 300 feet or more above sea level and is composed of the Pliocene-age Cypresshead Formation, reworked sediments of the Cypresshead Formation, and Quaternary-age sands and dunes. Many geologists think the Lake Wales Ridge was once much broader and longer and extended farther north, along with the Mount Dora Ridge, to a connection with Trail Ridge. The narrow, lake-filled valley running down the middle of the southern half of the Lake Wales Ridge, called the Intra-Ridge Valley, is sand hill karst that formed as the limestone beneath it dissolved.

One can have a panoramic view of the northern Lake Wales Ridge and vicinity from the top of the Citrus Tower, located on US 27 in Clermont about 1 mile north of Florida 50. Built in 1956, the tower rises about 500 feet above sea level, offering one of the highest views in the state. The tower is a relic of an earlier golden age of citrus. In the 1950s and 1960s, the site included a reptile exhibit and a glass-blowing shop. The view from the top reveals miles of even rows of orange trees. Now

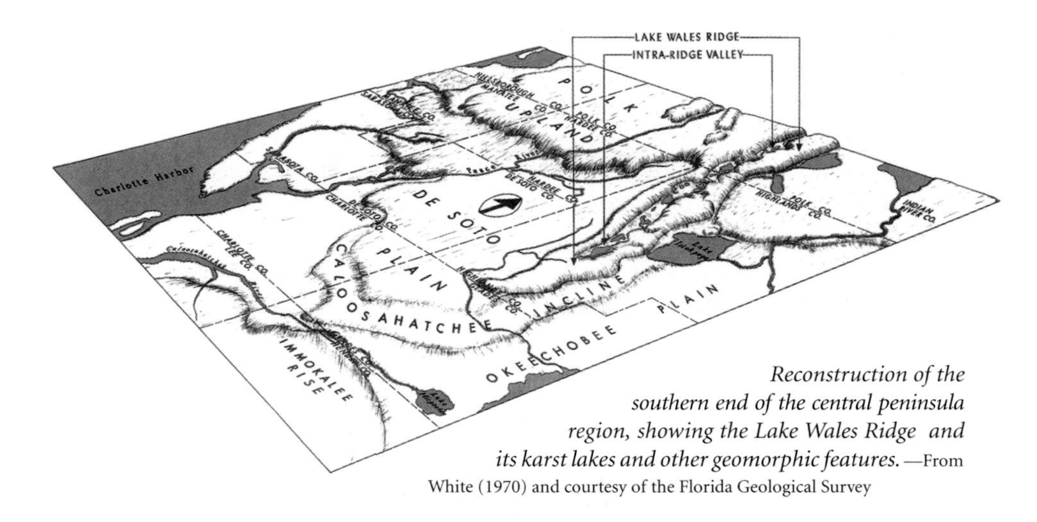

Reconstruction of the southern end of the central peninsula region, showing the Lake Wales Ridge and its karst lakes and other geomorphic features. —From White (1970) and courtesy of the Florida Geological Survey

the hilly terrain of the Lake Wales Ridge and numerous karst lakes that are very evident in the Clermont area are increasingly being crowded with subdivisions. To the northeast, Lake Apopka is especially visible.

In his slim book *Oranges*, John McPhee refers to the Lake Wales Ridge simply as "The Ridge" and describes it in these entertaining terms:

> *The Ridge is the Florida Divide, the peninsular watershed, and, to hear Floridians describe it, the world's most stupendous mountain range after the Himalayas and the Andes. Soaring two hundred and forty feet into the subtropical sky, the Ridge is difficult to distinguish from the surrounding lowlands, but it differs more in soil conditions than in altitude, and citrus trees cover it like a long streamer, sometimes as little as a mile and never more than twenty-five miles wide, running south, from Leesburg to Sebring, for roughly a hundred miles. It is the most intense concentration of citrus in the world.*

McPhee's description was written in the 1960s, long before the freezes of 1983, 1985, and 1989 systematically decimated nearly all of the citrus trees in this region. Except for such periodic, critically low temperature drops, the climate and sandy soils of the Lake Wales and similar ridges are ideal for citrus, which require cold (but not freezing) nights and wet conditions (but not standing water). The well-drained, sandy soils of the central peninsula ridges are just right. Several cities are located on the crest of the ridge, including Haines City, Lake Wales, Frostproof, Avon Park, and Sebring.

East of US 27 and running somewhat parallel to it is Florida's Turnpike, which offers one of the most panoramic highway views in the entire state. Along the turnpike about 20 miles south of its starting point at I-75, the karst sand hills of the Lake Wales Ridge are very evident, with many open valleys, lakes, and hills. The turnpike passes through the axis of several hills at about 28 and 32 miles southeast of I-75. At the county line between Lake County and Orange County, you descend from the Lake Wales Ridge onto the Orlando Ridge. There is still an abundance of lakes on the Orlando Ridge, but the relief is much less marked. Orange groves once covered these hills in perfect rows until the freezes of the 1980s, and citrus is making a limited comeback here. During winter months, oranges are especially evident, even from the highway. However, development pressures have resulted in many former orange groves becoming subdivisions. This is especially evident along the turnpike and US 27.

Sugarloaf Mountain

The highest point of the Florida Peninsula is Sugarloaf Mountain, at 312 feet. To see the "mountain," and more of the impressive topography of the Lake Wales Ridge and karst lakes, proceed north on US 27 from Florida 50 in Clermont for about 5 miles, then turn north (right) on County Road 561. Travel for 2.5 miles to Sugarloaf Mountain Road. Take this road up to the top of the ridge for a truly spectacular view. At the top, the road is about 300 feet above sea level. Lake Apopka is prominent to the east.

View from the top of Sugarloaf Mountain.

Between Clermont and Haines City, US 27 follows the western flank of the Lake Wales Ridge, bordering the Green Swamp to the west, which is a beach ridge plain of Pliocene age. West of US 27, between Haines City and as far as Lakeland, US 92 and I-4 pass over outliers of the Cypresshead Formation called the Winter Haven and Lakeland ridges. The sand hill karst is very evident, and there are numerous lakes.

In Lake Wales, be sure to visit the beautiful Historic Bok Sanctuary, perched upon Iron Mountain, which is part of the central Lake Wales Ridge. At this sandy exposure of the Cypresshead Formation, the ridge reaches about 298 feet, though a benchmark on top of the hill reads 295 feet. Some of the sands here are cemented with iron oxide, making them iron-cemented sandstones. The 205-foot-tall carillon tower, completed in 1929 by Dutch-American Edward Bok, is a national historic landmark. It was partly built with Cambrian-age Etowah Marble from Georgia and coquina of the Pleistocene-age Anastasia Formation from the east coast of Florida. Both of these building stones are visible on the tower's exterior. From US 27 north of Lake Wales, take the Mountain Lake Cut Off Road (County Road 17A) east for almost 1 mile, then turn south on Florida 17 and drive 0.8 mile. Turn east back onto Florida 17A and proceed about 1 mile to the entrance. While in Lake Wales, you can also experience your car "rolling uphill" at Spook Hill, located about 0.1 mile west of the Bok Tower entrance on County Road 17A.

The Lake Wales Ridge is associated with dune fields of Pleistocene age, which are scattered from the southern end of the ridge in southernmost Highlands County to at least the Ocala National Forest in Marion County, a distance of about 170 miles. The elongated dunes developed along the eastern side of the ridge, particularly between Haines City and Lake Placid, amongst the numerous lakes in this area. Scenic US 27A/ Florida 17 between Haines City, Dundee, Lake Wales, and Frostproof passes through some pronounced dune topography with numerous sand hill karst lakes.

The Lake Wales Ridge is of enormous biological significance, hosting one of the highest concentrations of endemic species in North America. This is thought to be related to the geological origins of the ridge. During high sea levels of Pliocene and Pleistocene time, the ridge was repeatedly isolated, or nearly isolated, from the North American mainland as a group of islands or a peninsula. This may have resulted in an exceptional number of species developing here in isolation from most of the larger continent, such as the scrub oak (*Quercus inopina*), scrub palmetto (*Sabal etonia*), and even a bird—the Florida scrub jay (*Aphelocoma coerulescens*).

Although the more ancient botanical communities associated with the Lake Wales Ridge are rapidly dwindling, they may still be explored

Bok Tower, partially constructed with coquina from the Pleistocene-age Anastasia Formation.

in several areas. Highlands Hammock State Park, 4 miles west of Sebring on County Road 634, preserves one of the few remaining temperate hardwood forests in Florida; it is also one of the most southern and ancient. In addition, the park hosts sand-dune scrub vegetation, which may have persisted here since Pleistocene time.

Lake June-in-Winter Scrub State Park preserves over 800 acres of sand-dune scrub habitat. About 12 miles south of Sebring, take County Road 621 west for 4 miles to Daffodil Road. Proceed 2 miles to the park. Lake Annie, which preserves an exceptional pollen record that reaches back over 40,000 years into late Pleistocene time, is just west of Bairs Den, at the intersection of US 27 and Florida 70. Very dry (xeric) scrub and grass savannah conditions persisted here until about 8,000 years ago, when the rising seas of the Holocene Transgression created higher water tables and more moist (mesic) conditions and vegetation. A short distance to the south, on the southern edge of the Lake Wales Ridge, is the Archbold Biological Station, an independent research facility founded in 1941 and devoted to ecological conservation, especially of the environments associated with the ridge. Archbold manages more than 19,000 acres of land in the region. US 27 travels off the southern tip of the Lake Wales Ridge south of Lake Placid, near Venus.

The "Mountains of Florida"— Dune Fields Past and Present

Humid, subtropical Florida is situated between latitudes 24 and 31 degrees north, where atmospheric high pressure prevails and dry air descends. This is the climatic setting that creates many of the northern hemisphere's deserts. Were it not for the enormous, moisturizing influence of the Gulf of Mexico, Florida would be more desert than orange grove. Florida was indeed drier, but not a desert, during Pleistocene time, allowing for the growth of a common arid-climate landform, the sand dune.

Unlike the great dunes of the Sahara and other deserts, Florida's dunes are primarily coastal in origin. They were, and continue to be, created by strong onshore winds blowing over beaches well-endowed with quartz sand. Salt-tolerant vegetation serves as a sediment trap that captures windblown sand and builds the dune. As the dune grows, it blocks sand very effectively and thus continues to grow. Many of Florida's dunes are associated with terrace escarpments or the long central peninsular ridges—all ancient shorelines. Dunes normally do not occur in isolation, but rather are high points of a much larger blanket of windblown sand called a *sand sheet.* Much of the surface of Florida is covered with such relict sand and dunes. So Florida clearly experienced periodic drier, windier climates during late Pleistocene and early Holocene time, up to 8,000 years ago. Some late Pleistocene dunes in the panhandle are now 70 feet or more in elevation and may have originally been much larger.

The dunes associated with the Central Highlands, such as those near the Lake Wales Ridge, host a unique plant community called *scrub.* It consists of evergreen shrubs, such as rosemary (several species), sand pine (*Pinus clausa*),

slash pine (*Pinus elliottii*), and scrub oak (*Quercus inopina*), plants especially adapted to the well-drained, sandy soils of old beach ridges and dunes. The scrub ecosystem is also adapted to fire. The association of scrub vegetation and dunes has been in place in Florida since Pleistocene time and persists on modern coastal dunes and interior ancient dunes and ridges.

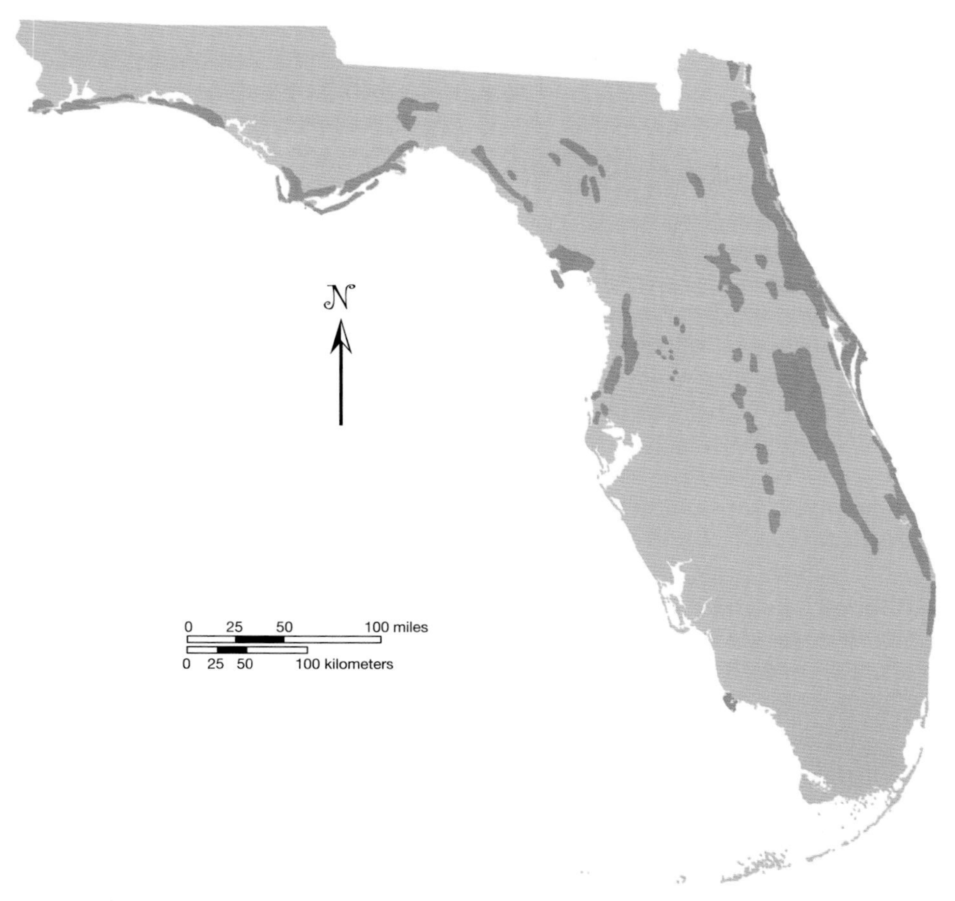

Dune fields (dark brown areas) of Florida. Most coastal dunes are entirely Holocene in age, whereas interior dunes may be late Pleistocene or early Holocene in age. This illustration shows dunes that have been mapped, but it is a conservative estimate of the total amount of dunes in the state. Vast areas of the state are covered with fine sand that is designated on the state geologic map as undifferentiated Quaternary deposits, and some sand deposits are not mapped at all because they are too thin (less than 20 feet thick). Undoubtedly, many of these sands represent windblown dunes or sand sheets. If these were included in this illustration, the coverage of dunes and windblown sediment would increase significantly. —Illustration by Amanda Kosche

Interstate 95, US 1, and Florida A1A

New Smyrna Beach—
Cape Canaveral—Fort Pierce

118 miles

I-95, US 1, and Florida A1A roughly parallel the straight Atlantic Coastal Ridge of Florida's eastern seaboard between New Smyrna Beach and Fort Pierce. Florida's east coast barrier island and coastal inlet coastline is the longest of its kind in the United States, extending some 200 miles. But unlike some barrier island coastlines, this barrier system has a discontinuous rocky surface beneath the sand—the Anastasia Formation. This rigid backbone of the Atlantic barrier coast has created some interesting surface features. There are no large drowned river valley estuaries, as there are on the west coast and panhandle coast of Florida. Instead, there are narrow coastal inlets and several shore-parallel rivers that are confined by the ridge, which acts as a coastal barrier, preventing easy entry into the Atlantic. For example, the St. Johns River flows far north, unable to enter the Atlantic until it is effectively past the rocky ridge at Jacksonville.

In the middle of this system, a very prominent coastal feature projects seaward: the cuspate foreland of Cape Canaveral, which extends over 12 miles into the Atlantic. During late Pleistocene time, when sea level was lower, the cape extended farther seaward. But as sea level rose during Holocene time, the cusp retreated in a landward direction as rising waters eroded the older cape. Offshore of the cape is a very complex series of shoals that formed as waves eroded and reworked older cape sediments. Behind the cape is Indian River lagoon, a unique estuary consisting of Mosquito Lagoon, the Banana River, and the Indian River. This estuary formed as the Holocene Transgression flooded a former beach ridge plain.

Most geologists believe that cuspate forelands, such as Cape Canaveral, form where longshore currents converge and collectively deposit sediments from two directions. Seafloor ridges or shoals may also control the location of the cusp, since these commonly form a nucleus for sediment deposition. Several coastal circulation cells may create areas where longshore currents converge and diverge, resulting in sand-rich beaches and beach ridges in the areas of convergence (depositional environments, such as Cape Canaveral), and pebble beaches and rocky outcrops in areas of divergence (erosive environments). This pattern has been documented up much of Florida's east coast.

The cuspate foreland of Cape Canaveral has been in place for most of the last 2 million years. The cape is constructed of sets of beach ridges. As seen from aerial photographs and satellite images, the beach ridge system of Cape Canaveral and vicinity is spectacularly developed. And these beach ridges continue, with exactly the same trend, far inland, forming a ruffled surface with a triangular shape. These beach ridges represent former shorelines. Just to the north of Cape Canaveral is a less obvious prominence called False Cape, where many of the older beach ridges converge, probably having formed an earlier cuspate foreland during Holocene time. Modern Cape Canaveral appears to represent the southern growth of False Cape.

Aerial photos clearly show a succession of former cusps inland along the same trend as False Cape. In fact, geologists postulate that this trend

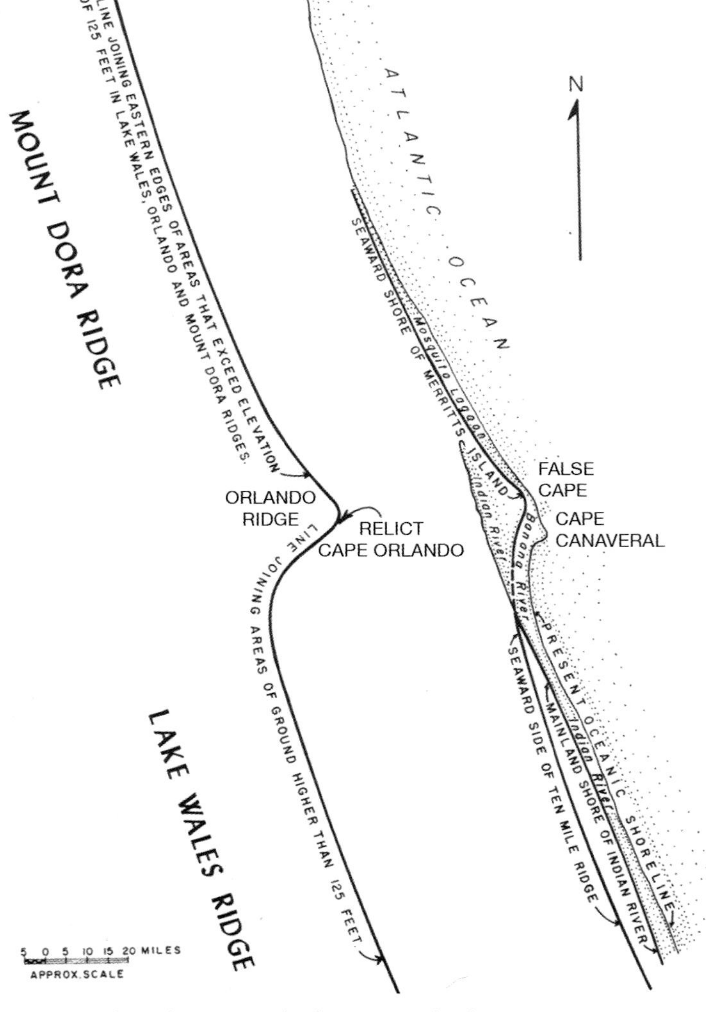

The False Cape and relict Cape Orlando. —From White (1970) and courtesy of the Florida Geological Survey

Coastal Development— Florida's Beach Ridge Plains

One of the most remarkable aspects of Florida's coastal geology—the surprising extent of its beach ridges—can best be seen in aerial photographs and satellite images. These coastal ridges are especially well-developed on St. Vincent Island along the panhandle, Cape Canaveral, and the Osceola Plain in Osceola County, but they are common along many coasts and barrier islands across the state. They may occur as thin, coastal strips, or in broader, washboardlike plains called *strand plains* or *beach ridge plains.* The ridges that are near sea level are Holocene in age, but the interior ridges, such as those across Osceola County, date to Pleistocene or even Pliocene time.

Beach ridges usually form as sand is successively accreted, or welded, to a beach by incoming surf, but they can also begin as dunes, shoals, and other sandy features. Each ridge is a former berm, or linear mound, that represents a previous shoreline (see **Anatomy of a Beach** in the **Sculpting a Land from the Sea**). Beach ridges form when the coastline has a great abundance of spare quartz sand and sea level is relatively stable. Ridges do not appear to be actively forming in Florida today, probably because dams and coastal construction trap sand, resulting in a drastic decrease in the overall sediment supply. Affected beaches are said to be "starved" of their normal sand supply, and their shorelines are retreating. This means there is net erosion occurring along most of the Florida coast.

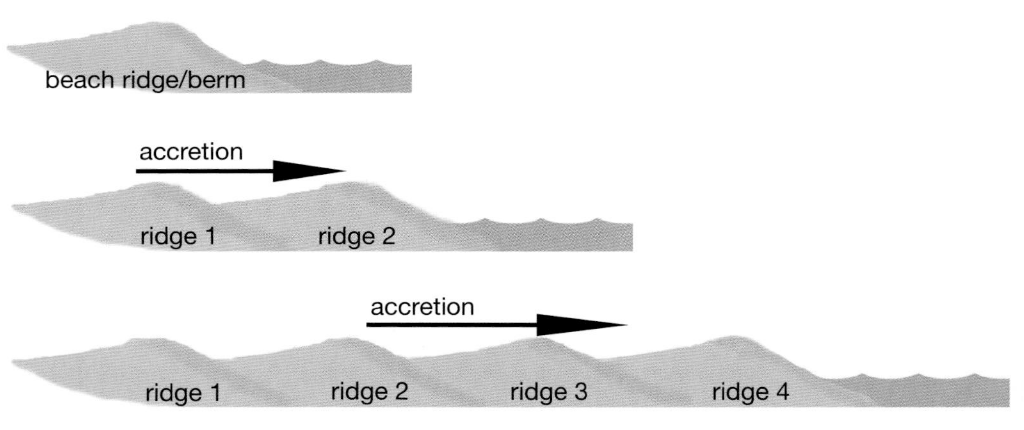

The origin of beach ridges. Probably the most common way beach ridges form is the accretion of successive berms as waves sweep sand to the shoreline. Ridges may be accreted even when sea level remains constant. If sea level falls, ridges may form at slightly different elevations. As the beach advances and a beach ridge plain forms, windblown sand and dunes may be draped over the older ridges. —Illustration by Amanda Kosche

continues as far west as the Orlando Ridge, some 25 miles or more inland, where a relict cuspate foreland called Cape Orlando formed a pointed east coast shoreline much like the modern one during Pleistocene time. The trend of the eastern margins of the Mount Dora, Lake Wales, and Orlando ridges is remarkably similar to, and parallel with, the modern western shore of Mosquito Lagoon, the former shoreline before the modern barrier island chain developed.

Cape Canaveral is of obvious, recent historic significance. The first test rocket launched here, a German V-2, was fired from the tip of the cape in 1950, and NASA was created in 1958. The Apollo Program was developed to fulfill President Kennedy's challenge to put a man on the moon, which it did with the mighty, thirty-six-story Saturn V rocket launched on July 16, 1969. The last Apollo flight, and humankind's last visit to the moon, was Apollo 17. It landed on the moon on December 11, 1972, carrying the first and only geologist to walk on the moon, Harrison "Jack" Schmitt. Numerous other rockets, satellites, and probes have been launched from Cape Canaveral, including the space shuttles.

Paleontological and archaeological sites are regularly uncovered in the Cape Canaveral region, especially when canals or borrow pits are dug. In 1982, a backhoe operator working near Titusville, about 5 miles from the cape and on the edge of Atlantic Coastal Ridge, inadvertently uncovered human skeletal remains from a layer of peat. Researchers have since uncovered 168 human bodies from what became the Windover Pond archaeological site, which was a burial pond. Archaeologists have dated the site as early Archaic Period, between 7,000 and 8,000 years old. Radiocarbon dates of human bone and tissue, wooden stakes, and peat indicate they are about 7,410 years old. The people who used the site wrapped their dead in a fabric and then staked them to the peat at the bottom of the lake. This peat burial resulted in extraordinary preservation, including soft tissue (ninety-one individuals even have preserved brains), remains of last meals, and DNA. A variety of artifacts made of stone, bone, and wood were preserved, as well as sophisticated fabrics and straw weavings. Plant and animal remains are also present. The Brevard Museum of History and Natural Science, located at 2201 Michigan Avenue in Cocoa, off US 1, displays artifacts taken from the site. However, most of the site has remained intentionally unexcavated—a common practice among archaeologists who await the development of more advanced research techniques.

In southern Brevard County, near the Indian River County line, shell beds with invertebrate and vertebrate fossils of early to middle Pleistocene age were uncovered in a borrow pit called the Tucker Borrow Pit. Similar fossils were recovered from strata along Sebastian Canal, about 12 miles east of the borrow pit near Roseland. Similar sites, which represent marsh and nearshore marine settings, are fairly common along the Atlantic Coast. They have been found as far south as Jupiter and Palm Beach.

An especially famous paleontological locality in this region is at Vero Beach. As early as 1913, dredging operations uncovered bones of many vertebrate animals of late Pleistocene age, including wolves, bears, saber-toothed cats, horses, sloths, elephants, and many smaller species. In 1915, human remains—a partial skeleton of a female, apparently a Paleoindian—were found with these animals. This startling discovery aroused the interest of many geologists, paleontologists, and anthropologists, who visited the site and disputed the age and stratigraphic association of the human bones and other fossils. In 1924, human remains associated with vertebrates of late Pleistocene age were recovered in similar strata near Melbourne Beach, sparking further debate over the age of the bones from Vero Beach, and now Melbourne Beach. Later anatomical work on the human bones indicated that they were

We Serve Anybody—Fossil Crabs of Florida

One especially interesting aspect of the paleontological record of Florida is the occurrence of well-preserved specimens of decapod crustaceans—crabs and shrimps. Although extremely abundant in modern marine waters, their fossil record appears to be rather meager. But careful collecting has shown that the record is better than it might at first appear. Approximately fifty species have been recovered from rocks of Eocene to Pleistocene age in Florida. Because carapaces (external skeletons) of crabs and shrimps usually break apart when the animals die, it is common to find only certain parts as fossils, such as individual pincers. Many small specimens can be found only by carefully washing sediments through a screen. Some whole-body fossils that

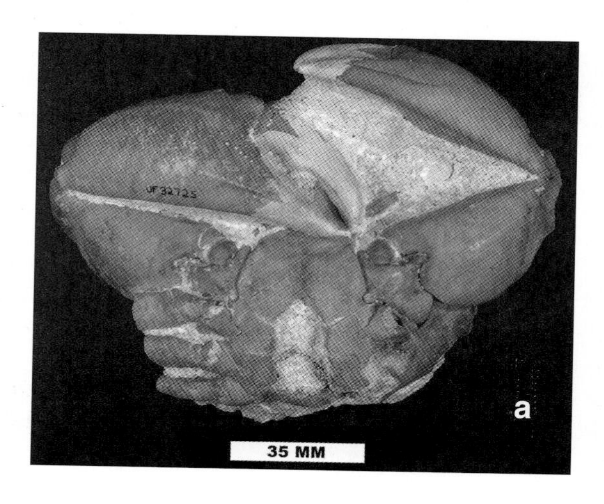

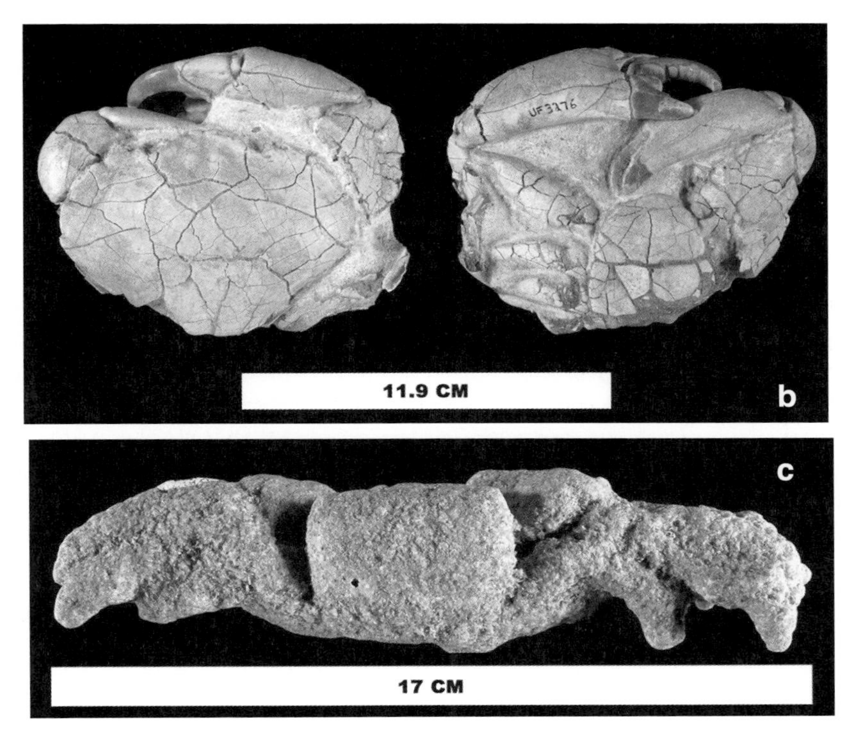

*Fossil crabs of Florida: (a) mud crab (*Ocalina floridana*) from the Ocala Limestone of Late Eocene age; (b) stone crab (*Menippe mercenaria*) from the Caloosahatchee and Fort Thompson formations of Pliocene and Pleistocene age; (c) ghost crab (*Ocypode quadrata*) from the Anastasia Formation of Pleistocene age. (These are frequently found along Brevard County beaches after nor'easters. Try Spessard Holland South Beach Park at Melbourne Beach.)* —Courtesy of the Invertebrate Paleontology Division, Florida Museum of Natural History, University of Florida

have been found may have been molts that animals shed during growth rather than a dead animal. A crustacean may shed many molts in its life, making the molts more likely to be fossilized.

Parts of the lower Ocala Limestone have abundant fossils of small pincers (usually only 0.375 inch or less in length) of the shrimp *Callianassa inglisestris*—so abundant that geologists have referred to this layer as the "shrimp claw limestone." *Callianassa* is a genus that reaches back to Cretaceous time and still thrives today. Because of its burrowing habit, it is one of the most common decapod crustaceans in the fossil record (burrowing greatly increases the likelihood of fossilization). Decapod crustaceans are present in the geologic record by Triassic time, more than 200 million years ago. They expanded and diversified continually throughout Mesozoic time, and especially during Cenozoic time, when decapods began to take on a distinctly modern appearance. Some decapods, such as shrimp, often leave behind their burrows as trace fossils, and their walking trails—impressions left in soft sediment—are sometimes found preserved in rock.

indeed Paleoindian remains. Other research concluded that the artifacts found with them were no more than 4,000 years old, but that the human remains were most likely Pleistocene in age, 10,000 years or older.

The dispute over the true ages of "Vero Man" and "Melbourne Man" remains unresolved. Unfortunately, the human bones were lost, misplaced, or stolen at some point during transfers between Florida and the Smithsonian Institution. Radiocarbon dating of the bones could have easily settled the controversy, but the bones went missing before this technique was developed in the 1950s. As a result, these immensely important anthropological finds have become an enormous embarrassment to Florida scientists and others. But all hope is not lost. Given the fact that human remains have been found in several areas along the east coast, paleontologists and anthropologists expect that similar deposits will be discovered in the future.

Offshore of most of the Atlantic coast are many rocky ledges that are heavily encrusted with marine life. These ridges are present from Jacksonville to Daytona, resume south of Cape Canaveral, and are very common from the Vero Beach area to Miami, where the Florida Reef Tract begins. The geologic identity of the ridges is not fully understood. The northern ridges are probably the Anastasia Formation or similar sediments. Farther south, the ridges are fossilized reefs of early Holocene age.

Southern Peninsula

A River of Grass Flows Over It

The water is timeless, forever new and eternal.
Only the rock, which time shaped and will outlast,
records unimaginable ages.
Yet, as time goes, this limestone is recent.
—Marjory Stoneman Douglas, *The Everglades: River of Grass*

Nature writer Sandra Friend attempted to summarize the geology of Florida in just two words: "ocean floor." This is not far off the mark, and it is particularly true for the southern peninsula. Gazing across the open expanse of the River of Grass—the Everglades—there is virtually no topographical variation evident. It takes little geological imagination to envision this flat terrain as a sea bottom. Much of the area south of and including Lake Okeechobee is part of the Pamlico Terrace, the ocean floor of the last interglacial interval of Pleistocene time, which occurred about 100,000 to 150,000 years ago. As ice masses melted in the north, sea level rose, submerging the southern peninsula. In the shallow Pamlico Sea, sea level was at least 25 feet higher than today.

All of the surface rocks of the southern peninsula formed during Pliocene and Pleistocene time, as the peninsula was repeatedly submerged beneath shallow seas. Two formations dominate: the Pliocene-age Tamiami Formation, composed of sandy, phosphatic limestones and fossiliferous sands and clays, crops out in much of the southwestern corner of the peninsula; and the Pleistocene-age Miami Limestone is exposed in the southeastern portion of the peninsula. Undifferentiated shelly sands of Pliocene and Pleistocene age floor much of the Everglades, and unnamed Holocene-age sands cover the southwestern coast. The phosphatic sands, clays, and limestones of the Peace River Formation are well exposed in Hardee and DeSoto counties, and the Key Largo Limestone is exposed in and around Biscayne Bay.

Subsurface strata in southern Florida are much thicker than strata of the rest of the peninsula. Over 16,000 feet of sediments have accumulated over volcanic rocks of early Mesozoic age in the South Florida Basin. While this great thickness of strata was being deposited, the southern peninsula was slowly subsiding, as evidenced by the strata. The Avon Park Formation, for example, which certainly formed in shallow marine and coastal waters, is found at the surface in Citrus and Levy counties in the northern peninsula but occurs more than 1,100 feet below the

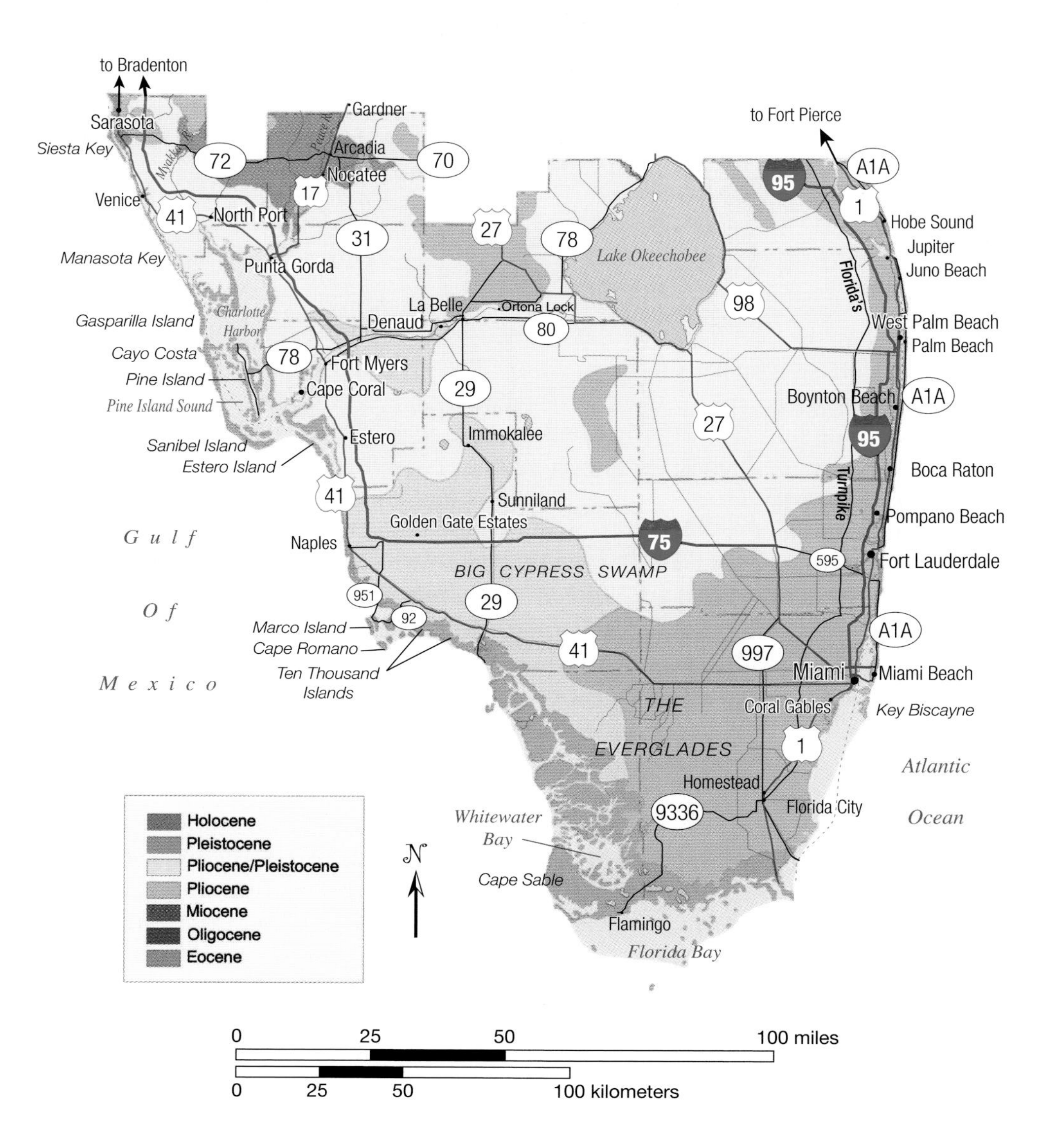

Road log map of the southern peninsula. —Map by Mountain Press

An agricultural field near Homestead showing the reddish soils known as terra rossa. The reddish color comes from fine-grained sediments that were transported by wind from the Sahara region of Africa. They were deposited in shallow-marine carbonate environments in southern Florida during late Pleistocene time. The carbonate sediment has slowly dissolved away, leaving the terra rosa behind.

surface in Miami-Dade County. This continual, steady sinking, which essentially began as Pangaea rifted apart, has resulted in the accumulation of a very complete sedimentary archive of the geological history of southern Florida from its inception in Middle Jurassic time, about 170 million years ago, to the present day.

Carbonate deposition in southernmost Florida began in Middle Jurassic time and continued with only minor influxes of quartz-rich sediments into Middle Miocene time, about 12 million years ago. Deposition of quartz-rich sediments replaced carbonate deposition in Late Miocene and Pliocene time, but in Late Pliocene and Pleistocene time carbonate deposition resumed and continues to the present day. This important sedimentary transition reflects both a decrease in a quartz-rich sediment supply and a change in climate, from subtropical to tropical. Carbonate sediments are deposited where the climate is warmer, and quartz-rich sediments where the climate is slightly cooler. In addition, mangrove forests, a tropical habitat, replace the salt marsh vegetation so common along the more northern Florida coasts.

Dust from dust storms in the Sahara Desert in Africa has been blowing across the Atlantic for thousands of years. Some of the dust settled in the water where the Miami Limestone was forming and was incorporated into the sediment. As a result of the dust content, weathering of this limestone creates terra rossa, or red soil. The plowed fields of the Redlands, just west of Florida City along Florida 9336, have a reddish hue. Interestingly, the Saharan dust contains a very minor radioactive component. The radioactive elements undergo natural radioactive decay and produce radon gas, a radioactive gas that can be a health hazard to people in confined, poorly ventilated areas.

The well-known and intensely studied marine paleontological record of southern Florida seems inexhaustible. It is no overstatement to say that nearly all of southern Florida is densely fossiliferous. The shell beds of the Tamiami and Caloosahatchee formations, along with other rocks and sediments of Pliocene and Pleistocene age, are second to none for their invertebrate fossil record, and they have yielded some surprising vertebrate fossils as well.

Interstate 75 and US 41

Bradenton—Fort Myers—Naples

128 miles

I-75 and US 41 run roughly parallel between Bradenton and Naples, just inland of most of the southern end of Florida's west coast barrier island chain, from Anna Maria Island to Marco Island and Cape Romano. Like the islands to the north in Pinellas County, these barriers have been very dynamic in historic times. Tidal passes open and close regularly, especially in response to hurricanes. Several of the islands in this chain are adorned with beach ridges and dunes. Most of them are also heavily developed and armored by seawalls and other structures in an attempt to retard the severe erosion that is prevalent here. Communities have replenished many beaches by pumping offshore sand onto them. Considering the area's high population and the economic importance of tourism, this coastline presents a fine case study and field laboratory for environmental geologists.

Fortunately, some of these coastal islands are accessible only by boat or are otherwise protected, so they are in near pristine condition. Shell collecting on these islands is good. Some of the best beaches are Stump Pass Beach State Park, Don Pedro Island State Park (accessible by boat), Gasparilla Island State Park, Cayo Costa State Park (accessible by ferry or boat), and Delnor-Wiggins Pass State Park (near Naples). Holocene-age beach ridges are visible on Siesta Key, Gasparilla Island, Cayo Costa,

Sanibel Island, and Marco Island. Exposures of coquina rock can be found on Lido Key and at Point of Rocks in the middle of Siesta Key. The geologic age of this coquina has not been established.

In the geological world, the Sarasota area is perhaps most famous for its Pliocene- to Pleistocene-age paleontological bounty in the incomparable shelly Pinecrest Beds of the Tamiami Formation and in the Caloosahatchee Formation. These fossil-rich strata have been well exposed at the APAC Mine (also called the Newburn Shell Pit or Mac-Asphalt Quarry) and the nearby Quality Aggregates Shell Pit (across from APAC on I-75). Some 60 feet of fossiliferous strata have been exposed at APAC, which is now a lake.

In the past, the APAC Mine was open for fossil collecting, but today, like many other quarries in Florida, it is closed to the public. Liability issues, past misuses, and unauthorized entry by collectors have understandably restricted access. This closure is an example of why it is so important to not abuse the privilege of access to a private quarry or any other potentially dangerous locality.

Pliocene and Pleistocene Mollusks

Some of the most spectacular and densely concentrated deposits of fossil shells in the world are found in the Pliocene- and Pleistocene-age formations of southwestern Florida. The sheer volume of molluscan shells in these deposits is staggering, and the preservation is excellent. The shell beds and related strata formed in nearshore marine environments. In the Pliocene-age Pinecrest Beds of the Tamiami Formation, there are well over two hundred species of bivalve mollusks. The total number of gastropod species is not known, but paleontologists estimate it may be three times that of the bivalves. Estimates of the total number of molluscan species in the Pinecrest Beds range from eight hundred to twelve hundred, making it comparable to the Miocene-age Chipola Formation of northern Florida. Both tropical- and temperate-water molluscan species are present in the Pinecrest Beds. Florida's Pliocene-age mollusks are part of a unique biogeographical association called the Caloosahatchian Province, which extended around the Gulf of Mexico and north up the eastern seaboard of the United States.

The Caloosahatchee Formation, which spans the boundary between Pliocene and Pleistocene time, is also mollusk rich. Although it is not as diverse as the Pinecrest Beds, paleontologists estimate that it may contain about seven hundred species. Unlike the Pinecrest mollusks, the Caloosahatchee fauna is distinctly tropical in composition. Other Pleistocene-age units also contain rich molluscan assemblages that are similar to the modern molluscan fauna of Florida.

The formation of the Isthmus of Panama in Late Pliocene time, connecting North and South America, had a profound effect on marine life. The tectonic rising of this land bridge severely changed ocean circulation, cutting off exchange between the Caribbean and the Pacific and intensifying the north-flowing Gulf Stream. These changes may have facilitated the buildup of northern-hemisphere ice at the beginning of Pleistocene time by forcing warm water and moist air toward the arctic, increasing snowfall. The molluscan fauna that was restricted to the Gulf of Mexico and Caribbean suffered many more extinctions than the Pacific fauna.

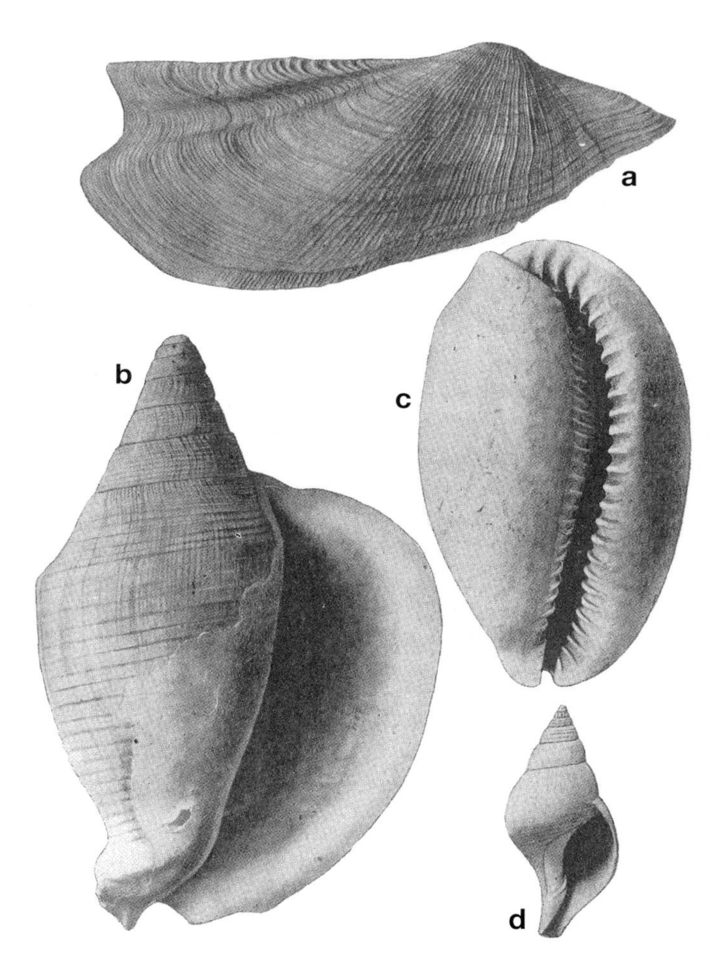

Some mollusks of the Caloosahatchee Formation: a) Arca wagneriana*; b)* Strombus leidyi*; c)* Siphocypraea problematica*; d)* Fasciolaria apicina.
—From Cooke and Mossom (1929) and courtesy of the Florida Geological Survey

An amazing assemblage of fossil birds, up to forty-five species of mostly coastal forms, such as grebes and rails, has been recovered from sediments of Late Pliocene age at the APAC and Quality Aggregates quarries. The avian assemblage is dominated by an extinct species of cormorant (*Phalacrocorax filyawi*), represented by many thousands of bones and complete skulls, including 137 partially articulated or whole skeletons. Because of the great volume of individuals of a single species of shorebird, paleontologists hypothesize that this is a mass death assemblage, perhaps related to a red tide event.

A red tide is a type of harmful algal bloom—a temporary population explosion of certain phytoplankton (microscopic, one-celled algae). During a bloom, seawater can turn a brownish red color, and a natural toxin produced by the algae can become so highly concentrated that it is lethal to many fish and some marine invertebrates. Even large animals such as sea birds, manatees, and humans may be affected. In 1996, 149 manatees, roughly 10 percent of Florida's population, died in southwest Florida during a red tide.

Spanish explorers reported fish die-offs in Florida, possibly related to red tides, in the 1530s, and coastal surveyors reported die-offs in the late 1800s. Human-related pollutants such as nitrogen and phosphorus products (from fertilizers and detergents) provide nutrients for algae, making them the probable cause of some harmful algal blooms. It is very likely that some fossil deposits, such as the cormorants mentioned above, are related to ancient blooms. Mass die-offs can drastically increase the probability that the affected animals will be fossilized simply because of the sheer number of specimens.

About 18 miles south of Sarasota is Venice Beach, which is famous among fossil collectors for its abundance of fossil shark and ray teeth of Miocene and Pliocene age, which have washed onto the beach from offshore rocks. Nearby Caspersen Beach also has many fossils. Teeth and other small vertebrate remains wash up on the beach in this area from Casey Key to the north down to Englewood Beach on Manasota Key. To collect, you can simply scan the shelly gravels on the beach or use a mesh screen to sift away smaller material. Local stores sell long-handled, screened scoops for just this purpose.

According to popular legend, the long-rumored "fountain of youth" sought by Juan Ponce de Leon is Warm Mineral Springs, located in the town of North Port, about 11 miles east of Venice. US 41 passes through North Port, and I-75 passes north of the town. This third magnitude spring is very warm—up to 87 degrees Fahrenheit—and brackish, about 19 parts per thousand salts, or 1.9 percent. Normal seawater is about 35 parts per thousand salts (3.5 percent). The salty water also has a sulfurous odor, which is hydrogen sulfide. A small industry has been built

Fossil and Living Sharks and Rays

It seems to be the prize of every fossil collector, whether novice or advanced—shark teeth! And like many states along the Gulf and Atlantic coastal plains, Florida is loaded with fossil shark teeth. Because of their phosphatic mineral composition, and thanks to the fact that sharks continually shed and replace their teeth, shark teeth are very common in some deposits. And in Florida's phosphate deposits of Miocene and Pliocene age, they can be especially abundant. These deposits accumulated very slowly on the seafloor, allowing ample time for falling shark teeth to become concentrated in condensed sediment layers.

Sharks and rays belong to the Chondrichthyes class of fish, which have a cartilaginous internal skeleton. Among sharks, only their teeth and the tiny, toothlike dermal denticles, which are embedded in their skin and act as an external skeleton, are normally fossilized. Among stingrays, their teeth and barbed tail spine are fossilized. Most shark species can be identified from teeth alone. But shark teeth are highly variable throughout the jaw, so the range of variation for each species must be understood before accurate identifications can be made from individual teeth. There are over 350 living species of shark and some 430 species of rays worldwide. About 75 species have been found in Florida's formations, representing most major groups.

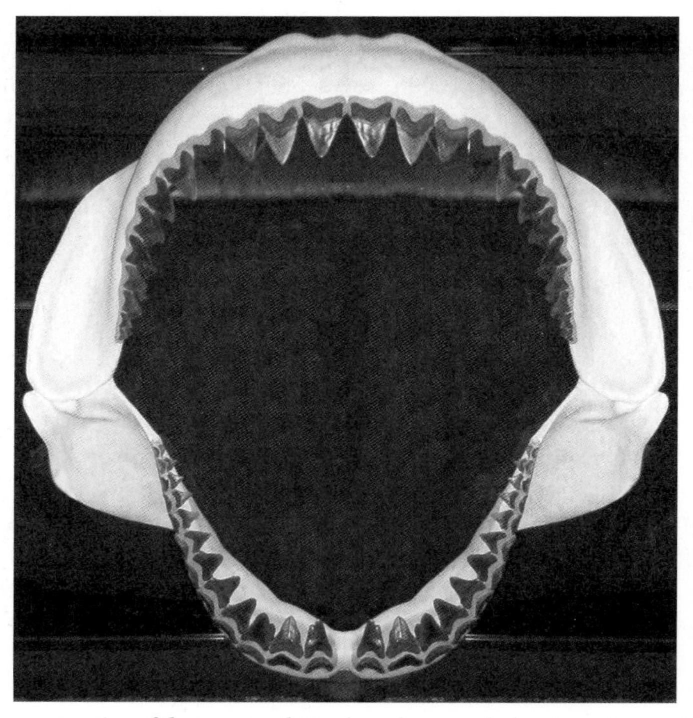

Jaw reconstruction of the megatooth Carcharodon megalodon *(gape is over 6 feet).*

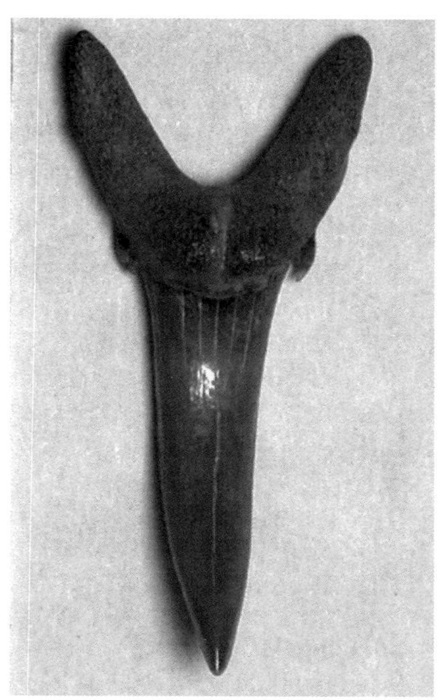

Sand tiger shark (Carcharias taurus*),*
3 centimeters long.

Extinct snaggletooth shark (Hemipristis serra*),*
3 centimeters long.

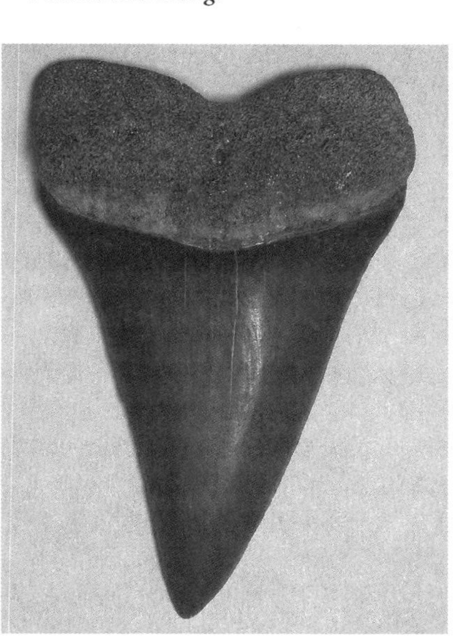

Extinct mako shark (Isurus hastalis*),*
4 centimeters long.

Stingray barb, 14 centimeters long.

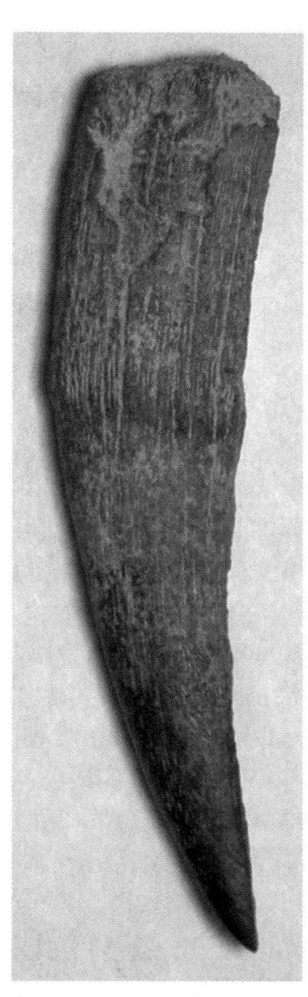

Sawfish tooth (Pristis *species), 5.5 centimeters long.*

—Courtesy of the Vertebrate Paleontology Division, Florida Museum of Natural History, University of Florida

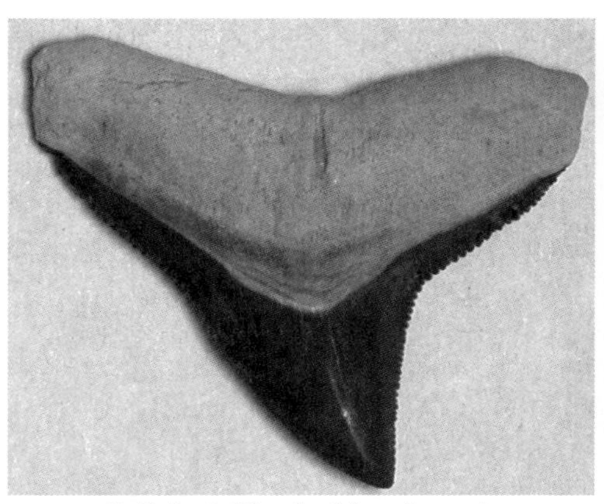

*Bull shark (*Carcharhinus leucas*), 2 centimeters long.*

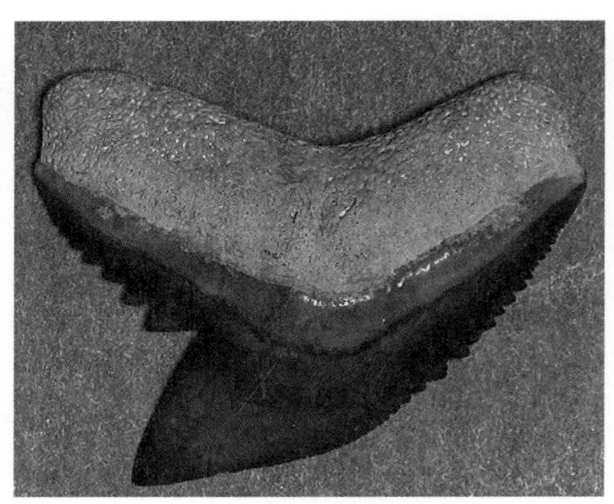

*Tiger shark (*Galeocerdo cuvier*), roughly 2 centimeters long.*

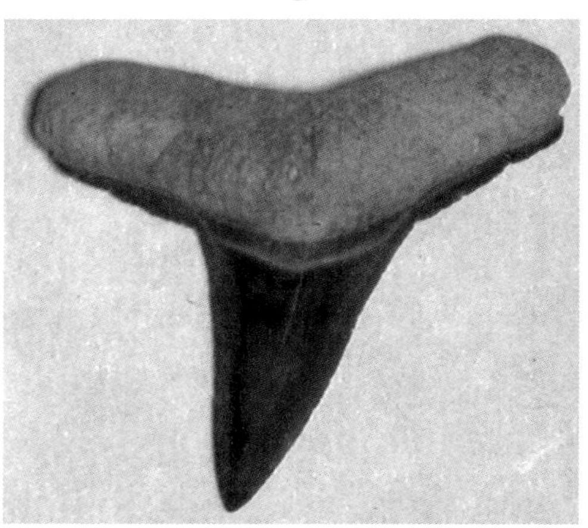

*Lemon shark (*Negaprion brevirostris*), 2 centimeters long.*

*Seafloor pavement composed of the flat shell-crushing teeth of Myliobatid stingrays. The individual teeth, which are about 3.5 centimeters long (*left to right*), often separate from the pavement.*

—Courtesy of the Vertebrate Paleontology Division, Florida Museum of Natural History, University of Florida

Sharks and rays belong to ancient and very successful lineages. Some modern species, such as the sand tiger shark (*Carcharias taurus*), have been around since Cretaceous time. Fossil teeth of the genus *Carcharias* have been found in sediments of Eocene to Holocene age in Florida. Remains of the unusual and distinctive snaggletooth shark (*Hemipristis serra*) have been recovered from Oligocene- to Pleistocene-age sediments in Florida. Only one species of this genus—*Hemipristis elongatus*—lives today, in the Red Sea and Indian Ocean.

The most famous shark, fossil or living, is the great white, a common name that generally refers to the genus *Carcharodon*. Some early forms of the great white that were exceptionally large are called *megatooth sharks*. Three mega-

tooth species have been found in Florida's formations: *Carcharodon sokolowi* (Middle Eocene to Early Oligocene), *Carcharodon subauriculatus* (Late Oligocene to Early Miocene), and *Carcharodon megalodon* (Middle Miocene to Early Pliocene). The largest teeth of the gigantic *Carcharodon megalodon* are nearly 7 inches long, though most are smaller. The teeth of *Carcharodon megalodon* are common in Miocene- and Pliocene-age formations of Florida and along the Atlantic Coastal Plain. Paleontologists estimate that this species grew more than 55 feet long and weighed over 60 tons. The modern great white, *Carcharodon carcharias*, and *Carcharodon megalodon* both lived during Miocene time; however, fossils of the modern great white are more common in Pleistocene-age deposits, which accumulated after *Carcharodon megalodon* became extinct.

Stingrays and other cartilaginous fish are also well represented in Florida's fossil record and are found in Eocene- to Pleistocene-age sediments. The unusual sawfish (genus *Pristis*), which has a long, flat snout lined with teeth on both sides (resembling a timber saw), is usually represented by isolated saw teeth. Rays of the genera *Aetobatus*, *Aetomylaeus*, *Dasyatis*, *Myliobatis*, and *Rhinoptera* are represented by dental plates (used for crushing crustaceans and other prey), isolated teeth, and barbed tail spines.

around the purported healing properties of the warm mineral water, which is said to offer relief from a variety of ailments, including arthritis and skin disorders. The spring has been developed into a healing and wellness swimming park.

The source of the warm, sulfur-rich, salty water appears to be the Boulder Zone of the Floridan aquifer system, an area of porous and fractured rocks of Paleocene and Early Eocene age located 2,300 to more than 3,500 feet below the surface. These strata are known to be rich in evaporite minerals, including gypsum and anhydrite, which are good sources of both salt and sulfur. The warm temperatures may result from both bacterial activity and the great depth of the spring's water source, since ground temperature increases with depth.

The spring's sinkhole developed in the Arcadia Formation, which crops out at the edge of the pool. The spring is about 230 feet deep, and at the bottom is a large sediment cone of limestone, dolostone, and sand standing 100 feet above the bottom. Two ledges with recessed areas occur at depths of 43 and 65 feet. They contain stalactites, which indicates the sinkhole was dry at some point at both depths. Archaeologists have focused their attention on Warm Mineral Springs because of the discovery of Paleoindian artifacts in the spring, including bone and shell tools, some of which indicate they used a spear-throwing device.

About 3 miles south of Warm Mineral Springs, Little Salt Spring is also known for Paleoindian remains. Two species of giant Pleistocene-age tortoises have been recovered in Little Salt Spring: *Hesperotestudo crassiscutata*, which was as large as the giant modern Galapagos tortoise (*Geochelone nigra*, which can grow up to 5 feet in length), and *Hesperotestudo incisa*, which was associated with human artifacts, including a wooden stake that may have been used to kill the tortoise. The tortoise and wooden stake are between 12,030 and 13,450 years old. Little Salt Spring was a sinkhole during the Paleoindian and early Archaic periods. Artifacts of the Archaic Period recovered here are 9,500 years old.

The Peace River, which has its headwaters in the Green Swamp in Polk County and empties into Charlotte Harbor near Port Charlotte, has historically been the source of many fossils of Miocene to Pleistocene age. As early as 1850, fossils from the Peace River were mentioned in scientific literature, and in 1889 Joseph Leidy described vertebrate fossils found in the river. The early finds were largely associated with the discovery of economically valuable river-pebble phosphate in 1886 and 1887. The river concentrates fossils from the Peace River Formation of the Hawthorn Group, including the famous Bone Valley Member—so well known for its profuse fossils in the phosphate district.

The Peace River offers great opportunities for fossil collecting. We recommend accessing the river by canoe between October and July. Much of the river is shallow enough for mask and snorkel. Some of the best collecting is off US 17 between Zolfo Springs and Nocatee. Canoe access points are in Zolfo Springs (Pioneer Park, at the corner of Florida 64 and US 17, and a good fossil site), Gardner, Brownville, Arcadia, and Nocatee. There are canoe outfitters and even commercial fossil excursions available in the area. Similar fossils can be collected in nearby Horse Creek, a tributary of the Peace River. Take Florida 72 about 10 miles west of Arcadia to the creek. Shell Creek, another Peace River tributary, is also a good fossil locality. It has good exposures of the Caloosahatchee Formation. Near Punta Gorda, take US 17 north from either I-75 or US 41. About 5 miles from I-75, turn right on County Road 764 and proceed 4.4 miles to Shell Creek Park.

Like its counterpart Tampa Bay, Charlotte Harbor is a classic drowned river valley that was flooded during the Holocene Transgression. The Myakka and Peace rivers are at the head of this estuary, and a small bayhead delta has formed at the mouth of the Peace River. Both Charlotte Harbor and Tampa Bay have been enlarged by karst dissolution of the underlying limestone.

It is said that the sport of tarpon fishing had its beginnings at the mouth of Charlotte Harbor and at Pine Island Sound in the 1880s. Boca Grande Pass, between Cayo Costa and Gasparilla islands, is still called the "tarpon fishing capital of the world." The magnificent tarpon *Megalops atlanticus*, also known as the silver king because of its large silver

Fossil Sea Stars of Florida—The Panamanian Connection

Although sea stars, also called *starfish*, are common in the shallow coastal waters of Florida today and were almost certainly so in the past, their fossil remains are extremely rare and fragmentary. Sea stars do have a fossilizable skeleton composed of thick calcium carbonate ossicles and spines, but the ossicles are held together by an organic material that disintegrates after death, so the skeleton rapidly breaks apart. Individual ossicles, which often look like miniature loaves of bread (usually less than 0.25 inch in size), are commonly found with other microfossils in fossil-rich sediments, but whole skeletons are extremely rare. For a whole skeleton to be fossilized, it must be buried rapidly. Because sea stars live on top of the seafloor rather than burrowing beneath it, whole specimens are even less likely to be fossilized.

But in 1986, a fossil deposit rich in whole sea stars was discovered at El Jobean, in Charlotte County, in the Pliocene-age Tamiami Formation. Hundreds of individuals were recovered from what was probably a storm deposit that buried the sea stars. Although it is an amazing example of rare fossilization, what is most fascinating about this deposit is that it consists entirely of *Heliaster microbrachius*, a tropical-subtropical species that today lives in the eastern Pacific Ocean, from the Gulf of California to Central America and along most

Heliaster microbrachius *from the Tamiami Formation of Pliocene age in Charlotte County.*
—Courtesy of the Invertebrate Paleontology Division, Florida Museum of Natural History, University of Florida

Modern Heliaster microbrachius *from Antofagasta, Chile. Specimen is 7.5 centimeters wide.* —Courtesy of the Invertebrate Paleontology Division, Florida Museum of Natural History, University of Florida

of the western coast of South America. It no longer lives in the Gulf of Mexico. This species is one of the so-called sun stars because it contains far more than the usual five arms, or rays. *Heliaster microbrachius* can have as many as fifty rays. The individual fossils had an average of thirty rays.

Prior to about 3 million years ago, the Isthmus of Panama, that narrow land connection between North and South America, was submerged, and the waters of the Atlantic Ocean, Caribbean Sea, and Pacific Ocean circulated together freely. So prior to Pliocene time, invertebrate faunas were essentially the same in the tropical eastern Pacific and tropical western Atlantic. But after the tectonic emergence of the isthmus, many species became extinct in the gulf, perhaps due to the radical changes in ocean circulation and climate. *Heliaster microbrachius* was one of those species.

scales, ranges across the southern Gulf of Mexico and the northern Caribbean. Seasonally, it swims to the northern gulf. It is a prized game fish because of its high and frequent leaps. It is primarily an inshore species that frequents shallow coastal waters, including bays, estuaries, and lagoons. Tarpon are another example of a living fossil, belonging to a lineage that reaches back to Late Jurassic time. There are two species living today, one in the Atlantic and one in the Pacific. The genus *Megalops* first appears in Eocene-age rocks, but in Florida it has been found only in younger fossil localities, from Middle Miocene to Pleistocene in age.

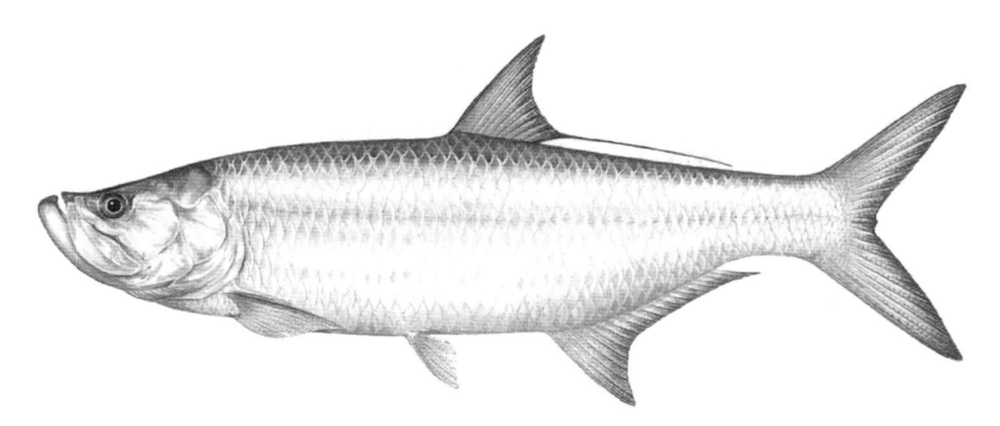

The tarpon, or silver king, Megalops atlanticus. —Illustrated by Diane Rome Peebles and courtesy of the Florida Fish and Wildlife Conservation Commission, Division of Marine Fisheries Management

Fossils of this genus are easily recognizable by their large, thin scales. Modern tarpon average 50 to 75 pounds and may grow up to 8 feet long. The world record is just over 283 pounds, caught in Africa. The Florida record is 243 pounds, caught at Key West.

Florida has one of the best records of fossil mollusks in the world. And for collectors of modern shells, Florida remains a prime destination, and Sanibel Island, southwest of Fort Myers, must be considered the conchological headquarters of the state. The great variety of marine mollusks along the Florida shore is nearly unmatched anywhere in the United States, and the extensive coastline provides almost unlimited opportunities for indulging in the pleasurable and inexpensive pastime of beachcombing. Some popular beaches may get picked over, but they are regularly resupplied by spring tides and storms. One of the reasons for the great variety of Florida's mollusks is that the state extends through several climatic zones and oceanic regions. Temperate, subtropical, and tropical influences are evident in its mollusk species, and the shoreline makes contact with the Atlantic Ocean, the Gulf of Mexico, and the Florida Current with its Caribbean Sea influence. Sanibel Island hosts an annual Shell Fair in March and has done so since 1937. Cobbles of coquina, which wash up from offshore, may also be found along the shores of Sanibel Island. Geologists are uncertain about their exact age or what rock formation they erode from.

In Estero Bay, south of Fort Myers and accessible only by boat, is Mound Key Archaeological State Park. It preserves large and complex shell and earth mounds built by the Calusa Indians, who did not farm but lived by the bounty of the sea. The Calusa densely populated the area around Charlotte Harbor, Pine Island Sound, and Estero Bay up to the period of

Common mollusks of the Florida shore. Rulers are in centimeters.

southern quahog
(Mercenaria campechiensis)

giant Atlantic cockle
(Dinocardium robustum)

yellow cockle
(Trachycardium muricatum)

disk dosinia (Dosinia discus)

sunray venus (Macrocallista nimbosa)

Atlantic bay scallop
(Argopecten irradians)

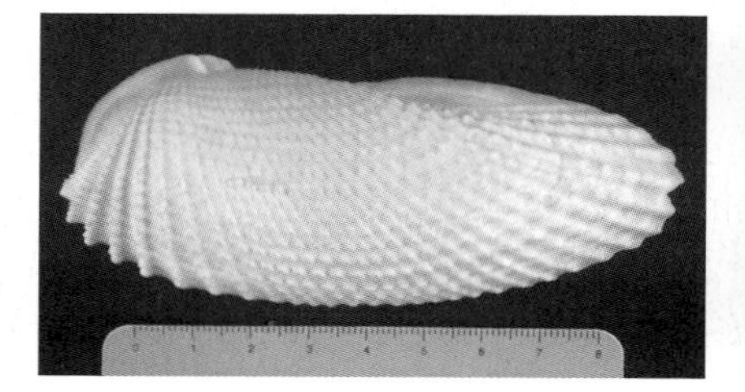

angel wing (Cyrtopleura costata)

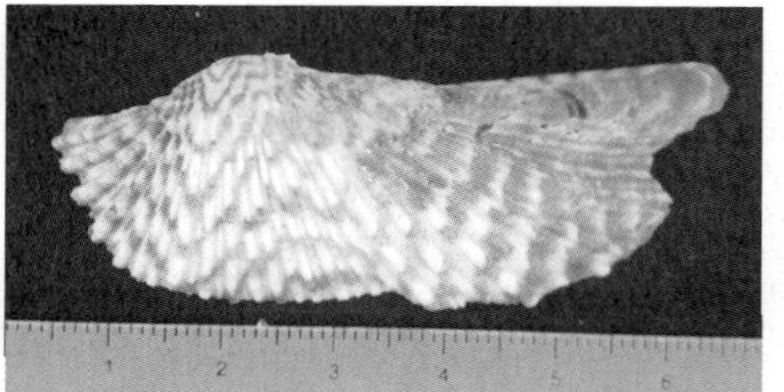

turkey wing (Arca zebra)

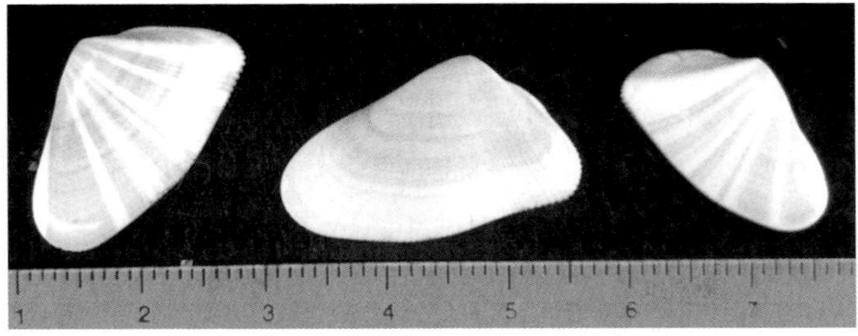

*coquina shell (*Donax variabilis*)*

*from left to right, cross-barred venus (*Chione cancellata*), lady-in-waiting venus (*Chione intapurpurea*), and imperial venus (*Chione latilirata*)*

*ponderous ark (*Noetia ponderosa*)*

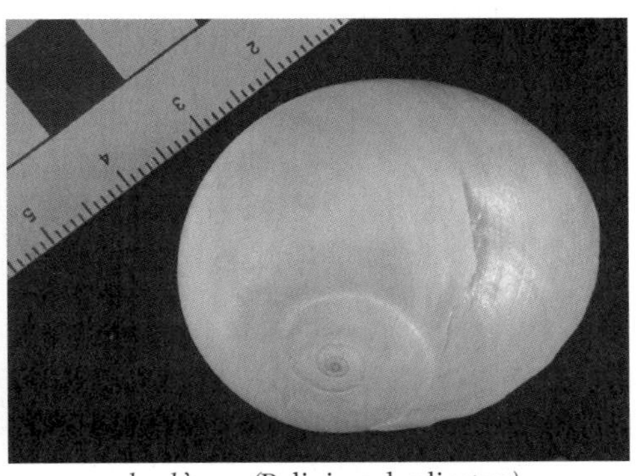

*shark's eye (*Polinices duplicatus*)*

apple murex
*(*Murex pomum*)*

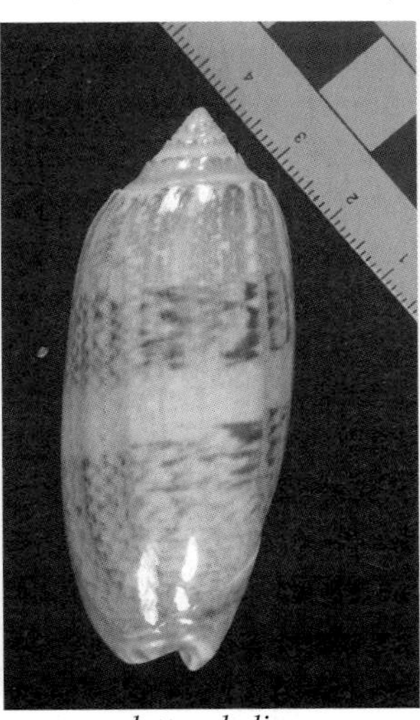

lettered olive
*(*Oliva sayana*)*

banded tulip
*(*Fasciolaria lilium*)*

Common Atlantic auger (Terebra dislocata)

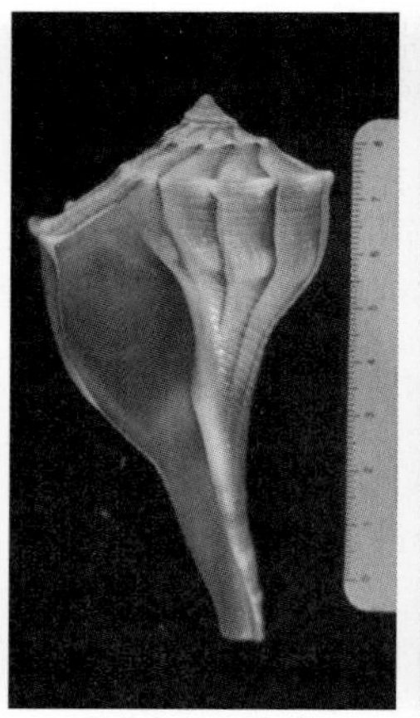

lightning whelk (Busycon contrarium)

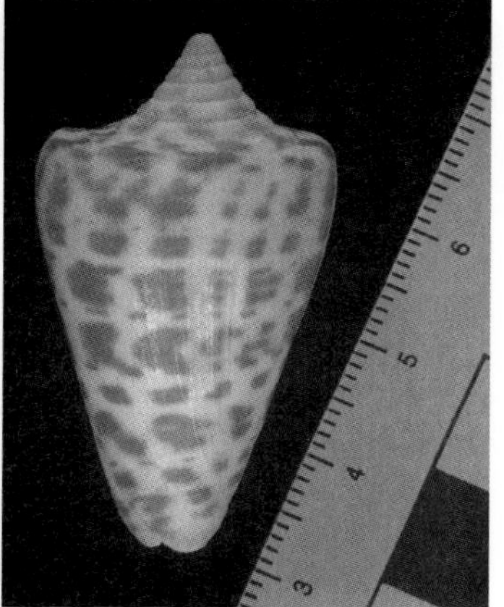

alphabet cone (Conus spurius)

scotch bonnet (Phalium granulatum)

common spirula (Spirula spirula)

European contact. Mound Key is thought to have been the ceremonial center of the Calusa nation. In the past one hundred years, erosion has been very severe on the southern end of Estero Island, the shore having retreated some 400 feet, exposing some rock of uncertain identity.

On the mainland just south of Estero is Koreshan State Historic Site, which preserves a village built by the followers of Cyrus Reed Teed. Teed attempted to establish a utopian community here beginning in 1894. His followers, the Koreshians, believed the Earth was hollow, a notion once considered a scientific possibility by some. Edmond Halley (Halley's Comet) proposed in 1691 that Earth was composed of concentric layers and was partly hollow. The Koreshians also believed that they were living *inside* a hollow sphere! The hollow Earth theory inspired much science fiction, most notably Jules Verne's 1864 novel *Journey to the Center of the Earth.*

The Estuarine Environment

Estuaries are partially enclosed bodies of coastal water that are diluted by river water. They are typically brackish, less than about 30 parts per thousand salts (30 grams of salt per 1,000 grams of water). Normal marine water averages 34 to 35 parts per thousand salts. There is a great variety of estuaries in Florida, and nearly all of them formed during the Holocene Transgression. Many, such as Pensacola Bay, Tampa Bay, and Charlotte Harbor, are coastal embayments that reach far inland. They formed as rising sea level flooded large river valleys. Indian River Lagoon, on the east coast near Cape Canaveral, consists of three elongated bodies of water that developed over beach ridge plains as sea level rose. Apalachicola Bay is a bar-built estuary, created by the development of sand spits and barrier islands. Apalachee Bay is a broad, indented, shallow coastline that is fed by sixteen rivers.

The ecological importance of estuaries cannot be overstated. They have been called the "cradle of the oceans" because so many marine species spend part of their life cycle in these relatively protected environments. Well over 90 percent of commercially important fish and shellfish depend on estuaries for some part of their life cycle, and some species depend on them entirely. Estuaries are also of enormous geological significance. They are good sediment

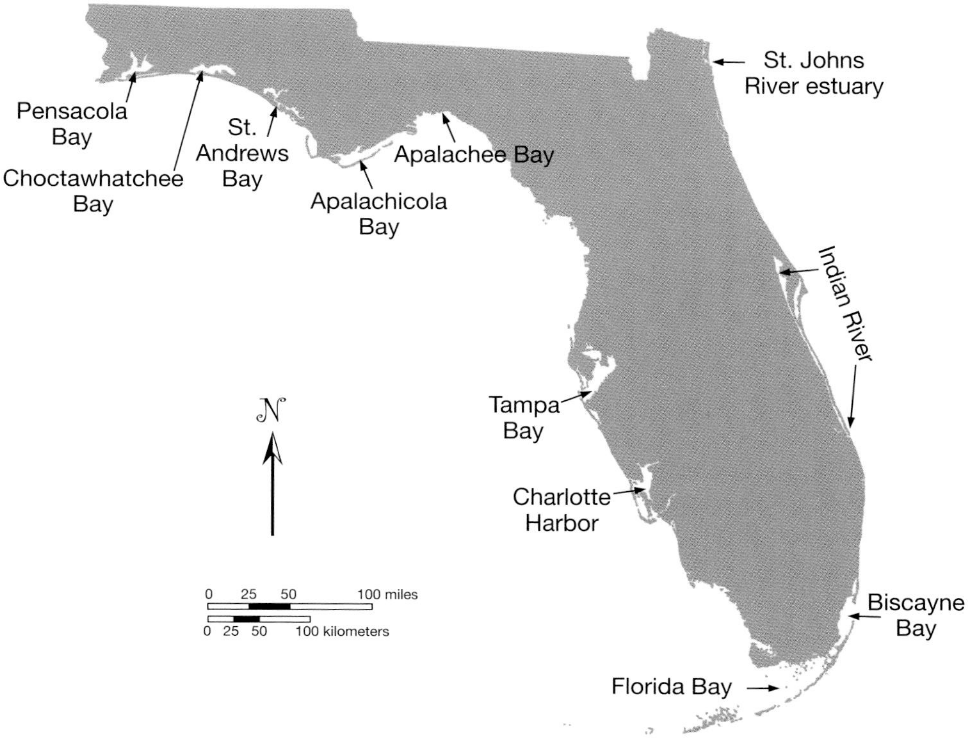

Major estuaries of Florida. —Illustration by Amanda Kosche

traps and rapidly fill with both river and marine sediment. Such deposits are abundant in the geological record, and many fossil deposits in Florida are estuarine in origin. Because estuaries are transitional environments between the terrestrial and marine realms, it is not uncommon for estuarine deposits to preserve marine, brackish, freshwater, and terrestrial fossil species. Such a mix of marine and terrestrial species in a single deposit allows geologists to correlate the ages of rock strata in marine and terrestrial realms, which is otherwise very difficult to do.

US 41 (Tamiami Trail)

Naples—Miami

107 miles

Much of the western segment of US 41, the Tamiami Trail, traverses surface exposures of the Pliocene-age Tamiami Formation, a densely fossiliferous, carbonate, clay, and quartz sand formation. Construction of the Tamiami Trail began in the early 1900s as workers dredged canals, and almost any roadside ditch, canal, or spoil pile along the road (or along the parallel I-75 to the north, called Alligator Alley) can be the source of numerous Pliocene- and Pleistocene-age fossils. The Tamiami Formation is exposed in various places in southern Lee and western Collier counties. Mines and exposures occur along County Road 846; County Road 951, especially between County Road 886, the Golden Gate Parkway, and County Road 856, also called Radio Road; Florida 29 from Immokalee to US 41; and US 41 between Florida 29 and Monroe Station. Expect to find abundant fossils of oysters, scallops, barnacles, corals, and echinoids.

In the area of Golden Gate Estates, east of Naples and off I-75, canals have been dug every 0.25 mile or so into the Tamiami Formation, producing fossil-rich spoil piles. A canal runs next to County Road 951, and spoil piles here may contain coral patch reef remains, part of the Golden Gate Reef Member of the Tamiami Formation. Patch reefs in this area, extending from at least Naples north to Estero, are composed of nearly fifty species of coral. Pliocene-age reef corals have been found across parts of Collier and Lee counties and as far north as Punta Gorda, in Charlotte County.

Located off Florida 951, 1 mile south of US 41, is Rookery Bay National Estuarine Research Preserve. Rookery Bay and Ten Thousand Islands are part of a subtropical mangrove forest and estuary—one of the few essentially undisturbed ecosystems of this type in North America. The preserve is active in a number of initiatives in research, education, and community outreach.

From US 41 about 8 miles southeast of its intersection with Florida 951, County Road 92 (San Marco Road) heads west to Marco Island. Although it is heavily developed, there is an obvious dune field of Holocene age along the road. The dunes rise about 25 feet or more above sea level and encircle Barfield Bay at the southeastern edge of the island. This dune field has the highest elevations in the state south of Lake Okeechobee. To the south of Marco Island, and accessible only by boat, is Cape Romano Island. This is the southernmost extent of quartz-rich sediment along Florida's Gulf Coast. Beyond this, the islands and coastline of the southern peninsula are limestone and calcium carbonate sediments. The pointed, southern end of Cape Romano Island is a cuspate foreland, which waves have sculpted over the past 3,000 years. Offshore of Cape Romano is a bank, or shoal, composed of several parallel sand ridges.

Southeast of Marco Island, from Cape Romano to Key McLaughlin, is the archipelago of Ten Thousand Islands. This maze of mangrove trees, oyster shoals, and tidal channels is 2 to 3 miles wide and 20 miles long. The tidal channel network is better developed in the northern half of the island group, where tidal influence is more pronounced. The southern segment has more freshwater influence, from Big Cypress Swamp and the Everglades, and the islands are more spread out. A shore-parallel series of lagoons, called the Chokoloskee Bay system, lies between most of the islands and the mainland.

The islands are part of a very interesting sequence of strata deposited between 7,000 and 3,000 years ago, at the very end of, and following, the Holocene Transgression. A basal layer of peat, about 6.5 to 16 feet below sea level, overlies eroded limestone of Pliocene and Pleistocene age. This peat is 3,500 to 6,500 years old. The peat is older offshore and becomes younger near the modern coast, showing the landward movement of both sea level and coastal vegetation during Holocene time. Quartz sands and shelly sands overlie the peat, having been deposited as sea level rose to its present position. On top of the sands of the inner islands are Holocene-age oyster reef beds overlain by modern peat and living mangroves.

Many of the outer islands have, in addition to oyster beds, a layer (up to 10 feet thick) of the unusual mollusk *Petaloconchus varians*. This wormlike mollusk, a snail of the Vermetidae family, is often confused for a worm tube. The *Petaloconchus* snails actually grow together, attaching their elongated shells to form thick clusters called "worm rock." Peat and living mangroves overlie the worm rock, just as they overlie the oyster beds on the inner islands. These *Petaloconchus* clusters actually formed a

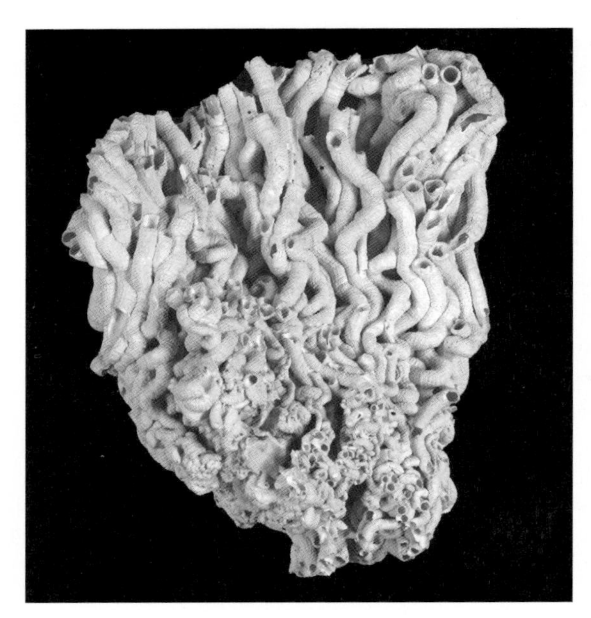

"Worm rock," composed of many shells of the vermetid snail Petaloconchus. *Specimen is 6.2 centimeters wide.* —Courtesy of the Invertebrate Paleontology Division, Florida Museum of Natural History, University of Florida

small barrier reef system from about 3,000 years ago until fairly recently. Oddly, the decline of vermetid gastropods over the past 100 years or so is a global phenomenon, and scientists have not been able to adequately explain it. *Petaloconchus* snails are still present in the region but are not forming reefs. Vermetid reefs are similar in many respects to some Cretaceous-age rudist bivalve reefs (see **When Rudists Ruled the Earth—Cretaceous Time** in **Geological and Paleontological History**).

Mangrove Swamps

Mangrove swamps, or forests, are coastal wetlands found along tropical to subtropical coasts. In Florida, mangrove coasts cover the southern half of the peninsula, and they are especially dense in the Ten Thousand Islands region and the coastal tracts of Big Cypress National Preserve and Everglades National Park. Mangroves become increasingly rare north of Tampa on the west coast and north of Merritt Island on the east coast, at roughly 28 degrees north latitude. Although isolated mangrove trees may be found as far north as Cedar Key and St. Augustine, salt marsh vegetation replaces mangrove communities along higher-latitude coastlines. It is estimated that mangroves cover up to 500,000 acres of Florida coastline.

Although there are about forty mangrove species along Pacific Ocean coastlines, there are only about eight in the western hemisphere, and only

a few of these are common. Only four species of mangrove grow in Florida. The red mangrove (*Rhizopora mangle*) is sometimes called the "walking tree" because of its distinctive prop roots that elevate the tree above the mud. The black mangrove (*Avicennia germinans*) is known by its unique pneumatophores, breathing structures that raise part of the root system out of the mud. Sometimes the structures are called "dead man's fingers." The white mangrove (*Laguncularia racemosa*) usually occurs at higher elevations than the red or black and does not have a visible root system. Buttonwood or button mangrove (*Conocarpus erectus*) also grows at slightly higher elevations. Generally, red mangroves occur in the most seaward position, with the other species growing landward of them. Mangroves are uniquely adapted to live in seawater, and the strange root systems of the red and black mangroves allow them to live in oxygen-poor black mud. Mangroves take up saltwater but have the ability to exude salt through large pores on the underside of their leaves, where it is common to see fairly large salt crystals.

Mangrove swamps are an important coastal ecosystem. Bacteria decompose much of the leaf litter and organic detritus from the mangroves and pass nutrients up the food chain. Mangroves are also nursery grounds for numerous invertebrate species and fish. The elevated root structures provide a maze of habitat for numerous marine invertebrates, such as sponges, hydroids, barnacles, mollusks, crabs, and many others. Geologically, mangroves have influenced the preservation of several fossil deposits in southern Florida, and mangrove root casts and even wood are occasionally fossilized. Because they require fairly warm water, above 72 degrees Fahrenheit, mangroves are usually associated with coral reefs along tropical coastlines.

*The red mangrove (*Rhizopora mangle*) in Florida Bay. Sometimes called the* walking tree, *the red mangrove has distinctive prop roots that elevate the tree above the mud.*

Sunniland Oil Field

On Florida 29 about 25 miles north of Carnestown is Sunniland, the location of the Sunniland oil field. It was here that the Humble Oil and Refining Company drilled the first commercial oil well in Florida in 1943. They encountered oil at 11,500 feet in the Sunniland Formation of Early Cretaceous age (about 115 million years old) and drilled seventeen additional wells near the first one. A total of fourteen oil fields have since been discovered in the surrounding area, all of them tapping into the rudist reefs and other fossiliferous limestones of the Sunniland Formation.

In 1955, a very curious fossil discovery was made in Early Cretaceous strata in a deep exploratory oil well near Lake Okeechobee. At a depth of 9,210 feet, the 4.4-inch-diameter core recovered the anterior portion, but not the skull, of a sea turtle, the specific identity of which has not yet been determined. Cretaceous-age sea turtles, as well as some dinosaurs, are found in the famous Selma Group strata of Alabama and elsewhere along the Gulf Coastal Plain where Cretaceous-age strata are exposed at the surface. Such a rare and fortuitous find in Florida's subsurface seems very unlikely to be repeated.

Big Cypress Swamp and the Everglades

There are no other Everglades in the world.
—Marjory Stoneman Douglas

East of Naples, I-75 and US 41 cross Big Cypress National Preserve, and US 41 travels alongside a portion of Everglades National Park, both of which are part of the larger Big Cypress Swamp and Everglades ecosystem. Big Cypress occupies 728,000 acres, and the Everglades National Park spans 1.5 million acres. Before it was altered by urban and agricultural development and the redirection of surface water through drainage canals, the Everglades ecosystem encompassed 2.24 million acres (3,500 square miles). Big Cypress is so named because of its numerous cypress strands, which are water-filled channels with cypress trees growing in them. Corkscrew Swamp, between Estero and Immokalee, has some virgin strands, with trees up to 100 feet tall and 700 years old. Most of the remaining Florida panthers (*Puma concolor coryi*) live in Big Cypress Swamp.

The Everglades, also called the River of Grass, is a vast plain and marsh dominated by Jamaica swamp sawgrass (*Cladium jamaicense*), which covers about 70 percent of this ecosystem. The rapidly growing sawgrass, which can grow up to 1 foot or more every two weeks, is a fire-adapted species that grows in shallow, flowing water. All water in the Everglades is surface runoff. There is no contribution from groundwater seepage. Most of the water that flows through this region does so during the wet season, May through September. The southernmost portion of the Atlantic Coastal Ridge is largely responsible for the formation of the Everglades. The coastal ridge forms a bucketlike rim along the east coast, confining fresh surface water inland and causing it to flow southward from Lake Okeechobee, through the Everglades, and to the tip of the peninsula.

Dome-shaped tree islands are common in the Everglades. Some islands are limestone pinnacles with tropical hardwood hammocks. There are thousands of these islands, each elongated in a teardrop shape with the more pointed end facing southward, the direction the surface water flows. The hammocks tend to not burn when seasonal fires sweep through the Everglades because of the karstic moats that surround them. The moats form as acidic leaf litter dissolves the limestone immediately around each hammock. Each island typically has a similar solution hole in its center, as well. The islands are nearly impenetrable, and their canopy-covered interior is shady and cool. Distinct populations of the colorful tree snail *Liguus fasciatus* occupy each island. Each of these snail populations has a unique color pattern.

There are three distinctive deposits of biological origin in the Everglades that produce specific types of rock commonly found in the geological record: diatomaceous earth, marls, and peat. Often associated with peat and muck is a silica-rich sediment called *diatomaceous earth.* In general, diatomaceous earths are cream or white in color, but the Florida deposits are generally dark brown or black due to the abundance of carbon-rich organic matter. The silica comes from the microscopic shells, or tests, of freshwater species of single-celled

River of Grass—the Everglades.

algae called *diatoms.* Diatomaceous earth has many commercial applications and is commonly used as a filtration medium and an additive in abrasives. It has not been commercially mined in Florida for many years.

Marls are gray to white, calcium carbonate–rich clays that also develop with peat. Some marls contain abundant molluscan shells of freshwater origin. The Lake Flirt Marl underlies much of the peat of the Everglades. The calcium carbonate in marl is derived from several unusual freshwater organisms. Periphyton is an association of filamentous cyanobacteria, sometimes called *blue-green algae,* and microalgae. Periphyton is not a genus or species name; it refers to the stringy, matlike clusters of these bacteria and algae. Periphyton concentrates calcium carbonate to form thick mucks. In places, the muck has accumulated up to 15 feet thick, but so much surface water has been drained from the Everglades that much of this marl has dried out, oxidized, and eroded away. It now rarely exceeds 5 feet thick, and dried, curled, and cracked periphyton crusts are very evident across the Everglades during the dry season.

Charophytes, a class of common freshwater algae, are also primary contributors to the marls of the Everglades. Sometimes called *stoneworts,* these algae produce tiny calcareous capsules that encase their female reproductive structures. About 50 percent of the dry weight of charophyte algae can be calcium carbonate sediment, and they are major contributors to the marl in the Everglades. Numerous freshwater mollusks also contribute their shells to the calcium carbonate content of the marls.

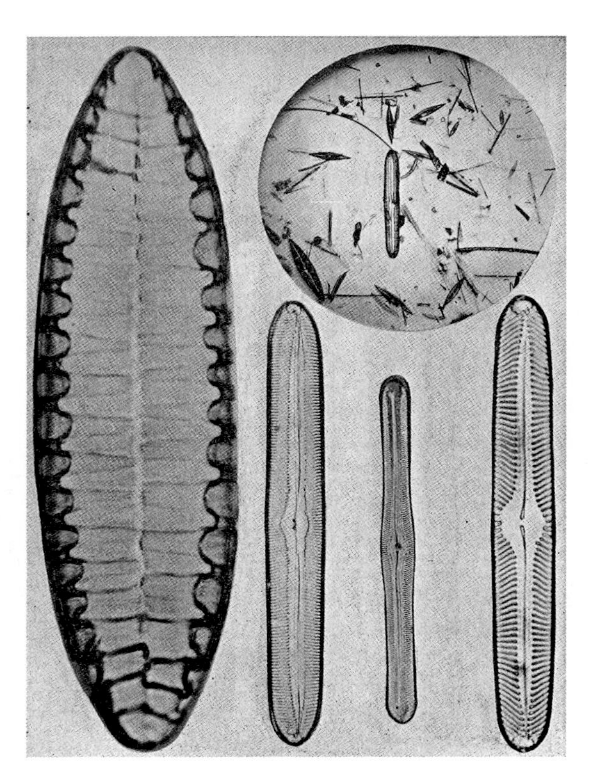

Freshwater diatoms from Florida peat. Their tests accumulate and form diatomaceous earth. —From Davis (1946) and courtesy of the Florida Geological Survey

Peat is a compacted accumulation of plant and woody debris that forms in lakes, ponds, river floodplains, freshwater marshes, salt marshes, cypress swamps, mangrove swamps, bogs, and several other soggy-bottomed environments. While peat is found in many wetland environments in Florida, the southern peninsula preserves the most, and the Everglades contains more than all of the other areas combined. For many years, Florida has been among the leading states in the economic production of peat, but this resource continues to be lost to drainage, exposure, and erosion.

Peat can convert to coal when it is buried and compacted by overlying sediments. Geologists have studied peat formation in southern Florida extensively to understand the early stages of coal formation. There are no coal deposits in Florida's geologic record, so it is a curious fact that coal-forming environments are presently active here. The great Appalachian coal beds of Pennsylvanian time (318 to 299 million years ago), created by very different types of vegetation, nonetheless formed in environments that were similar to the Everglades. These coal beds commonly overlie marine limestone, just as peat in the Everglades overlies the Miami Limestone and other Pleistocene-age marine beds. Most Everglades peat is Holocene in age, younger than 10,000 years. Although peat accumulation rates in south Florida are highly variable, it may be deposited as fast as 4 inches every 100 years.

To reach Everglades National Park, follow US 1 or the Florida Turnpike to Florida City or take Florida 997 south from US 41 to Florida City, then proceed west on Florida 9336 to the visitor center. Most of the Everglades is underlain by the Miami Limestone, which is well exposed along canals on Florida 997 en route to the park. And within the park, the heavily karstified Miami Limestone can be seen almost everywhere. The karst development is on a smaller scale than elsewhere in Florida and is called *microkarst.* It is especially well exposed during the dry season along the Pah-hay-okee Overlook Trail, Rock Reef Pass, and the Long Pine Key Trail. The Anhinga Trail traverses a freshwater slough, which is also a classic example of a modern peat-producing environment in the Everglades. The West Lake Trail passes through a mangrove ecosystem that surrounds a brackish lake. Mangrove and sawgrass peat is accumulating here in abundance. Farther south in the park, the Mahogany Hammock Trail explores one of the dome-shaped tree islands.

Boat tours of the Everglades are available in Flamingo, at the park's southern end. Nearby Cape Sable is a wave-sculpted cuspate foreland of Holocene age with several beach cusps that are clearly visible on maps. Their carbonate sand has abundant molluscan debris. Cape Sable also has many beach ridges that accreted to the coastline, forming much of the cape between 3,000 and 1,500 years ago.

Finally, just as northern Florida has its swamp legends of Tates Hell and the Wakulla Volcano, southern Florida has some swampy tall tales of its own. Although rarely seen or photographed, the 7-foot-tall, red-haired orangutan-like Florida bigfoot reportedly prowls the wetlands of the Everglades, Big Cypress Swamp, and other areas. A flurry of "sightings" occurred in the 1970s, curiously following a widely seen film that depicted a bigfoot walking in a northern

California forest. But the Floridan species is easily distinguished from other North American bigfoots—it has a chronic case of BO! It is said to have an appalling smell, likened to rotten eggs, manure, or a dirty, wet dog. Hence the true name of the Florida Bigfoot—Skunk Ape. If you visit the Everglades, keep that camera ready!

Alligators and Crocodiles

The first alligators appeared in Late Cretaceous time, more than 65 million years ago, and the genus *Alligator* appeared in North America during Late Eocene time, more than 38 million years ago. Almost since it first emerged from the sea, Florida has been infested with alligators! They first appeared in Florida in Early Miocene time, about 18 million years ago, as the small, 7- to 8-foot *Alligator olseni,* which has been found at Thomas Farm and other fossil localities. The modern *Alligator mississippiensis* appeared as early as Late Miocene time, at least 6 million years ago, and its fossils are common in formations of Pliocene, Pleistocene, and Holocene age. The most common fossils of alligators (and crocodiles) are their blunt, conical teeth and the oval to squarish bony plates, called *osteoderms,* that are embedded in their skin and serve as body armor.

Alligators can be seen almost anywhere in the state in lakes, streams, and marshes. They are especially abundant in southern Florida in the Everglades and along the appropriately named Alligator Alley (I-75), where they are top

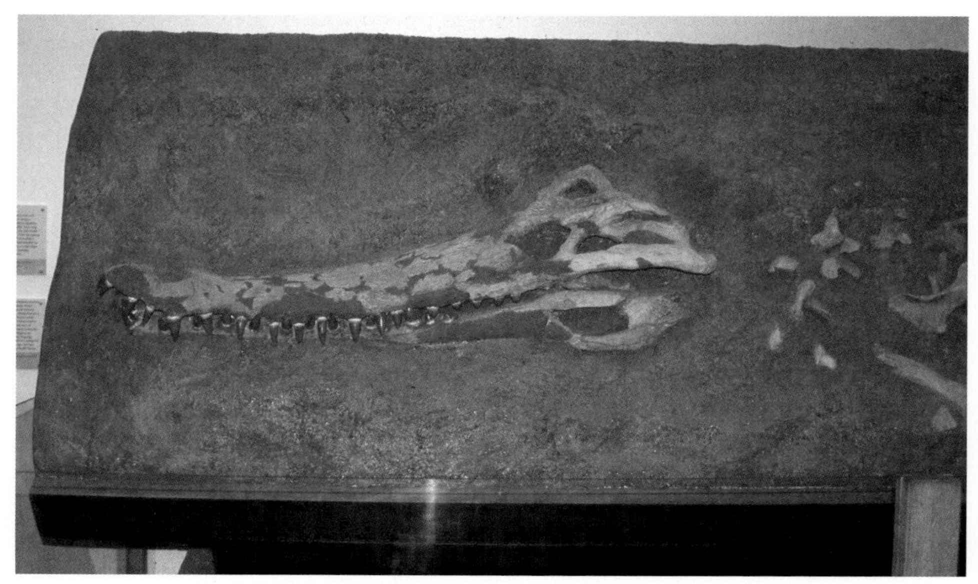

Gavialosuchus americanus *of Miocene age from the Haile Quarry complex.*

predators in this freshwater wetland ecosystem. Their large wallow holes, called "gator holes," are also commonly seen throughout the wetlands. The American alligator can be found throughout the southeastern United States, from North Carolina to Texas, and may reach as far into the interior as Arkansas. The largest populations are in Louisiana, southern Georgia, and Florida. The only other modern species of *Alligator* lives in China (*Alligator sinensis*).

Florida has also long been the home of crocodiles. Crocodiles first appeared in Middle Triassic time, more than 230 million years ago. The earliest Florida crocodiles appeared in Miocene time. *Gavialosuchus americanus* was a long-snouted, estuary-dwelling crocodile that grew up to 39 feet long. It is found only in Middle- to Late-Miocene deposits in Florida but is related to a modern crocodile (*Tomistoma schlegelii*) living in Thailand, Malaysia, and Indonesia. The modern American crocodile (*Crocodylus acutus*), a threatened species, is not found in Florida's fossil record, but three breeding populations live in the mangrove swamps of the Everglades and Florida Keys. One population is on the north side of Key Largo at Crocodile Lake National Wildlife Refuge. Southern Florida is the northern range of the American crocodile, a tropical species found in coastal wetlands bordering the entire Caribbean Sea, and on the Pacific Coast from southern Mexico to Ecuador. Only four hundred to five hundred individuals live in Floridan waters. Another crocodilian, the spectacled caiman (*Caiman crocodilus*), is also found in Florida, but it is a nonnative species that was evidently released by pet owners.

Lake Okeechobee

Perhaps more than any other area of the state, the flat terrain north of Lake Okeechobee is classic Florida Cracker country, where early cowboys drove their herds on horseback, "cracking" their rawhide whips. The Florida Cracker Trail, which includes sections of Florida 64, Florida 66, and US 98, traverses the entire peninsula from Bradenton to Fort Pierce. It was established, in part, to provide a greenway that preserves this cultural heritage.

Lake Okeechobee is the largest lake in Florida, covering 730 square miles, and one of the largest in the United States. It is very shallow, with depths averaging just 9 feet, and receives much of its water from the Kissimmee River, which has its headwaters just south of Orlando and Walt Disney World. Water flows south from Lake Okeechobee as a broad sheet of freshwater, forming the headwaters of the Everglades. When viewed from Herbert Hoover Dike, which surrounds the lake, Lake Okeechobee seems like a vast inland sea. The lake basin is, in fact, a former lagoon or estuary, and the modern lake is a simply the freshwater-filled portion of this elongated, Pleistocene-age embayment.

There have been other proposals for the origin of the lake. One hypothesis was that the lake formed in a depression that resulted from downward faulting. A more colorful suggestion was that the lake is actually a meteorite crater!

Most geologists favor the lagoon-estuary origin theory. Karst dissolution may also have lowered the area to some extent.

Devastating hurricanes struck the Lake Okeechobee area in 1926 (category 4) and 1928 (category 5). The 1928 storm, also called the San Felipe Hurricane, first pummeled Puerto Rico, killing fifteen hundred people. As it approached southern Florida, the storm was strengthened by warm Gulf Stream waters. The wind-driven storm surge flooded the farming communities along the southern edge of Lake Okeechobee, killing more than three thousand people. Although little known, this was the second-worst hurricane disaster in U.S. history, the first being the Galveston storm of 1900, and the second- or third-worst natural disaster of any kind in the United States, on par with the 1889 Johnstown Flood. This devastating loss of life, and the economic impact of the storms, prompted the federal government to construct the Herbert Hoover Dike and its associated canals to prevent such catastrophic flooding. Today, the Lake Okeechobee Scenic Trail sits atop the dike and extends 110 miles around the lake, with many access points.

The surface geology of the southern peninsula is largely a record of rising and falling sea level during Pleistocene time. Geologists recognize five distinct episodes of rise and fall, which occurred during the latter half of Pleistocene time. During each phase, sea level was 20 feet or more higher than at present, and the shoreline may have reached as much as 150 miles inland of Florida's present southern shoreline. Each of these five phases is separated by rocky crusts of brown, thinly laminated calcium carbonate called *duricrust, calcrete, caliche,* or *soilstone.* These crusts formed as acidic groundwater dissolved some of the limestone bedrock, and then the calcium-carbonate-laden water evaporated beneath the soil, creating a hard crust. Soil particles may also be incorporated into the hard duricrust. Each surface may be associated with fossil casts of roots, such as those of mangrove trees, and soils, microkarst, and freshwater limestone. So each crust represents a depositional hiatus, a period when sea level had dropped, the peninsula was exposed, limestone dissolved, soils formed, and coastal and freshwater sediments formed over the marine limestones. Crusts are found in all of the Pleistocene-age formations in southern Florida, including the Caloosahatchee, Fort Thompson, and Anastasia formations and the Key Largo and Miami limestones.

During high sea levels, Lake Okeechobee was a restricted bay or estuary, meaning it had a limited connection to the sea. It formed as the ocean transported quartz-rich sediments southward along the east and west shores of what was then a narrower peninsula. Deposition along the southern end of the peninsula isolated and enclosed the bay. Shell pits near Lake Okeechobee yield abundant mollusks of late Pleistocene time, including oysters and thick beds of the common rangia clam (*Rangia cuneata*). These clams are good indicators of brackish water, so it is likely that Lake Okeechobee was an estuary. Water depths in southern Florida ranged from a maximum of 100 feet in early Pleistocene time to 20 feet during late Pleistocene time. The modern surface of the southern peninsula is essentially the seafloor—the Pamlico Terrace—of the last interglacial period (the Sangamonian), about 125,000 to 100,000 years ago.

During higher sea levels, more-open marine conditions developed farther south, over the area now occupied by the Everglades and Florida Bay. Water depths were approximately 30 feet in late Pleistocene time, creating conditions similar to those presently found on the Bahama Banks. East of Lake Okeechobee was a shelly complex of barrier island beaches, dunes, and lagoons that formed the Anastasia Formation. This complex graded southward into an extensive sandbar belt that formed the oolitic Miami Limestone. Farther south, a reef system became the Key Largo Limestone, and to the west, in the Lower Keys, another sandbar complex became the Miami Limestone. See **No Frozen Mammoths—Pleistocene Time** in **Geological and Paleontological History** for more information about southern Florida during Pleistocene time.

Some of the best exposures of the Pleistocene-age formations of the southern peninsula crop out along the banks of the Caloosahatchee River, west of Lake Okeechobee. Exposures along the Caloosahatchee, like those along the Peace River, were the subject of early geological and paleontological investigations in Florida. There are boat access points all along the river. To explore the entire Caloosahatchee by boat, we recommend you put in near Fort Myers. From I-75 in Fort Myers, proceed east on Florida 78 about 3 miles to Florida 31. Turn right (south) and follow Florida 31 to the river at Franklin Lock.

Notable Pleistocene-age outcrops along the Caloosahatchee River	
Denaud	The Caloosahatchee and Fort Thompson formations are visible along the north and south banks of the river near the town's bridge. From Fort Myers, take Florida 80 to County Road 78A. Turn north and follow County Road 78A to Denaud and the river.
La Belle	The Caloosahatchee Formation is visible at the picnic area along the river on the north side of the Florida 29 bridge. Exposures are also visible along the river east of the bridge for 4 miles.
Ortona Lock	The Caloosahatchee Formation is visible along the north bank of the river, west of the lock. From La Belle, take Florida 80 east to Goodno, then drive north on Goodno Road SW to Ortona Lock.

Florida A1A (I-95 and US 1)
Fort Pierce—Miami
119 miles

From Fort Pierce to Miami, barrier islands continue with Hutchinson Island, Jupiter Island, and various coastal spits and beaches. These are largely quartz sand beaches, supplemented with a rich mix of molluscan sand. The molluscan debris is derived from both recent shells and erosion of coastal outcrops of the Anastasia Formation, a molluscan coquina limestone. At the southern end of this route, the sands of Miami Beach, Virginia Key, and Key Biscayne are primarily calcium carbonate with some quartz sand mixed in. This change approximates the shift from subtropical to tropical conditions.

It is evident when driving on Florida A1A that Hutchinson Island (as well as Orchid Island to the north) is significantly higher than sea level. As with the northern section, this segment of the Atlantic Coastal Ridge consists of the Anastasia Formation, which is covered with well-vegetated coastal dunes and barrier islands. Like the barrier islands to the north, Hutchinson is a perched barrier island because it consists of a drape of sediment "perched" on top of a narrow coastal exposure of hard coquina. Dunes further increase the elevation of these islands.

One of the most curious features of Florida's east coast is the tube worm reefs that occur from Cape Canaveral to Key Biscayne in depths up to 30 feet. These reefs are constructed by the marine worm *Phragmatopoma caudata*, popularly known as the bristle worm. Adult bristle worms are about 0.125 inch in diameter and can grow up to 2 inches long, though most are less than 1 inch. They construct colonies of tubes by cementing sand particles together into a tough sheath, in which they live. They can live in densities ranging from fifteen thousand to sixty thousand individuals per square meter, forming massive, wave-resistant mounds. The worms thrive in turbulent, plankton-rich coastal waters that flush food to them. Turbulent water also suspends sediments, which the worms trap to form their tubes. At low tide the worms may be exposed, but they seal their tubes, which can hold sufficient water to keep them until the next high tide. Like coral reefs, the worm reefs form an important marine habitat for a wide variety of marine life. Snorkeling along the reef front is a popular and educational experience.

The tube worm reefs are best developed off the coast between Fort Pierce and Jupiter, where they grow over exposures of the Anastasia Formation. Probably the best exposure is at Bathtub Reef, at the southern end of Hutchinson Island. This amazing coastal reef clearly illustrates the convergence of geology with biological oceanography. Worm reefs are found in many locations around the world, but exposures as excellent as those at Bathtub Reef are unique to Florida. To see the worm reefs up close, go straight (south) on MacArthur Road instead of following the sharp turn at the southern end of Hutchinson Island, where Florida A1A turns sharply right and crosses the Indian river. Proceed about 2.5 miles to Bathtub Reef. The headquarters of the Florida Oceanographic Society is on A1A just west of MacArthur Road. Stop in for information on the worm reefs and other interesting oceanographic features. Just north of Bathtub Reef on MacArthur Road is Gilbert's Bar House of Refuge, a private maritime museum that sits on top of a beautiful exposure of the Anastasia Formation. The Anastasia is up to 15 feet high here, and outcrops extend for 2 miles north along the shore.

South of Hobe Sound off US 1/Florida A1A is Jonathan Dickinson State Park, named for an English Quaker. In 1696, Dickinson and a crew of twenty-four were en route from Jamaica to Philadelphia when they were shipwrecked near the mouth of the Loxahatchee River, possibly on worm reefs. After being captured and eventually released by the Jobe Indians, the party worked its way north to St. Augustine. Dickinson recorded the story of the shipwreck and journey up the coast in *God's Protecting Providence*, written in 1699. As one of the few records of this time, place, and culture, it is a classic of Florida's historic literature.

Numerous parks and beach access points along this entire route offer opportunities to view Florida's coastal geology. A prominent, stable,

Worms reefs of Phragmatopoma caudata *growing over the Anastasia Formation at Bathtub Reef.*

A large mound of Phragmatopoma caudata *at Bathtub Reef.*

vegetated dune ridge guards the coast, and many of the beaches have rocky outcrops of the Anastasia Formation along the shore. On Jupiter Island, one of the best and most picturesque exposures of the Anastasia can be seen at Blowing Rocks Preserve. This coquina limestone crops out for perhaps 0.5 mile or more along this coast, but the best exposures are at the southern end of the preserve. This is one of the few rocky coastlines in Florida. Small sea cliffs, some nearly 15 feet in height, sea arches, and caves can be seen here, especially at low tide. At high tide, crashing waves can force water through "blowholes" in the rock, blasting water over 50 feet in the air. The preserve is just north of the town of Jupiter off County Road 707. Other excellent coastal exposures of the Anastasia Formation occur at Coral Cove Park, south of Blowing Rocks on County Road 707. Many beaches in Palm Beach County also have Anastasia exposures, including Carlin Park, Juno Beach, Ocean Reef Park, Phipps Ocean Park, Gulfstream Park, Jap Rock, and South Beach Park.

South of Jupiter, in an older, affluent area of Palm Beach, a narrow roadcut reveals a truly exceptional exposure of the Anastasia Formation. The east-west-trending outcrop, which is more than 20 feet tall and approximately 600 feet long, displays sediments from many different depositional environments. On the western end of the exposure, it is a barrier island beach deposit composed of fine-grained shell hash, and on the eastern end, it is a dune deposit composed of sand-sized grains layered in large cross-beds inclined at 30 degrees or more. The barrier island rock is older than the dune rock, and the two are separated by a fossil soil horizon. Few places preserve such a well-defined "petrified" beach and dune deposit that is so easily accessible and visible. In West Palm Beach, Florida A1A crosses the Flagler Memorial Bridge. Once across the bridge, on the east side, proceed to North County Road. Turn north and drive about 1.6 miles to Country Club Road. Turn west on Country Club Road and find somewhere to park. The exposure is obvious. Note that this is a narrow road with little space between the road and the outcrop.

Beginning near Fort Pierce and extending to the Miami area is a series of offshore rocky ridges in depths of 10 to 150 feet or more. They rise 5 to 30 feet above the seafloor. There are three or four of these ridges,

The Anastasia Formation at the Palm Beach roadcut.

and they are roughly parallel to the modern shoreline. Well known to divers and often referred to as "reef ridges," they are well documented in dive books and on charts. The ridges host a number of marine invertebrates and other marine life, creating reeflike conditions. Sea life that is associated with reefs is found as far north as Jupiter, where the warm Gulf Stream passes near the shoreline before continuing northward, away from the Florida coast and into the Atlantic. Nonetheless, the sea creatures on these ridges are not actually constructing a reef but are growing on an older fossil barrier reef, which was dominated by elkhorn coral (*Acropora palmata*), a major Caribbean reef constructor. Geologists have dated these reefs as being between 4,000 and 7,000 years old. The reefs died during the Holocene Transgression for reasons not fully understood, although some geologists hypothesize that the rising sea eroded coastal soils, loading the reefs with excess nutrients. Nutrient loading is a well-documented cause of reef death today. See **Holocene History of the Florida Reef Tract** in the **Florida Keys** chapter for more information about these ancient reefs.

The continental shelf offshore of the Miami area is called the Miami Terrace. The upper terrace of this narrow, stair-stepped, Miocene-age surface is in depths of 650 to 1,230 feet, and the lower terrace is in depths of 1,970 to 2,290 feet. The upper terrace is primarily karstified shallow-water limestone. The karst features formed when the shelf was above sea level. Strong currents that intensified during Middle Miocene time, about 16 to 11 million years ago, scoured and eroded the lower terrace. Both surfaces of the Miami Terrace are rich in phosphatic rock, and the upper surface is littered with bones of marine mammals and shark teeth. A similar but larger terraced continental shelf of Miocene age, the Pourtales Terrace, is located offshore of the Florida Keys in depths of about 650 to 1,300 feet.

Another interesting oceanographic phenomenon off the east coast of Florida occurs in water depths of 1,640 to 2,296 feet in the northeastern Straits of Florida and near the Little Bahama Bank. Here, numerous deepwater mounds, or lithoherms, up to 984 feet long and 164 feet high are constructed by soft corals (sea fans, sea whips), crinoids, sponges, hydrocorals (colonial animals that resemble stony corals but belong to another class of animals called hydrozoa), and other skeletal invertebrates. They are teardrop-shaped: high on the up-current side but tapering off on the down-current end. Similar structures are commonly found in the geologic record, especially in rocks from Paleozoic time (hundreds of millions of years ago), but they are rare today.

Although the city of Miami may not seem a likely location to conduct a geology field trip, there are several significant exposures of the Miami Limestone in this city. Formerly known as the Miami Oolite, this unique limestone formation underlies much of the southeastern tip of Florida, including all of Miami-Dade and portions of Collier, Monroe, and Broward counties. To the south, the Miami Limestone extends under Florida Bay and is replaced in the Upper Keys by the Key Largo Limestone. But in the Lower Keys, from Big Pine Key to the Marquesas Keys, the Miami Limestone appears again and overlies the Key Largo Limestone.

The Miami Limestone forms much of the southern end of the prominent Atlantic Coastal Ridge, upon which the cities of Miami, Fort Lauderdale, and surrounding towns were built. Near the border of Broward and Palm Beach counties, the Miami Limestone grades into the Anastasia Formation, which then underlies the Atlantic Coastal Ridge up to St. Augustine. The Miami Limestone also forms the bulk of the Biscayne aquifer, which is part of the surficial aquifer system. This highly permeable aquifer, which can be up to 200 feet thick, supplies potable water to residents of the southeastern coast of Florida.

The Miami Limestone consists of two sedimentary rock types that formed in distinct depositional environments. Geologists refer to sedimentary depositional environments as *facies*. The oolite facies is exposed primarily in the Miami area and the Lower Keys. It consists of cross-stratified, oolitic limestone that was deposited in high-energy, shallow-water shoals that were influenced by strong tidal currents. Ooids are tiny (0.3 to 1 millimeter in diameter), polished spheres of calcium carbonate that precipitate from shallow, warm, turbulent, tropical seawater. Ooids have a tiny nucleus consisting of a sand grain, shell fragment, or other particle that is coated by thin (1 to 3 microns thick), concentric layers of calcium carbonate in the form of the mineral aragonite. Ooids are not forming today in southern Florida waters, but they are actively forming across the Florida Straits on the Great Bahama Bank. In fact, a comparison of the modern Great Bahama Bank with the Pleistocene-age record of southern Florida reveals an undeniable mirror image. Clearly, the shallow marine conditions of the Great Bahama Bank are like those that existed in southern Florida over 100,000 years ago. The Great Bahama Bank is a very shallow (generally less than 30 feet), flat-topped, limestone-producing plateau with coral reefs on the eastern margin, off Andros Island. But there are extensive ooid sand shoals on the western margin. Southern Florida was very much like this during Pleistocene time. See **No Frozen Mammoths—Pleistocene Time** in **Geological and Paleontological History**.

The bryozoan facies of the Miami Limestone extends west of the Miami area and underlies much of the Everglades. A fine-grained, muddy limestone that contains large, knobby, branching colonies of the bryozoan *Schizoporella floridana*, it formed in a broad quiet-water lagoon behind the ooid shoals. This facies is surprisingly vast, extending over 2,000 square miles from Fort Lauderdale through the Everglades. It is well developed at the southernmost end of the peninsula at Flamingo. Bryozoans, or so-called moss animals because of the mosslike appearance of some modern forms, are colonial marine animals. Each individual, called a *zooid*, is quite small, usually less than 1 millimeter in length, and superficially resembles a tiny coral polyp. But the bryozoan has a more complex anatomy than the coral, and its tentacles don't have stinging cells. Despite the small size of a single zooid, colonies of *Schizoporella floridana* can be large, up to a foot or more in diameter. Brownish to yellowish colonies can be found today living on the lagoon side of the Florida Reef Tract, and they commonly grow within sea grass beds. They are also common in the lagoon areas behind the modern ooid shoals of the Bahamas.

Perhaps the best exposure of the Miami Limestone in the Miami area is the well-known outcrop at Coral Gables Canal. This outcrop, composed of about 18 feet of cross-stratified oolitic limestone, is located at the intersection of Lejeune Road and Sunset Drive. Cross-stratification indicates the oolitic sand was deposited as a wave- and tide-influenced shoal. From US 1 in South Miami, take Sunset Drive (Marjory Stoneman Douglas Boulevard) east to a roundabout at Lejeune Road. There is a Coral Gables Waterway historical marker here, and good exposures of the oolite along the canal.

On Bayshore Boulevard between Mercy Hospital and Coconut Grove, there are several outcrops of the Miami Limestone on the west side of the road (on private property). The vertical faces of these exposures are a wave-cut escarpment that the sea cut into the Silver Bluff Terrace. The escarpment formed when sea level was about 8 feet higher than it is now. Also near Coconut Grove, the beautiful Mediterranean villa at Vizcaya Museum and Gardens is largely constructed of the oolitic Miami Limestone and the Key Largo Limestone. Several older houses in Miami Beach were also constructed of the oolite.

Key Biscayne is the southernmost extent of the long Atlantic Coast barrier chain. On the northeastern shore of the island, there is an exposure of the Miami Limestone that is best seen at low tide. It preserves fossilized casts of mangrove roots that closely resemble the elaborate root systems of the black mangrove (*Avicennia germinans*). Preservation of ancient tree root systems is not uncommon. They are frequently found in the sedimentary rock formations of the great coal-forming swamps of Pennsylvanian time (318 to 299 million years ago), such as those of the Appalachians, and have also been fossilized elsewhere in Florida. However, geologists think the root casts of Key Biscayne are early Holocene

The Miami Limestone along Bayshore Boulevard, with prominent cross-stratified oolitic limestone at the base.

in age, substantially younger than the Miami Limestone in which the roots were embedded. Some have interpreted the structures as root casts of turtle grass (*Thalassia testudinum*), rather than mangrove. Whatever the species, its roots penetrated the relatively soft limestone.

The geology of Biscayne Bay is of interest because the Key Largo and Miami limestones are juxtaposed at the surface here. The oolitic Miami Limestone forms the Atlantic Coastal Ridge on the mainland, but the Key Largo Limestone forms an offshore ridge, represented by Elliot Key to the south. Shallow Biscayne Bay, which rarely exceeds 12 feet in depth, separates these two Pleistocene-age ridges—one of oolite, the other of coral. From US 1, take toll road 913 (Rickenbacker Causeway) to Key Biscayne. The road crosses Virginia Key and Bear Cut Channel.

These geologic features in the Miami area tell us much about sea level fluctuations during the waning stages of Pleistocene time. About 125,000 to 100,000 years ago, during the Sangamonian interglacial period, sea level was an estimated 25 feet higher than at present, and the Key Largo reef was forming across most of the Keys, including Elliot Key. West of Elliot Key, on what is now the mainland, large ooid shoals of the Miami Limestone formed. Between the reef and the ooid shoals was a lagoon.

During the ensuing glacial stage, the Wisconsinan, sea level dropped, exposing the ooid shoals and reefs, which were cemented into limestone under the influence of freshwater, which both dissolves and reprecipitates calcium carbonate within the pore space of limestone. When the sea was approximately 8 feet above the present level, its waves cut a cliff into the Miami Limestone, forming the Silver Bluff Terrace (noted back along Bayshore Boulevard). At the height of the Wisconsinan glacial stage, sea level was near the edge of the modern continental shelf, nearly 390 feet lower than at present. During the Holocene Transgression, early Holocene reefs (today's "reef ridges"—mentioned earlier in this section) developed offshore of the modern reef tract, then reef development shifted to its modern position, slightly offshore of Key Biscayne, as the sea approached its present level about 6,000 years ago, flooding Biscayne and Florida bays.

Although vertebrate fossils are generally rare in this part of Florida, one site is worth mentioning. The Cutler Hammock locality, a sinkhole and cave deposit of late Pleistocene age in southern Miami-Dade County, preserved forty-two individuals of the extinct dire wolf (*Canis dirus*). In addition to the dire wolf, the site also preserved the Florida cave bear, black bear, saber-toothed cat, horse, mammoth, mastodon, bison, jaguar, American lion, Florida panther, and numerous other, smaller vertebrate species. The sinkhole-cave deposit, which developed in the Miami Limestone near Cutler Ridge, is about 16 by 20 feet across and nearly 10 feet deep.

Paleontologists have interpreted this site as a dire wolf den due to the abundance of carnivore remains and splintered, gnawed bones of prey species. Many of the smaller fossils were probably the prey of predatory raptors. The site may have also been a natural trap for some animals. Human remains, radiocarbon dated as being about 9,670 years old, were found here as well, and many artifacts, such as points and scrapers, have also been recovered. Humans may have used this site in latest Pleistocene time as well, and they may have contributed some of the animal remains, including bones that were obviously burned.

Florida Keys
Pleistocene Republic

Probably no other area in the United States has been as over-romanticized as the place "to get away from it all" as the narrow archipelago of tropical islands called the Florida Keys. But the reputation of the "Conch Republic" is at least partly deserved. The Florida Keys are a beautiful and exotic natural place. When famed naturalist John James Audubon visited the Keys in 1832 to study their avian fauna, he was elated by the natural wonders he saw and referred to the place as part of "the Divine conception."

Because of their location along the Gulf of Mexico, the Florida Straits, and the Caribbean, the Keys have a long, colorful, and tragic maritime history of marauding pirates, shipwrecks with sunken treasure, and disastrous hurricanes. On one day alone in 1733 a hurricane dashed twenty-one Spanish galleons, many of them carrying gold, on the reefs between Key Largo and Key Vaca in the Upper Keys. And about 4 miles from the Marquesas Keys lies the famous treasure-laden *Atocha*, which sank during a hurricane in 1622. The tropical combination of hurricanes and coral reefs is an ancient and unforgiving nautical hazard.

Geologically, the Florida Keys and the environments associated with them are unique. You need look no further for evidence of ancient sea levels, or to see the ongoing, massive production of limestone that has characterized nearly all of Florida's geological past. Nowhere is the construction of coral reefs, both past and present, more easily observed. The Florida Keys, which never rise more than 20 feet above sea level, lie along a semicontinuous, exposed ridge of Pleistocene-age limestone. The Upper Keys, from Key Largo to Bahia Honda Key, are composed of the coralline Key Largo Limestone. The Lower Keys, from Big Pine Key to Key West and the Dry Tortugas, are composed of the oolitic Miami Limestone. The Key Largo and Miami limestones can be seen in numerous spoil piles and road and canal cuts, and along some shorelines. There are numerous small sinkholes in the Keys, many of which are now ponds filled with peat and vegetation.

Offshore of the Keys is the largest tropical coral reef ecosystem on the continental shelf of the United States—the Florida Reef Tract. It is truly an exhilarating experience to put on fins, snorkel, and mask and jump into the sea near a living reef. The scene is dazzling—a riot of marine life displayed in a full spectrum of colors. The Keys, the reef tract, and

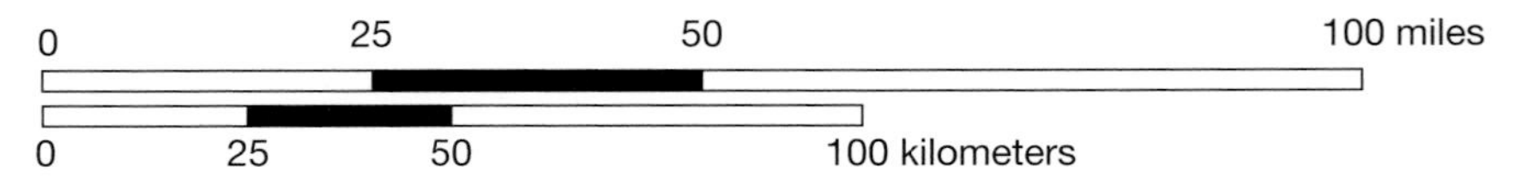

Road log map of the Florida Keys. —Map by Mountain Press

surrounding environments compose one of the most studied continental shelf margins in the world. It is as necessary for a geologist to see the Florida Keys and the Florida Reef Tract as it is to see the Grand Canyon.

The Florida Reef Tract

The Florida Reef Tract is a discontinuous band, 3 to 5 miles wide, of living coral reefs growing along the southern edge of the Florida Platform. The reefs grow in water depths of 0 to 80 feet. The reef tract is about 150 miles long and extends from near Miami to the Dry Tortugas. Biscayne National Park and the Florida Keys National Marine Sanctuary encompass the Florida Reef Tract. Within the bounds of the national marine sanctuary are also John Pennekamp Coral Reef State Park, Key Largo National Marine Sanctuary, Looe Key National Marine Sanctuary, and Dry Tortugas National Park. At around 25 degrees north latitude, the Florida Reef Tract lies in the northern latitudinal range of most tropical reefs. Waters are of normal marine salinity (34 to 37 parts per thousand salts, or 3.4 to 3.7 percent) with average water temperatures of 72 degrees Fahrenheit in winter and 82 to 86 degrees Fahrenheit in summer. Adjacent to the reef tract is the Florida Straits, through which the Florida Current flows. The Straits of Florida contain the most diverse concen-

Snorkeling along the Florida Reef Tract.

tration of marine life in the entire Atlantic Ocean. Once through the straits, the Florida Current joins the Caribbean Antilles Current off Florida's Atlantic coast, forming the famous Gulf Stream.

In his geological classic, *The Structure and Distribution of Coral Reefs*, Charles Darwin classified the Florida Reef Tract as a fringing reef complex. Others, such as Joseph LeConte, Louis Agassiz, and T. W. Vaughan, considered the Florida reefs a barrier reef. Some recent popular literature boasts that the Florida Reef Tract is the third-longest barrier reef in the world, behind the Belize Barrier Reef and the Great Barrier Reef. However, the discontinuous nature of the Florida Reef Tract technically precludes it from being a true barrier reef. Individual reefs in the Florida Reef Tract do not exceed 1 to 2 miles in length, whereas barrier reefs form a more continuous rim. But like true barrier reefs, the Florida Reef Tract has a rimmed shelf—a shallow, flat continental shelf with a marginal ridge, or rim, composed of carbonate sediments of organic origin. However, the Florida Reef Tract rim is not constructed solely of coral reefs; it also incorporates mounds of coral debris, shoals of sand derived from skeletal material, and other limestone. Most geologists classify the living portion of the Florida Reef Tract as being of two basic reef types: platform margin reefs and patch reefs.

Platform margin reefs, also called *bank margin reefs* or *bank reefs*, grow near the edge of the carbonate platform, 4.5 to 8 miles offshore of the Keys, adjacent to the Straits of Florida. It is estimated that there are over one hundred platform

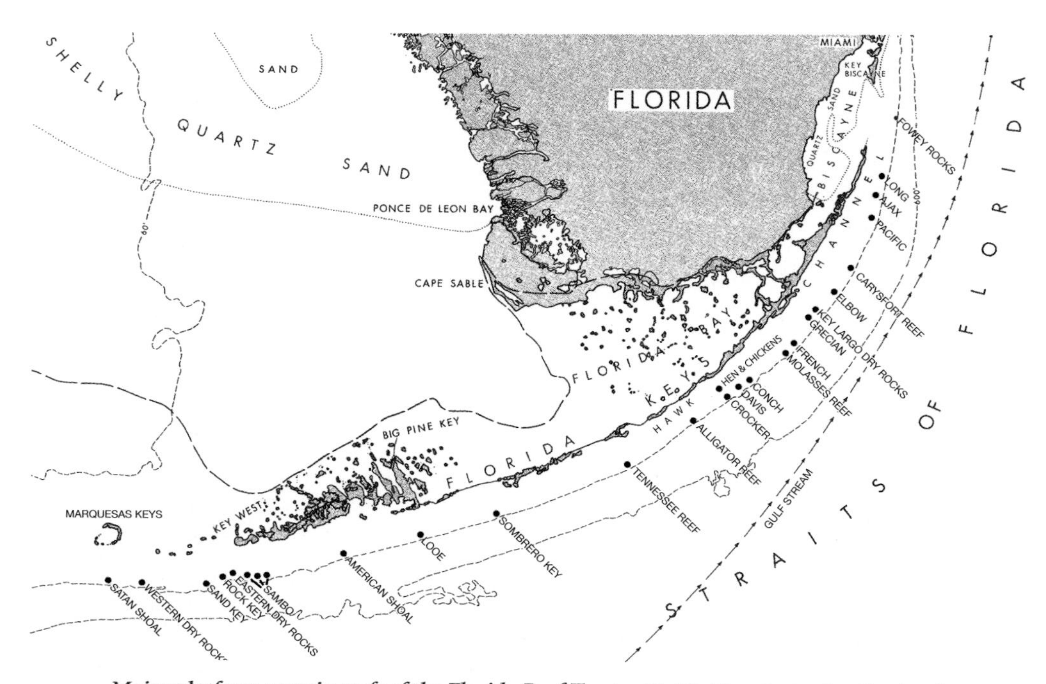

Major platform margin reefs of the Florida Reef Tract. —Modified from Enos and Perkins (1977)

margin reefs along the Florida Reef Tract from Fowey Rocks, south of Key Biscayne, to the Dry Tortugas. The more prominent and well-known platform margin reefs are named, oftentimes after the first ship that ran aground on the reef. As platform margin reefs grow, they create many subenvironments and niches for a wide variety of marine life. Three major, distinctive zones are easily discernable in a lateral profile of most platform margin reefs: the reef crest, the fore reef, and the back reef.

The reef crest is the highest part of the reef and takes the brunt of the waves. The corals of the reef crest are forms with thick branches, such as elkhorn coral (*Acropora palmata*), that can withstand and disperse much wave energy without breaking. In some cases, the reef crest also has abundant coralline red algae growing over the corals and reinforcing the ridge—an algal ridge. Reef-crest corals may grow right up to the spring low tide level.

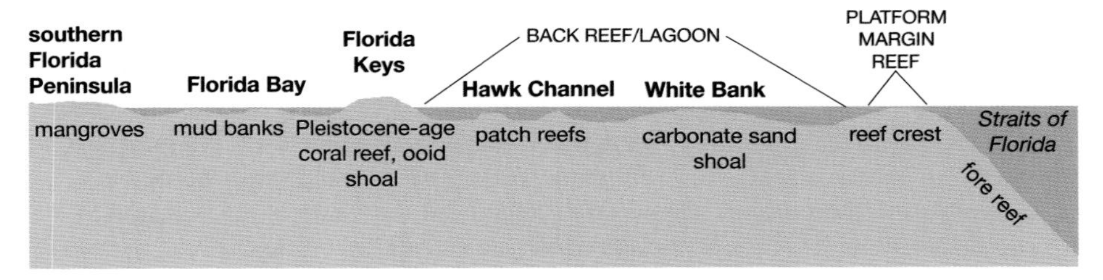

Ecological zonation across the Florida Reef Tract, Keys, and Florida Bay. —Illustration by Amanda Kosche

The elkhorn coral (Acropora palmata)*, a reef-crest builder.*

The fore-reef zone is seaward (and windward) of the reef crest, in depths of about 16 to 80 feet, and primarily composed of large head corals, such as star coral (*Montastrea annularis*) and brain coral (*Diploria labyrinthiformis*). On a few reefs, the fire coral *Millepora alcicornis* is common on the fore reef. In the deeper fore reef, where light begins to fade, thin, platelike corals of the genus *Agaricia* commonly grow outward from the reef wall. Reef growth drops off dramatically below 80 to 100 feet of water depth. On many reefs, the upper fore reef has distinctive ridges and channels called *spur-and-groove* structures. The spurs are constructed of corals, mostly elkhorn coral, and grow perpendicular to the reef crest. They may extend many hundreds of feet in depths of 3 to over 30 feet. The grooves are erosional in nature, acting as channels through which reef debris may wash downslope during major storms.

The back-reef zone is landward (and leeward) of the reef crest, in the lagoon. The water here is generally calm because the reef crest absorbs much of the incoming wave energy. Patch coral reefs, small coral mounds, soft corals (sea fans, sea whips), and meadows of sea grass and bottom-dwelling algae are common here. The back-reef zone and lagoon of much of the Florida Reef Tract, called Hawk Channel, has a sandy or muddy bottom covered with sea grasses near the mainland. Closer to the platform margin reefs, the channel has patch reefs, bare limestone hardbottom, and ripple-marked sand shoals composed of skeletal material. The largest skeletal shoal is White Bank. Located in the Upper Keys, it is about 25 miles in length and about 1 mile wide.

The brain coral Diploria labyrinthiformis.

Little Africa patch reef as seen from the top of the lighthouse at Loggerhead Key in the Dry Tortugas. Note the white "halo" of sand around the reef. This area has been heavily grazed by fish and invertebrates living in the reef.

Patch reefs typically grow in the protected waters of the lagoon between the platform margin reefs and the Keys in depths of about 30 feet or less. There are at least five thousand patch reefs across the Florida Reef Tract, and many have been named. Patch reefs are usually smaller than platform margin reefs and do not have the same ecological zonation. They rise as high as 15 feet off the seafloor and range in diameter from several feet to as much as 2,300 feet. They are usually round, oblong, or some similar shape. Patch reefs are constructed of brain corals and other species that form massive coral colonies, various delicate coral species, and a variety of soft corals, which don't build a calcareous exterior. Patch reefs are almost always surrounded by a "halo" of white carbonate sand. Beyond the sand is sea grass. The halo develops as fish and sea urchins that live in the patch reef graze the sea grass around the reef. Extensive thickets of staghorn coral (*Acropora cervicornis*) are also common in the back-reef area. See **Florida Land Developer—The Coral Animal** later in this chapter for more information about corals.

The Florida Reef Tract and its associated environments are part of a well-defined geobiological system. In this tropical climate, carbon dioxide from Earth's atmosphere (gaseous Earth) is dissolved and saturated in the hydrosphere (water Earth), where it is taken up by shelled marine life in the biosphere (living Earth) who use it to construct their calcareous skeletons. The skeletal material is then transformed into lithosphere (solid Earth) as strata of shelly sediment and limestone. This process is part of the global carbon cycle. The cycling of carbon dioxide gas, a major factor affecting the Earth's greenhouse effect, is one of the most important factors in global climate change through time.

US 1 (Overseas Highway)
Biscayne Bay and Upper Keys (Key Largo to Bahia Honda Key)
70 miles

Conveniently, the Keys are marked by mile markers, which begin at 0 in Key West and end at 106 in Key Largo, where US 1 turns north to the mainland. Since most highway travelers will necessarily begin in Key Largo, this road log begins there. Throughout, we cite points of interest using mile markers and other highway indicators. We include Biscayne National Park in this road log because of its connection to the Florida Reef Tract.

Where do the reefs of the Florida Reef Tract actually begin? Although reef-associated fauna such as sponges and soft corals (sea fans, sea whips) are found as far north as Jupiter (27 degrees north latitude), true construction of coral reefs begins just south of Cape Florida, near Soldier Key and within Biscayne National Park. The visitor center for the park is located on the mainland. From US 1 (Dixie Highway) at Homestead, drive about 9 miles east on SW 328th Street (North Canal Drive). Exposures of the Miami Limestone are visible along the canal that parallels Canal Drive, especially along the eastern end near the park headquarters. The Dante Fascell Visitor Center has a facade of tile stones of the coralline Key Largo Limestone. Snorkeling and scuba trips, glass-bottom boat tours, and canoe and kayak rentals can be arranged from the visitor center. We highly recommend Biscayne National Park if you want to examine reefs and reef environments. It is far less congested than areas within the Keys proper, where reef-seeking visitors usually dash as a first instinct, bypassing the park altogether.

Within the park, Soldier Key marks the northernmost exposure of the Key Largo Limestone, which extends from Miami to at least the Dry Tortugas and seaward into the Straits of Florida. The limestone is in the subsurface in the Lower Keys. The Key Largo varies in thickness from about 69 feet to over 200 feet and is composed of reef-forming scleractinian (stony) corals and other coarse skeletal sediment. Due east of Soldier Key is Fowey Rocks, which represents the northernmost occurrence of modern reef development in the Florida Reef Tract. This important northern ecological boundary is conveniently marked by Fowey Rocks Lighthouse, which in 1878 replaced the mainland Cape Florida Lighthouse as a more effective marker of the reef tract. There are six ocean lighthouses across the Florida Reef Tract, posted about 30 nautical miles apart. Most are solar powered. The Key Largo Limestone is also exposed on Elliott Key. Offshore of Elliot Key is Triumph Reef,

Long Reef (which at more than 2 miles is one of the longest unbroken reefs in the Florida Reef Tract), and Pacific Reef.

Two roads lead from the mainland peninsula to the Keys: US 1 and Old County Road 905A (also called Dixie Highway and Card Sound Toll Road), both of which lead to Key Largo. At about 30 miles in length, Key Largo is the longest Florida key. Once on Key Largo by way of County Road 905A, it is an additional 9 miles southwest on County Road 905 to rejoin US 1.

At the intersection of County Road 905 and US 1 (mile marker 106), there are good exposures of the Key Largo Limestone on the north side of US 1. Samples of the Key Largo Limestone can be picked up almost anywhere in the Upper Keys. On County Road 905 just 0.25 mile north of its intersection with US 1, Dagny Johnson Key Largo Hammock Botanical State Park preserves one of the largest remaining tracts of West Indian hardwood hammock. Approximately 2.5 miles south of that same intersection, US 1 crosses over a canal that extends from Blackwater Sound to the Atlantic side of Key Largo. The canal cuts through the Key Largo Limestone and exposes a beautiful cross section of the upper portion of the formation. Caution! Dangerous tidal currents move through this canal.

The entrance to John Pennekamp Coral Reef State Park is at mile marker 102.5. Founded in 1960, it was the first undersea park in the United States. It covers about 70 square nautical miles (93 square miles). Nearshore tropical marine environments, including sea grass beds and mangroves, are on full display and can be reached by kayak or canoe, or along designated shoreline areas. Snorkeling and scuba diving excursions are offered many times daily. The park also offers an outstanding marine aquarium and interpretive center.

There are several platform margin reefs within the boundaries of John Pennekamp Coral Reef State Park. Carysfort Reef has excellent reef zonation with elkhorn coral (*Acropora palmata*) and an algal ridge across the reef crest. The Carysfort Reef Lighthouse marks the reef. Built in 1824, it was the first ocean lighthouse in the United States. The creaking, groaning iron structure is said to be haunted by a former keeper who died at the lighthouse and returns nightly to see that all is well. Elbow Reef shows good spur-and-groove structures with fire coral (genus *Millepora*) growing over the spurs. Key Largo Dry Rocks, which also exhibits good zonation with an elkhorn coral reef crest, is the site of the 9-foot-tall bronze statue so familiar to snorkelers and reef divers, the famous Christ of the Deep. Grecian Rocks, averaging only 6 feet deep, is good for snorkeling. It is one of the most studied reefs and also has good zonation. The oldest corals in this reef deposit are approximately 6,000 years old. The reef has grown over carbonate sands and peat. French Reef

Holocene History of the Florida Reef Tract

Growth of the modern, living Florida Reef Tract began between 7,000 and 5,000 years ago, near the end of the Holocene Transgression. As much as 40 feet of sediment and rock have accumulated in some areas, consisting of coral rubble and the skeletal debris of countless other marine invertebrates and algae, notably green algae of the genus *Halimeda*. The modern growth phase of the Florida Reef Tract is very similar to a portion of one of the five growth phases identified during the Pleistocene history of the Key Largo Limestone reef complex (see **Pleistocene History of the Key Largo Limestone** below), and thus represents a sixth phase of reef growth in the region. In fact, most of the Florida Reef Tract reefs began growing as fringing reefs over the remnants of Pleistocene-age reefs or other limestone.

But even within Holocene time, a complex history of sea level rise and reef growth is evident. For example, drowned Holocene-age shorelines with cemented beach rock have been documented at depths from 196 feet to over 325 feet. And seaward of the modern Florida Reef Tract, especially in the keys southwest of Key Largo, there are several dead reefs in water depths of 23 to 44 feet. These extinct reefs, called *outlier reefs*, are of late Pleistocene to early Holocene age and are considered by some geologists to be part of the Florida Reef Tract. The Holocene-age outlier reefs are 5,000 to 9,000 years old. Similar extinct reef ridges (4,000 to 7,000 years old) exist offshore from Fort Pierce to Miami in water depths of 10 to 150 feet or more (see the **Florida A1A** road log in the **Southern Peninsula** chapter). In addition, many dead elkhorn coral (*Acropora palmata*) ridges, about 8,000 years old, are found in about 50 feet of water throughout much of the Caribbean.

The growth of reefs as far north as Fort Pierce in early Holocene time probably reflects the well-documented global warming interval that occurred between 9,000 and 5,000 years ago. But the existence of so many dead Caribbean and Floridan reefs of Holocene age in relatively deep water is perplexing, because most tropical reef-forming corals have fairly fast rates of growth and should be able to keep pace with normal sea level changes. Geologists believe the reefs must have died abruptly and have proposed two hypotheses to explain why.

First, since reef corals are extremely sensitive to temperature, nutrients, and suspended sediment, change in any of these three factors could instantly harm an otherwise healthy reef. As the Holocene Transgression progressed, the mainland was flooded. Radiocarbon dating of mangrove peats indicates that the initial shelf flooding occurred around 7,000 years ago. Hawk Channel (the lagoon of the modern reef tract), which was then an Everglades-like marsh environment, was flooded by about 6,000 years ago. By 4,000 years ago Florida Bay was flooded, and by 2,000 years ago the modern shoreline was fully established. The erosion of organic-rich soil during flooding could have increased the sediment and nutrient load on the earlier reefs, resulting in their death. One geologist described this process as reefs being "shot in the back by their own lagoon."

A second hypothesis is that the shallow-water reefs were flooded too rapidly to a water depth from which they could not recover. Essentially, they drowned. The rapid rise in sea level resulted from the catastrophic collapse of polar ice as gigantic fragments of continental glaciers broke off (calved) and slowly slid into the sea. Three such events, each resulting in rapid rises of several feet, have been documented in western Atlantic and Caribbean reefs.

Many are concerned about the future of the Florida Reef Tract. Although heroic and successful efforts have been made to conserve the reef system, there remain numerous stressors on the ecosystem.

- **Ecotourism and boat traffic.** The sheer number of snorkelers and scuba divers on the Florida Reef Tract poses a threat, and boat traffic damages sea grass beds, increases turbidity over the reefs, and results in numerous groundings. Several strategies are used to help mitigate potential damage: A mooring buoy system helps prevent anchor damage to the reef tract, and snorkelers are often taken to sites that are sufficiently deep to prevent them from standing on the reef or otherwise coming into physical contact with it.

- **Grounding of large vessels.** The federally protected status of waters around the Florida Reef Tract effectively discourages all but the most wayward of barges and tankers, but groundings still do occur.

- **Pollution.** This comes from a variety of sources, including boat fuels and oil spills (especially near marinas), runoff containing fertilizers and pesticides, septic tank effluent, and shallow injection wells that pump wastewater into the limestone. Daily tidal pumping through the porous limestone brings pollutants to the living reef.

- **Hurricanes.** While tropical storms and hurricanes may be beneficial to the long-term health of the reef, increased frequency or intensity of these storms may delay the recovery of the ecosystem and cause the reef to suffer disproportionately. The 2005 hurricane season was one of the most active and intense on record. Climate scientists inform us that we may experience more frequent or intense storms in the near future.

- **Global climate change.** Scientists think warming ocean water contributes to coral bleaching (see **Florida Land Developer—The Coral Animal** in this chapter). If prolonged, this can result in widespread coral death.

- **Saharan dust.** Scientists have correlated windblown dust from the Sahara Desert in Africa, in conjunction with climate change and other anthropogenic effects, with diseases in the reef tract. The dust may be a source of airborne pathogens that affect coral.

Dense growth of both stony and soft corals (sea fans) behind Christ of the Deep.

has many caves and rocky ledges and is a popular dive spot. Molasses Reef is visited by glass-bottom boats and has outstanding spur-and-groove development with fire coral growing over spurs.

On the outer shelf between French Reef and Molasses Reef, in depths of 115 to 213 feet and next to the Straits of Florida, are patches and mounds of calcium carbonate nodules called *rhodoliths*. Rhodoliths are rounded pebbles or cobbles constructed of concentric layers of coralline red algae (rhodophyte algae) and other skeletal organisms. These rhodoliths also have an abundance of the encrusting foraminiferan *Gypsina vesicularis*. Nodules range in size from small pebbles of 0.375 inch or so to small cobbles up to 3.5 inches in diameter. Similar rhodoliths in the Oligocene-age Bridgeboro Limestone in the eastern

Stony corals and soft corals (including many sea fans) growing over old coral rock at Key Largo Dry Rocks.

Florida Panhandle are associated with reefs that grew along the flanks of the ancient Gulf Trough.

Located along the southeastern end of Key Largo and north of Tavernier is Rodriguez Key, a very small island (and shoal) that is well known to geologists. Geologists refer to this structure as an *inner-shelf carbonate bank* because it is a buildup, or shoal, of calcium carbonate mud and skeletal sand that rises above the seafloor. It is not a true reef, but it is certainly reeflike, with distinct ecological zones. The bank has a narrow rim on the windward side that is constructed of the delicate branching finger coral (*Porites porites*) and the sticklike coralline red algae *Goniolithon strictum*. Most of the bank consists of carbonate mud covered with sea grasses like turtle grass (*Thalassia testudinum*), and calcareous green algae of the genera *Halimeda*, *Penicillus*, and *Acetabularia*. The exposed part of the bank, Rodriguez Key, consists of peat covered with mangroves: black mangroves in the center, and red mangroves along the margins. Carbonate banks of various types are very common in the geological record. Rodriguez Key is a modern example that clearly reveals the nature and origin of these structures.

At mile marker 85.5, near Islamorada, is an absolutely essential stop for every geology enthusiast—Windley Key Fossil Reef Geological State Park. This park preserves three abandoned quarries in the Key Largo Limestone. Windley Quarry has the best rock exposures. In Flagler Quarry the outcrops are a bit overgrown but have higher walls—about 10 feet high. Russell Quarry is largely overgrown. The vertical cuts in the quarry faces and parts of the quarry floors provide outstanding cross sections that show exactly how corals constructed this Pleistocene-age reef. These quarries supplied stone for the construction of Henry Flagler's Overseas Railroad between 1905 and 1912 and were later a source of a decorative building stone locally called *keystone.* An interpretive center explains various aspects of the geology of the Key Largo Limestone reef.

Other modern reefs lie off Plantation, Windley, and Upper and Lower Matecumbe keys. An abundance of large conch shells lies around Conch Reef. Davis Reef has a distinct terrace of dead elkhorn coral at water depths of up to 20 feet. These dead colonies, which drowned, composed a reef crest that was active when sea level was about 20 feet lower than at present. Crocker Reef is known for its 20-foot-high coral walls in the fore-reef zone. Hen and Chickens Reef has many large coral heads of star coral (*Montastrea annularis*). They rise up to 10 feet off the seafloor, but many of the corals on this reef are dead, having suffered 80 to 90 percent mortality due to cold weather during the winter of 1969–70. Like Davis Reef, Alligator Reef also has a distinct terrace of dead elkhorn coral in depths of almost 20 feet. It overlies a layer of peat. Alligator Reef was named for the U.S. Navy ship that wrecked on it after engaging pirates in 1822. The crew blew up the *Alligator* to prevent it from being plundered. Alligator Reef Lighthouse, completed in 1873, marks the reef.

The Florida Reef Tract is a tropical reef ecosystem and nearly the northernmost representative of Caribbean reef systems. The Flower Garden Banks on the continental shelf of Texas comprises lower-diversity coral reefs growing on top of salt domes, and the Florida Middle Ground is also in the northern gulf (see **Central Peninsula** chapter)—both are significantly farther north than the reef tract. Both the Flower Garden Banks and Florida Middle Ground have species of Caribbean aspect but are of lower diversity than fully tropical Caribbean reefs. The most northern of Atlantic Basin reefs grow around the island of Bermuda, at 32 degrees north latitude. These are also lower-diversity reefs of Caribbean species and are only able to extend their range so far north because of the influence of the warm Gulf Stream.

The Florida Reef Tract is bathed by the warm, eastward-flowing Florida Current, which connects the Gulf of Mexico Loop Current to the Gulf Stream, but it is also partially influenced by cooler waters of

The Pleistocene-age Key Largo Limestone at Windley Key Fossil Reef Geological State Park.

Pleistocene-age reef corals of the Key Largo Limestone exposed at Windley Key Fossil Reef Geological State Park.

Pleistocene History of the Key Largo Limestone

The Key Largo Limestone is a coral reef of late Pleistocene age that flourished sometime between 145,000 and 90,000 years ago, during the Sangamonian interglacial period, when sea level was 20 to 26 feet higher than at present. Some geologists refer to the sea of this time as the Pamlico Sea. The Key Largo Limestone runs continuously from under Miami Beach to the Dry Tortugas, a distance of at least 220 miles. In places, it approaches 200 feet in thickness. While it is not a single, contemporaneous reef (there were multiple stages of reef growth during Pleistocene time), it does represent continuous growth in a limited time interval. It was indeed an extensive and luxuriant coral reef system.

The ecological nature of the ancient Key Largo reef presents some interpretive challenges. The long, uninterrupted nature of the limestone's accumulation gives it the appearance of having been a continuous shelf-margin reef—a barrier reef. But the corals in the Key Largo are characteristic of low-energy patch reef environments or the deeper fore-reef zone of a platform margin reef. These corals include star coral (*Montastrea annularis*), by far the dominant species; the brain corals *Diploria strigosa, Diploria clivosa,* and *Diploria labyrinthiformis*; several species of finger corals (genus *Porites*), *Porites astreoides* being the most common; staghorn coral (*Acropora cervicornis*); great star coral (*Montastrea cavernosa*); starlet coral (*Siderastrea radians*); maze coral (*Meandrina meandrites*); and lettuce coral (*Agaricia agaricites*).

What is conspicuously absent in the Key Largo Limestone is the primary reef-crest builder in the Caribbean, elkhorn coral (*Acropora palmata*). Elkhorn coral is very temperature sensitive, so its absence could reflect that the ocean was slightly cooler while this reef was being built. It is also possible that, although it first appeared at the end of Pliocene time, the elkhorn coral was only fully established as a reef-crest builder in Holocene time. Also, elkhorn coral prefers shelf-margin habitats as opposed to inner-shelf settings, where the Key Largo reef may have developed. In any case, it appears the Key Largo reef was not exactly the same type of reef as modern shelf-margin reefs of the Caribbean.

Geologists have interpreted the Key Largo Limestone reef in several ways. Some consider it an outer reef, the equivalent of the fore-reef settings of Florida's modern reefs. These reefs are usually located in water depths of 20 to 50 feet, just below where the elkhorn coral builds reef crests but where star coral (*Montastrea annularis*)—the dominant species in the Key Largo Limestone—is most prolific in today's environments. Other corals, such as brain corals (*Diploria strigosa* and *Diploria labyrinthiformis*), finger corals (*Porites astreoides* and *Porites furcata*), and starlet coral (*Siderastrea siderea*), are also common both in today's fore-reef zone and in the Key Largo Limestone. The presence of other, typically deeper-water corals such as maze coral (*Meandrina meandrites*) and lettuce coral (*Agaricia agaricites*) also supports the outer-reef (fore-reef) hypothesis. The difficulty with the deeper-water interpretation is that the Key Largo Limestone is presently between about 3.5

and 20 feet above sea level. When the Key Largo was growing as a reef, sea level was about 23 feet higher than present. That means the reef would have been forming in shallow water.

Other geologists consider the Key Largo Limestone to represent a series of coalescing patch reefs deposited in water that didn't exceed 10 feet in depth. This view is supported by the present topographic elevation of the Keys and the extent of past sea level as explained above. It is also supported by the lateral stratigraphic relation of the Key Largo with the oolitic Miami Limestone, also a shallow-water deposit; the overall composition of the coral fauna, which includes many back-reef forms; the great abundance of the calcareous green algae *Halimeda*, which is very common in modern back-reef settings; and the rarity of fire coral (genus *Millepora*) and coralline red algae, both of which are common components of the modern reef-crest and fore-reef environments but rare in back-reef settings.

Some geologists agree that the Key Largo reef was a shallow-water reef, but they argue that its geographic and stratigraphic location clearly indicate that it was not a patch reef complex but a shelf-margin reef—essentially a barrier reef. The absence of the environmentally sensitive elkhorn coral, which might be expected to live in such a reef setting, was due to direct contact with the slightly cooler and nutrient-rich waters of the Gulf of Mexico. In the Florida Reef Tract today, elkhorn coral grows best where it is shielded from gulf waters by the Florida Keys. Where elkhorn coral is rare or absent (as in the Dry Tortugas), it is not uncommon for star coral, the dominant reef constructor of the Key Largo Limestone, to dominate the reef-crest environment.

the southeastern Gulf of Mexico, particularly during winter months. So while reefs are located across the entire reef tract, they are best developed seaward of the larger islands, such as the northern Elliott Key and Key Largo, and the Lower Keys from Big Pine Key to Key West, which block the cooler gulf waters. From Plantation to Vaca Key, there is generally less reef development due to the increased influence of gulf waters, but hardy soft corals, such as fire coral, and massive head corals are more common.

On the Gulf of Mexico side of the Keys is triangular Florida Bay, wedged between the southern peninsula and the northern keys. This unique carbonate-producing environment is composed of calcium carbonate mud mounds or banks that form an interconnected network of shoals. The mounds are only about 6.5 feet high and have gentle slopes (less than 1 degree). Where they break the ocean's surface, they form small muddy islands with mangroves and other vegetation. There are presently 237 such mud islands in the bay. Sea grass beds are also common in the bay.

Geologists have long debated the origin of the creamy gray mud, but the fine sediment appears to be entirely biogenic—formed by the decomposition of skeletal calcium carbonate from corals, mollusks, and especially the tiny needles of aragonite crystals precipitated by green algae of the genus *Penicillus.* Freshwater peat underlies many areas of the bay, indicating there were Everglades-like conditions prior to the Holocene Transgression. The entire bay is underlain by the karstic Miami Limestone coated with a calcrete crust, or duricrust, indicating the bay was once above sea level. The Florida Bay mud banks have been of enormous interest to many geologists. Mud banks are very common in the geological record, especially in strata from Paleozoic time (542 to 251 million years ago), but there are few modern examples. Study of the Florida Bay mud mounds has elucidated the origins of these ancient structures.

A mud mound in Florida Bay.

Florida Land Developer—The Coral Animal

The coral animal is fascinating. Corals belong to the phylum Cnidaria, which includes sea anemones, soft corals (sea fans and sea whips), jellyfish, hydroids (like fire coral), and many extinct groups. Corals belonging to the order Scleractinia (or Madreporaria) are known as stony corals because they secrete a hard calcium carbonate skeleton in which their polyps are housed. An individual calcareous cup—the skeleton—is called a *calyx*. Stony corals may be solitary or colonial, but most reef-forming corals are colonial, consisting of numerous individuals attached like conjoined twins. Each coral polyp, which resembles a sea anemone, is armed with tentacles full of stinging cells called *nematocysts* or *cnidoblasts*, which work like microscopic harpoons to capture prey, mostly small invertebrates. Corals are generally nocturnal feeders. They reproduce both sexually and asexually (by budding). Coral growth rates are highly variable, ranging from 1 centimeter per year for massive colonies, such as star coral (*Montastrea annularis*), to 10 centimeters per year for long-branched species, such as staghorn coral (*Acropora cervicornis*).

Most reef-forming corals are uniquely adapted to living in tropical, well-illuminated waters (the photic zone) and cannot grow if winter water temperatures drop below about 65 degrees Fahrenheit. Most reef-forming corals also have a unique symbiotic relationship with a group of microscopic, single-celled dinoflagellate algae called *zooxanthellae*, which live within a layer of cells located beneath the surface layer of cells of the corals. Corals that host these algae are called *zooxanthellate corals*, or *z-corals*. This coral-algal symbiosis is a mutualistic relationship that benefits both species. When zooxanthellae perform photosynthesis they remove carbon dioxide and generate oxygen. This facilitates the coral's precipitation of calcium carbonate. The algae also provide photosynthates (fats and sugars produced by photosynthesis) that the coral consumes. Some studies indicate that 90 percent or more of coral nutriment comes from photosynthates. The algae, in turn, benefit from the protection the coral provides. Zooxanthellae are "naked," having no shells as their free-living cousins do. Corals also provide carbon dioxide, nitrogen, and phosphorus—important nutrients for the zooxanthellae.

Coral-algal symbiosis is an important and successful strategy in nutrient-poor tropical waters, where most shallow-water reefs grow, and most reef-forming stony corals are z-corals. Some mollusks, sponges, and foraminifera enjoy similar symbiotic relationships with algae. This symbiosis is a crucially important adaptation for reef corals. It is a strange paradox that most tropical coral reefs actually thrive in waters that have very low concentrations of dissolved chemical nutrients, such as nitrate and phosphorus, that many types of marine life need. Such nutrient-poor waters account for the crystal clarity of ocean waters around reefs. It is the coral-algal symbiotic relationship that allows corals to use and recycle the limited nutrients that are available. If the amount of nutrients over a reef increases due to runoff or pollution, other marine species, such as fast-growing sponges and fleshy algae, may overgrow and kill the reef.

Besides runoff and pollution, various other maladies have threatened Florida corals in recent years, including black band disease, a bacterial infection that rapidly destroys coral tissue. Another concern is coral bleaching, which occurs when the water gets excessively warm (over 90 degrees Fahrenheit). This increases the photosynthetic activity of zooxanthellae, which raises the oxygen content in corals to toxic levels. Corals then eliminate some of their symbiotic algae by shedding thick streams of mucus. Coral bleaching may also be caused by stress induced by tropical storms. If prolonged, bleaching may cause corals to become susceptible to various harmful infections. If a reef is weakened by disease, it may succumb to bioerosion, wherein various sponges, worms, mollusks, and fish actively bore into or even consume the reef. If growth rates do not exceed normal bioerosion rates, the reef will rapidly deteriorate.

It is often said that Caribbean reefs are not as diverse as Indo-Pacific reefs, but this is only partly true. The Indo-Pacific represents a much larger area than the Caribbean. If compared to an area of comparable size, Caribbean

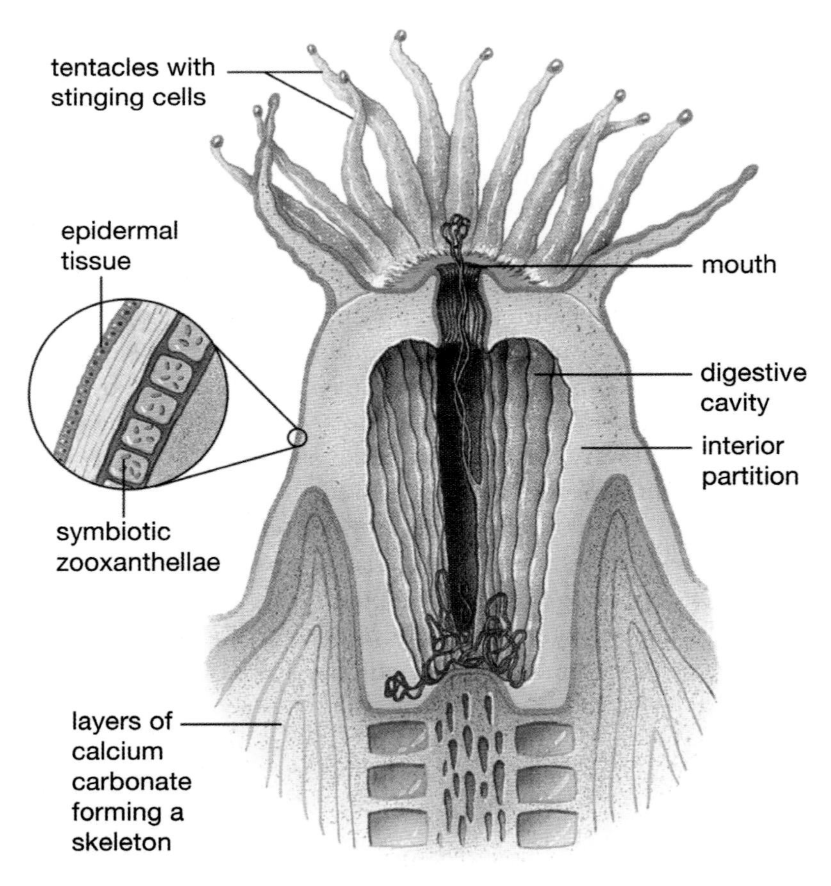

Anatomy of a coral polyp. —From *Oceanography, An Invitation to Marine Science* (with ThomsonNOWâ? ¢ Printed Access Card) 6th edition by GARRISON. 2007. Reprinted with permission of Brooks/Cole, a division of Thomson Learning: www.thomsonrights.com

reefs may contain just as many corals as their Pacific cousins. The modern Florida Reef Tract, located at the northern end of the Caribbean reef environment, contains nearly fifty species of stony corals, representing approximately 80 percent of the entire Caribbean reef coral fauna.

SOME COMMON CORAL SPECIES OF FLORIDA'S REEFS	
Elkhorn coral (*Acropora palmata*)	Found only on reef crests.
Staghorn coral (*Acropora cervicornis*)	Abundant in back-reef areas.
Star coral (*Montastrea annularis*)	A major reef constructor, and a dominant fossil in the Key Largo Limestone.
Brain corals (genus *Diploria*)	Major reef formers.
Starlet coral (*Siderastrea siderea*)	A hardy coral common in patch reefs.
Finger corals (genus *Porites*)	Common in quiet water in back-reef areas.
Lettuce, leaf, or plate corals (genus *Agaricia*)	Common in the deeper fore-reef zone.
Fire coral (*Millepora complanata*)	Common on reef crests and in shallow-water fore-reef zones; the name "fire coral" is derived from the sharp sting it may deliver to divers who swim too close. *Millepora alcicornis* is also widespread across the Florida Reef Tract. Fire corals are not actually scleractinian (stony) corals but belong to the related class Hydrozoa; they differ from true stony corals in many anatomical details.

Other Limestone Producers of the Florida Reef Tract

Most limestones are entirely organic in origin. The sediments from which they form are composed of the minerals aragonite or calcite, both of which are calcium carbonate but in different crystal forms. These minerals compose the skeletal remains of marine invertebrates or calcareous algae. Like stony corals, many marine invertebrate animals produce a calcareous shell. These invertebrates include foraminifera; polychaete worms, which produce calcareous worm tubes; mollusks (bivalves, gastropods, and tusk shells); barnacles; bryozoans (moss animals); and various echinoderms, including echinoids (urchins and sand dollars), holothurians (sea cucumbers), which produce tiny calcareous fragments called *sclerites*, and starfish, which produce small but thick calcareous fragments called *ossicles*. All of these animals produce enormous amounts of shell debris, but perhaps the most abundant producers of organic calcium carbonate are several types of bottom-dwelling algae that precipitate plates or crusts. Called *calcareous algae*, they belong to two major groups: green algae and red algae.

Among the green algae (chlorophytes) are several forms, including those of the genera *Penicillus*, *Rhizocephalus*, *Udotea*, and *Halimeda*. In Florida Bay, the tiny, fibrous needles of the merman's shaving brush (*Penicillus*) can be so abundant that they create a creamy, gray lime mud. And plates of *Halimeda* constitute 100 percent of the sediment in some areas of the Florida Reef Tract. Among the red algae (rhodophytes), also called *coralline algae* because of their superficial resemblance to corals, are the ubiquitous, sticklike species of the genus *Goniolithon*. Some coralline algae, such as those of the genera *Lithothamnium*, *Lithophyllum*, and *Melobesia*, also construct rhodoliths in the deep fore-reef zone. The coralline red algae are important reef constructors. Their crusty growth over coral and other skeletal debris helps bind the reef into a solid limestone structure.

Limestone producers on the Florida Reef Tract. The green algae Penicillus *(left) and* Halimeda *(upper middle); red alga* Goniolithon *(upper right); molluscan shell (bottom middle) and sea urchin test (bottom right).*

Where's the Reef? Reefs of Florida—Past and Present

As a land that has been submerged (completely or partially) beneath tropical and subtropical seas for nearly all of its post-Pangaean history, Florida has hosted many different types of reefs. Geologists usually define a reef as an organically built, rigid structure that rises above the surrounding seafloor. By this definition, there are many reef-forming marine organisms besides corals, including certain bacteria and algae, sponges, bryozoans, tube-building worms, oysters, and other mollusks. Only a small portion of Floridan strata can be directly attributed to reef growth. Most reefs are actually quite narrow and limited in extent. But reefs have an enormous influence on surrounding marine environments. Much of the limestone in Florida is clearly associated with past reef environments.

REEF DEPOSITS OF FLORIDA

Age	Formation and Location
Jurassic	Sponge-algal mounds in the Smackover Limestone in the western panhandle subsurface
Cretaceous	Rudist bivalve mounds in the Sunniland and Lawson formations of the peninsula
Paleocene	Reportedly, reef deposits in the Rebecca Shoals Dolomite of the peninsula and deposits of coralline red algae in the Salt Mountain Limestone of northern Florida
Eocene	Very small patch coral reefs in the Ocala Limestone, and probably shelf-edge reef buildups along the southern margin of the Florida Platform
Oligocene	Patch coral reefs and coral thickets common in the Bumpnose (subsurface), Suwannee, and Bridgeboro limestones and the Tampa Member of the Arcadia Formation
Miocene	Patch coral reefs and coral thickets in the Parachucla, Chattahoochee, St. Marks, and Chipola formations; offshore deposits on the West Florida Shelf
Pliocene	Coral reefs in the Tamiami Formation in the southern peninsula
Pleistocene	Coral reefs in the Key Largo Limestone, outlier reefs seaward of the modern reef tract, deepwater coral mounds on the West Florida Shelf, and various coral ridges on the outer continental shelf
Holocene	Drowned coral ridges and outlier reefs (early Holocene), the Florida Middle Ground, oyster reefs, worm reefs, vermetid gastropod ("worm rock") reefs, banks of *Oculina* coral and other deep coral banks, deepwater mounds (lithoherms) composed of many different invertebrates in the Straits of Florida, and the Florida Reef Tract

Between mile markers 81 and 82 is the Florida Keys Memorial, or Hurricane Monument, commemorating the hundreds of World War I veterans who perished during the category 5 Labor Day Hurricane of September 2, 1935. Islamorada, on Upper Matecumbe Key, experienced sustained winds of 200 miles per hour and a barometer reading of 26.35 inches for many hours, and the storm surge was between 15 and 20 feet. It was the most intense hurricane ever to affect the United States. Most of the buildings on Upper Matecumbe Key were destroyed. After the storm, 423 bodies were found, most of which were cremated. The veterans had been camped on Lower Matecumbe and Windley keys while they worked on US 1 for the Works Progress Administration. The train that was sent to rescue them was too late, and was itself blown off the tracks. In 1937, about three hundred cremated remains were placed in the crypt in front of the monument. The monument is constructed of the beautiful Key Largo Limestone, also called the *keystone rock.*

Lignumvitae Key Botanical State Park, accessible by boat, is one of the few places where the original tropical island vegetation of the Keys remains unspoiled. The island has a unique flora, a tropical hardwood hammock, with a strong West Indian influence. Some of the trees include West Indian mahogany; lignumvitae, also called the tree of life; Jamaica dogwood, also called Florida fishfuddle tree, which Caribbean natives used to make a fish poison that stupefied fish; gumbo-limbo, also called West Indian birch or the tourist tree (its peeling bark resembles the peeling skin of a sunburn); Florida poisontree, also called hog gum; mastic, also called jungleplum or wild olive; pigeon plum, also called pigeon seagrape; and Florida strangler fig, also called golden fig. There was a time when all of the Keys hosted this type of tropical vegetation, but it has been almost entirely obliterated since the late 1800s. Daily charters to Lignumvitae Key are available from local marinas at Mile Marker 78.5 on Upper Matecumbe Key, and tours are regularly given on the island.

In 1733 a powerful hurricane struck an unsuspecting Spanish flotilla of twenty-one galleons returning to Spain. The storm drove the ships across the Florida Reef Tract, sinking them all. Their remains are scattered offshore between Key Largo and Key Vaca. In most cases, all that remains of the ships are piles of ballast stones, but many artifacts are still retrieved by divers. One of the wreck sites, that of the Dutch-built *San Pedro,* is located about 1.25 nautical miles south of Indian Key, in 18 feet of water. It is protected in the San Pedro Underwater Archaeological Preserve State Park.

At mile marker 69, on Long Key, an easily accessible, abandoned quarry on the north side of US 1 exposes 2 to 3 feet of Key Largo Limestone and its fossilized coral boulders. It is about 2.2 miles south of the highest

Fossil coral of the Pleistocene-age Key Largo Limestone at an abandoned quarry at mile marker 69.

point of the Channel Five Bridge, and there is a small parking area. Other exposures of the Key Largo Limestone can be found along the northern end of Long Key. Much of the island is covered by a thin layer of calcium carbonate sand. Tennessee Reef is located offshore.

Throughout most of the Keys there are numerous rest stops and wayside areas along US 1 inviting you to stop and wade in the shallow water. You can see plenty of tropical marine life and limestone production in progress. Lime mudflats are well exposed at low tide, and there are some rocky limestone bottoms as well. The dark patches in the water that are so visible from the highway are sea grass beds. Offshore of Vaca Key is Sombrero Key Reef, another favored scuba diving location. The reef is marked by the tallest of the reef lights, Sombrero Key Lighthouse, completed in 1858 and standing 142 feet above sea level.

On the seaward (southern) side of Seven Mile Bridge are some extensive mudflats. At mile marker 37 is Bahia Honda State Park. Key Largo Limestone is exposed here, but unlike many of the rocky keys, Bahia Honda has a carbonate sand beach with coastal berm and even carbonate dune deposits, as on quartz-sand beaches. These features are visible along the park's Silver Palm Nature Trail. Bahia Honda is a favored snorkeling locality. The old Bahia Honda Bridge is in the park, a rusting reminder of Henry Flagler's Overseas Railroad. The southernmost exposure of the Key Largo Limestone is to the southwest, at Newfound Harbor Keys.

US 1 (Overseas Highway)

Lower Keys (Big Pine Key to Key West)

29 miles

The surface rock across the Lower Keys is the Miami Limestone, which is up to 35 feet thick in places. The Miami overlies the Key Largo Limestone, which extends throughout the Lower Keys not far below the surface. The obvious change in the shape and orientation of the Lower Keys compared with the Upper Keys has everything to do with the nature of the depositional environment in which the Miami Limestone was laid down. It was deposited as an ooid shoal (see the discussion of the Miami Limestone in the **Florida A1A** road log in the **Southern Peninsula** chapter). The Lower Keys are exposed Pleistocene-age ooid sandbars that are now cemented into a soft limestone. The islands were once strikingly similar to the modern ooid shoals and tidal channels that are common on the Great Bahama Bank. The northwest-southeast orientation of the Lower Keys primarily reflects the tidal current flow of Pleistocene time, but it also reflects modern tidal flow.

The reef systems of Pleistocene time had oolitic bars associated with them, but the modern reef tract in southern Florida does not. The lack of ooid development on the south Florida continental shelf is almost certainly related to the relatively low modern sea level and the narrowness of the modern shelf. If seas were higher and the tropical continental shelf was broader, as it is on the modern Bahama Banks and was during Pleistocene time in southern Florida, ooid sandbars would almost certainly be forming today.

On Big Pine Key near mile marker 30 is Blue Hole, an abandoned and flooded limestone quarry within the National Key Deer Wildlife Refuge. Depending on water level, 3 to 6 feet of the oolitic Miami Limestone is exposed and accessible. This is the best exposure of the Miami Limestone in the Lower Keys. Turn north onto County Road 940 (Key Deer Boulevard) and proceed 2.8 miles to Blue Hole. The contact between the Miami and Key Largo limestones was once exposed on Big Pine Key but is now covered by vegetation.

The National Key Deer Wildlife Refuge, which covers a portion of the Lower Keys, is home to the tiny, endangered key deer (*Odocoileus virginianus clavium*), a subspecies of the Virginia white-tailed deer. Key deer are the smallest subspecies of the white-tailed deer, only about the size of a German shepherd dog. (Some consider the key deer a distinct species.) They dwindled to as few as twenty-five individuals in the 1950s but have rebounded to approximately three hundred today thanks to the establishment of the refuge. An interesting biogeographic phenomenon, called *isolation*, occurred in the Lower Keys during the Holocene Transgression. As sea level rose at the beginning of Holocene time and flooded the south Florida continental shelf, the deer of the Lower Keys were isolated from the mainland. The isolation of a subpopulation from a larger stock is a common way new species evolve. The key deer are unique to the Florida Keys.

Looe Key Reef, named for the British frigate HMS *Looe*, which sank on this reef in 1744, lies within the Looe Key National Marine Sanctuary, which is off Big Pine, Ramrod, Summerland, and Cudjoe keys. The reef has exceptionally well-developed spur-and-groove structures. The spurs were originally constructed of elkhorn coral but are now primarily encrusted with fire coral. American Shoal Lighthouse is nearby. Sambo Reefs and several smaller reefs and shoals are offshore of Key West. The best reef development in the Lower Keys is between Looe Key Reef and the Sambo Reefs. Sand Key Lighthouse is situated about 8 miles southwest of Key West, next to an isolated shoal and island. Reefs at Sand Key have well-developed spur-and-groove structures covered with fire coral.

Marquesas Keys and Dry Tortugas

Between Key West and the Dry Tortugas is an area called the Quicksands, a thick (up to 40 feet) carbonate sand deposit that is constantly shifting because of strong currents, which generate giant sand waves that can be as much as 10 feet high. The sand is largely composed of thin plates of calcareous green algae of the genus *Halimeda*. These shifting, coarse carbonate sands can quickly bury shipwrecks.

The Marquesas Keys, located about 21 nautical miles west of Key West, are composed of about ten carbonate sand islands and spits arranged in a circular fashion, with mangroves growing on them. The carbonate sand is also composed of *Halimeda* plates. Numerous shoals and reefs surround the Marquesas, and north and northwest of the islands, between Ellis Rock and New Ground Shoal, are many linear patch reefs. About 6 miles due south of Marquesas are the Marquesas Reef Line and Cosgrove Shoal. West of the Marquesas Keys lie Halfmoon and Rebecca shoals, also composed of *Halimeda* sands, and about 4

*Conchs and other marine life are easily observed by snorkeling around Fort Jefferson. A living horse conch (*Pleuroploca gigantea*) is held on the left, and two juvenile queen conchs (*Strombus gigas*) on the right.*

miles west is the site where the famous treasure-laden Spanish galleon *Atocha* sank during a hurricane in 1622.

Dry Tortugas National Park, located 80 nautical miles west of Key West, is about the end of the line for the above-water tour of Florida geology. The islands were named "Tortugas" in 1513 by Ponce de Leon because of "the great amount of turtles which there do breed." The islands are "dry" because there is no freshwater to be found there. The Dry Tortugas are underlain by the Miami Limestone but covered with Holocene-age carbonate sediments at the surface. This is a great snorkeling location! One can easily observe coral reef fauna and flora growing in the water surrounding the islands and along the red brick walls around Fort Jefferson. Several charters leave from Key West daily, and camping is available on Garden Key.

Historic Fort Jefferson, Gibraltar of the Gulf, is on Garden Key. Construction of this fortress began in 1846 and continued sporadically for thirty years. It suffered many structural problems because the foundation was on a sand shoal, not coral rock. The fort was abandoned in 1874. There are numerous shipwrecks in the area. The fort walls are 8 feet thick and stand 45 feet high. They are not solid brick. The interiors of the walls are filled with coral debris taken from the surrounding shoals.

The interior of the walls of Fort Jefferson is filled with coral sand and rubble.

The reefs of the Dry Tortugas and the entire Florida Reef Tract have been the focus of much historic and ongoing geological research. The famous Swiss-American geologist and paleontologist Louis Agassiz conducted the first scientific survey of the Florida reefs in 1851. Agassiz was part of a crew commissioned to survey the area for the purpose of establishing several lighthouses along the coast of the newly acquired state of Florida in an effort to prevent shipwrecks along the reef tract. The reefs of the Dry Tortugas were later surveyed and mapped in great detail in 1881 by Louis's son, Alexander Agassiz. In 1904, the Carnegie Institution of Washington established a marine laboratory—the Tortugas Laboratory—on Loggerhead Key, which is 2.5 miles west of Garden Key. Much of the classic work on coral reef geology and biology was conducted here by such notable coral researchers as T. W. Vaughan, J. W. Wells, and many others. The lab burned in 1939.

The 151-foot-tall Dry Tortugas Lighthouse, completed in 1858 and repaired after a hurricane in 1873, is also on Loggerhead Key, which is surrounded by patch reefs. Loggerhead Key is known for extensive exposures of beach rock, a hardened coastal limestone that forms when carbonate beach sands are cemented at the surface without being buried.

T. W. Vaughan called the Dry Tortugas an atoll, a circular coral reef. The area consists of a discontinuous rim of coral and small coral sand islands and could properly be classified as a carbonate bank, but there is no official agreement on this. There are many well-developed patch reefs around the Tortugas, and they do form an atoll-like rim. Some of the reefs are up to 55 feet thick. Both Loggerhead and Garden keys are coral sand islands hosting a wide variety

Bush Key, connected to Garden Key by a carbonate sand ridge called a tombolo.

of environments, including bare sand and rubble, sea grass meadows, fleshy algae meadows, hardbottom encrusted primarily by soft corals, patch reefs dominated by star coral (*Montastrea annularis*), and staghorn coral (*Acropora cervicornis*) reefs.

About 50 miles west of the Dry Tortugas is a very important oceanographic site. Pulley Ridge Reef is roughly 30 miles by 15 miles and grows in water depths of 200 to 295 feet. It is one of the deepest photic-zone coral reefs known, and up to 60 percent of its surface is constructed of agariciid corals, such as those of the genera *Agaricia* and *Leptoseris*, which have a symbiotic relationship with zooxanthellae algae. Popularly known as "plate corals" because of their flat, dinner-plate-like form, agariciids are common in the deeper photic zone. Colonies on Pulley Ridge reach 1.6 feet or more in diameter. Several other coral species are present, as are many species of macroalgae, or seaweed. The occurrence of seaweeds here is unusual, as they normally thrive in much shallower water. Nonetheless, they are thriving in this environment where the intensity of the light is only 2 percent of that at the surface. Geologists think that nearby Howell Hook, in depths of 490 to 625 feet, is the remains of a similar reef system of Pliocene and Pleistocene age.

Of special geological interest is the fact that Pulley Ridge Reef developed over a north-south-trending barrier island complex. This complex formed during the Wisconsinan glacial period, when continental glaciers in North America achieved their maximum size. During this period, between 25,000 and 18,000 years ago, sea level was much lower, near the edge of the modern continental shelf. These barrier islands were completely flooded during the Holocene Transgression as the ice sheets melted, from 18,000 to 6,000 years ago. It is not uncommon for coral reefs to grow on top of seafloor bumps of various geologic origins, such as barrier islands, rock ridges, earlier reefs, or salt domes.

West of the Dry Tortugas and Pulley Ridge, an important oceanographic change occurs along the margin of the Florida Platform. Here, the continental shelf edge curves to the north into the Gulf of Mexico, where instead of being a shallow, flat-topped, rimmed shelf margin, as along the Florida Reef Tract, it becomes what is called a *carbonate ramp*—a broad continental shelf with a gradual slope down to its edge. Carbonate ramps are commonly associated with nontropical continental shelves. The West Florida Shelf slopes about 1 to 2 degrees to a depth of over 3,000 feet, where it drops off precipitously to the deep ocean floor. This deepwater ledge, called the West Florida Escarpment, is constructed of Cretaceous-age limestone and has been deeply eroded by slumping (submarine "landslides"). Many strange deep-sea creatures live on the outer West Florida Shelf and beyond. One of the most unusual, and prehistoric in appearance, is the giant deep-sea isopod *Bathynomus giganteus*, affectionately known as the sea roach. These creatures grow up to 1.5 feet in length and generally inhabit depths of 1,000 feet or more. They resemble the terrestrial pill bug, or roly-poly bug.

There is one final oceanographic phenomenon in Florida that should be mentioned because of its great geologic importance. At several locations along the base of the West Florida Escarpment, in water depths of 1.5 to 2 miles,

some very unusual associations of deep-sea marine life occur. Called *brine seep communities* or *chemosynthetic communities*, they consist of interesting clams, worms, and other invertebrates. These creatures thrive on briny waters rich in hydrogen sulfide that seep out of the Cretaceous-age limestone at the foot of the escarpment. These creatures do not depend on direct sunlight for food. Instead, the base of the food chain consists of bacteria that oxidize hydrogen sulfide and use the chemical energy released by that oxidation to synthesize food. Scientists consider these and similar deep-sea chemosynthetic communities some of the most ancient of marine ecosystems.

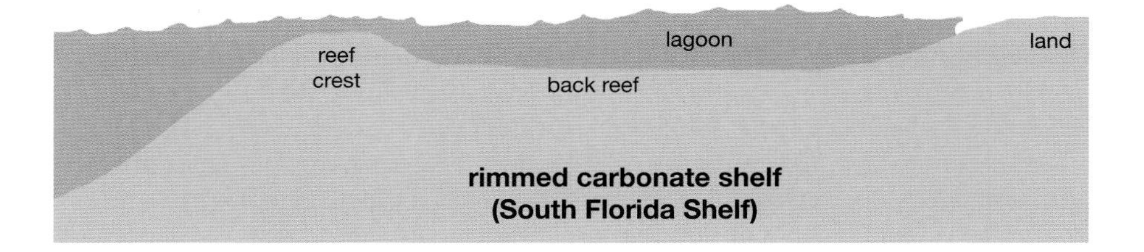

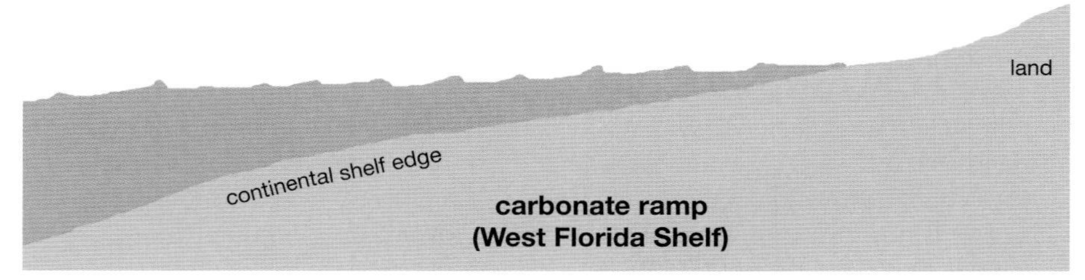

Rimmed carbonate shelf and carbonate ramp profiles. —Illustration by Amanda Kosche

*A denizen of the deep Gulf of Mexico—the sea roach (*Bathynomus giganteus*). Specimen is 9 inches long.*

Glossary

agate. A banded variety of chalcedony. Agate typically grows as layers of finely crystalline quartz along the interior of hollow fossils or other rock cavities, forming geodes.

algae. See **calcareous algae**.

alluvial stream. A stream primarily supplied by surface water. Alluvial streams typically have many tributaries.

aquifer. A large body of porous and permeable rock or sediment that is saturated with groundwater.

aragonite. A calcium carbonate mineral. In Florida, it most commonly occurs as the mineral composing the skeletons of corals and calcareous algae and the shells of most mollusks.

Archaic Period. An archaeological time period from about 8,000 to 1,000 BC, characterized by various complex artifacts and homesites. Large shell middens from this period are common, and clay pottery appeared in the homesites about 2,000 BC in Florida.

archipelago. A group or chain of many islands, such as Ten Thousand Islands in southwestern Florida or the Florida Keys.

artesian spring. A spring that flows under pressure rather than merely seeping out at the surface. The hydraulic pressure is created naturally as groundwater becomes confined beneath or between impermeable rock or sediment, and when the area that recharges the spring is higher than the spring itself.

Atlantic Coastal Ridge. A long, narrow complex of raised (above-water) Pleistocene- and Holocene-age dunes, berms, beaches, coquina rock, and oolitic limestone that extends from Amelia Island to Biscayne Bay along the east coast of Florida.

Bahama Banks. The complex of flat-topped carbonate platforms that compose the Bahama Islands and the shallow marine environments around them. Today, the Bahama Banks actively produce carbonate sediment and rock, including coral reefs, ooid bars, and the skeletal debris of marine invertebrates.

bar. A sediment deposit within a stream channel, usually consisting of coarse sand and gravel. It may also refer to a sandbar in shallow marine water.

bar-built estuary. An estuary enclosed by offshore barrier islands or spits. Apalachicola Bay is a good example.

barrier island. A sandy, usually elongated coastal island that parallels the shore and creates a protected bay between the island and the mainland. Barrier islands may form in a variety of ways, but many in Florida are created by the growth and modification of a spit, or from the shoaling of an offshore sandbar, also called a *longshore bar*. See also **perched barrier** and **drumstick barrier**.

basement (or basement rock). The deepest and oldest continental rock, which is usually buried beneath thick, younger sedimentary strata. Florida bedrock is Precambrian, Paleozoic, Triassic, and Early Jurassic in age.

bayhead delta. A river delta that enters an estuary. Also called an *estuarine delta.*

beach. A buildup of unconsolidated sand on an ocean shore or lake shore.

beach face. The inclined ramp of the berm over which breakers rush up onto the shore.

beach nourishment. The addition of dredged nearshore sand or trucking of river sand (or other beach sand) to a beach to restore what has eroded.

beach ridge. A relict coastal landform that parallels the shoreline. Normally occurring as stranded berms, they can also be composed of lines of beach dunes. A plain with many stranded beach ridges is called a *beach ridge plain* or *strand plain.* These form on shorelines that experience successive episodes of ridge growth.

beach rock. A crusty rock found on some tropical carbonate beaches that is calcareous sand and other materials, such as shell debris and quartz sand, cemented together.

berm. A mound of sand deposited by breaking waves. The berm is usually steeply inclined toward the water. Also called a *beach ridge.* Beaches may have both high and low tide berms, and perhaps a winter berm. See also **beach face**.

biostrome. A layered accumulation of fossil shell debris. It is similar to a reef, but in life the organisms did not build a rigid structure or one that rises above the surrounding seafloor. Many oyster beds are good examples.

blackwater stream. A stream primarily supplied by surface water and groundwater seepage from swamps and freshwater wetlands. The seepage water is rich in dissolved organic materials and tannic acid, creating water the color of iced tea.

Bone Valley. A name coined by Florida miners and fossil collectors for the Central Florida Phosphate District, which occurs in portions of Polk, Hillsborough, Manatee, Hardee, and DeSoto counties. Rich deposits of Miocene- and Pliocene-age marine and terrestrial mammals have been collected here and continue to be uncovered by mining operations.

boulder. A sediment particle greater than 256 millimeters (about 10 inches) in diameter.

brachiopod. A marine invertebrate that has a bilaterally symmetrical shell with two halves. They are commonly called *lampshells.* Although they superficially resemble clams, their anatomy and shell structure are entirely different.

bryozoan. A colonial marine animal, sometimes called a "moss animal." Most secrete a calcareous skeleton.

calcareous. Composed of calcium carbonate.

calcareous algae. Primarily a group of tropical, bottom-dwelling green (chlorophyte) algae that precipitate plates or needles of the mineral aragonite. These calcareous parts form, and have formed, enormous volumes of limestone-producing sediment. This term is also applied to coralline red algae.

calcite. A calcium carbonate mineral. It composes much of Florida's limestone. Oyster and scallop shells are also composed of calcite.

calcium carbonate. A solid compound that comprises the minerals calcite and aragonite, and forms the rocks limestone and marble.

calcrete. A thinly layered limestone that forms beneath a soil zone that rests on top of limestone. The deposit forms as water containing calcium carbonate moves upward. The water evaporates beneath the soil horizon and deposits the calcium carbonate, often integrating organic matter into the deposit. The deposit is often referred to as a *duricrust.*

carbonate. Refers to sediments or sedimentary rock composed of the minerals calcite or aragonite (calcium carbonate) or dolomite (calcium magnesium carbonate). Limestone and dolostone are the most abundant carbonate rocks.

carbonate bank. A buildup (also called a *shoal*) composed of calcium carbonate sediment (mud, skeletal particles, or other larger debris) that marine organisms produced.

carbonate platform. Large calcium carbonate–producing continental shelves, including carbonate ramps, rimmed shelves, and others.

carbonate ramp. A broad continental shelf with a gradual slope down to the edge of the shelf, primarily composed of calcareous sediment. The West Florida Shelf in the Gulf of Mexico is a carbonate ramp. Carbonate ramps are most commonly associated with nontropical continental shelves that lack coral reefs. See also **rimmed shelf**.

cast. A fossil consisting of the sediment filling of an external mold. See also **mold**.

Central American Isthmus. The narrow land connection between North and South America. It rose tectonically in Pliocene time, between 5 and 1.8 million years ago, resulting in restricted ocean circulation between the Caribbean Sea and the Pacific and the intensification of Gulf Stream flow in the Atlantic. It also allowed land animals to migrate between the two American continents. Also known as the *Isthmus of Panama* or the *Panamanian Land Bridge.*

Central Highlands. A series of long ridges in the middle of the Florida Peninsula that parallel it from Georgia to near Lake Okeechobee. Examples include Lake Wales Ridge, Brooksville Ridge, and Trail Ridge.

chalcedony. A very-fine-grained variety of the mineral quartz. It is more commonly known by its varietal names, such as *chert, flint,* and *jasper.*

chert. Generally a brown variety of chalcedony, but color may vary.

clay. A sediment particle less than 0.004 millimeter in diameter.

coastal plain. A broad, tectonically undisturbed wedge of sediments and sedimentary rock of Jurassic to Holocene age (the last 200 million years) that extends from continental bedrock or a mountain front to the edge of the continental shelf. Florida is part of both the Gulf of Mexico and Atlantic coastal plains.

cobble. A sediment particle between 64 and 256 millimeters (2.5 and 10 inches) in diameter.

Cody Scarp. In part, the escarpment of the Pamlico Terrace and approximate shoreline between 150,000 and 100,000 years ago. The base, or toe, of the scarp actually varies between 50 and 125 feet of elevation and reflects not only sea level change but also erosional (karstic) modification.

continental shelf. The flooded margin of a continent and the shallowest part of the ocean floor. The continental shelf is underlain by granitic bedrock and represents the true geological edge of a continent.

coral. A general term for a class of animals (Anthozoa) that includes soft corals (most polyp animals, such as sea anemones, sea fans, sea whips), which lack a hard skeleton, and stony corals, which make a hard, calcium carbonate skeleton in which the polyp lives. "Coral" is also used for the fire corals, which have a hard skeleton but actually belong to another group of polyp animals called *milleporids* (of the class Hydrozoa).

coral head. A large, usually rounded colony of stony corals such as a brain corals or star corals.

coralline algae. A group of bottom-dwelling red (rhodophyte) algae that precipitate thin crusts or sticks of the mineral aragonite. They are major reef-constructing organisms (but also grow in cool waters) and can produce great volumes of limestone-forming sediment. Their hard crusts can superficially resemble rock made by stony corals. Coralline algae commonly form nodules or algal balls called *rhodoliths.*

cover collapse sinkhole. A sinkhole that forms rapidly (in minutes to hours) as surface sediments fall into an underlying cavity that can no longer support the overburden. These are the largest of sinkholes.

cover subsidence sinkhole. A sinkhole that develops as loose, permeable sediment, usually quartz sand, gradually sinks, hourglass style, into an underlying, enlarging cavity within limestone. Large groupings of subsidence sinkholes create a landscape called *sand hill karst.*

crinoid. A class of echinoderm animals (which includes sea urchins and starfish) consisting of a crown (a calcareous cup with food-gathering arms) attached to a stalk, which attaches to the seafloor. Also called *sea lilies.*

cross-stratification (or **cross-bedding**). Thinly bedded, inclined layers of sediment that are deposited by currents or wind. Cross-strata are common in nearshore sands (such as longshore bars), beaches, river channels, and sand dunes. See also **trough cross-stratification**.

cuspate beach. A beach with regularly spaced sandy points, or cusps. These cusps occur along a single beach line and should not be confused with the much larger, pointed coastal form called a *cuspate foreland.*

cuspate foreland. A large-scale pointed coastline, such as Cape Canaveral.

cutoff. A short connecting stream that has bypassed a river meander, or bend.

delta. A sediment deposit at the mouth of a river where it enters a lake, estuary, or ocean. The shape of a delta may vary depending on the amount of sediment the river is carrying, the tidal range of the coast at the river mouth, and the strength of currents along the front of the delta. Deltas have many ancillary streams that flow out of the main river channel. They are called *distributaries* because they redistribute sand across the delta.

delta plain. The swampy, marshy environment of a delta where a river meanders and divides into smaller streams (called *distributaries*) before entering the ocean. Because it is frequently flooded, the delta plain has shifting and eroding river channels, many freshwater wetlands, and much sediment deposition.

dendritic pattern. A stream drainage pattern in which streams and their tributaries form a branching, bushlike pattern.

dip. The angle of inclination of rock strata. Most strata in Florida rarely exceed 10 feet per mile, about 0.1 degree of dip.

disappearing stream. See **swallet**.

dissolution. A weathering process that affects carbonate rocks (limestone, dolostone) and calcite-cemented sandstones at or near the Earth's surface. Weakly acidic groundwater dissolves the mineral calcite. The water becomes acidic by absorbing carbon dioxide and other compounds from the atmosphere (as rainfall) and more carbon dioxide as it percolates through vegetation-rich soil (as groundwater).

divide. A ridge or relatively high ground that separates or delineates drainage basins.

dolomite. A calcium carbonate mineral that has the element magnesium in its crystal structure.

dolostone. A sedimentary rock composed of the mineral dolomite. It is a common sedimentary rock in Florida (for example, in the Avon Park Formation).

drowned karst. A karst terrain that has been flooded by a rise in sea level. It is characterized by a rocky or salt marsh shore that usually lacks barrier islands. In Florida, the Big Bend coast from Ochlockonee Bay to Anclote Key is a classic example.

drowned river valley. A river valley flooded by the sea during the Holocene Transgression. These valleys are now long estuaries. Pensacola Bay, Tampa Bay, and Charlotte Harbor are examples.

drumstick barrier. A barrier island with a drumstick shape. They form as a longshore current erodes sand from an ebb tide delta and redeposits it, forming an oblong barrier that is wide where the sand is deposited but narrower at the other end of the island. Drumstick barriers are common along the Georgia and South Carolina coasts, and a few are found in Florida.

dugong. A manatee-like animal that was once widespread in the western hemisphere. Today only one species (*Dugong dugon*) remains, dwelling in the Indo-Pacific, Indian Ocean, and Red Sea. Dugongs were extremely abundant in Florida from Eocene to Pliocene time.

dune lake. A lake that forms when coastal sand dunes impound freshwater. The lake may exchange water with the ocean.

duricrust. See **calcrete**.

Eastern Valley. A topographically low area that extends down most of the eastern side of the Florida Peninsula and is situated between the Central Highlands and the Atlantic Coastal Ridge. The north-flowing St. Johns River flows in it. The valley is a relict Pleistocene-age beach ridge plain.

echinoderm. A major phylum of marine invertebrates that includes starfish, brittle stars, sea urchins, sand dollars, sea cucumbers, crinoids (sea lilies), and many extinct forms. They typically have a pentameral (five-fold) symmetry to their body and a bumpy and/or spiny surface.

echinoid. A class of echinoderm animals commonly known as *sea urchins, pencil urchins, sea biscuits,* or *sand dollars.* They have complex, calcareous skeletons (called *tests*) covered with many spines.

endemic species. A species that is very restricted in its geographic occurrence.

erosion. The transport of weathered minerals and rocks by water, wind, ice, or gravity.

escarpment. A relatively steep slope or cliff that separates two relatively flat levels of land. Florida's escarpments represent the landward edges of terraces, where waves cut into upland sediments in the past. Though they approximate former shorelines, erosion modified them after sea level fell, making them less continuous and less recognizable. Also called a *scarp.*

estuary. A nearly landlocked body of coastal water that is significantly diluted by river water.

evaporite. A mineral, such as halite and gypsum, that precipitates from saltwater brine. Evaporite minerals are found in the subsurface of Florida, especially in strata of Paleocene age.

fault. A fracture in rock along which there has been movement.

fauna. An assemblage of living or fossil animals or protozoans.

flint. A black variety of chalcedony.

floodplain. The broad, muddy surface surrounding a river channel that is flooded when the river overflows its natural channel.

flora. An assemblage of living or fossil plants or algae.

Florida Bay. A broad, shallow estuary on the west and northwest side of the Florida Keys. Florida Bay is of special geological interest because of the large amount of calcium carbonate mud produced there, primarily by calcareous green algae.

Florida Current. An ocean current that flows between Cuba and the southern end of the Florida Peninsula through the Straits of Florida. It joins the Gulf Stream on Florida's Atlantic coast.

Floridan aquifer system. Florida's largest aquifer. It underlies all of the state and consists of Eocene- to Miocene-age carbonate rocks that are over 2,000 feet thick in some cases. It is largely an artesian aquifer.

Florida Platform. A massive sedimentary rock structure that underlies, and includes, Florida. It has been primarily constructed of limestone deposited over Florida's basement rock over the last 200 million years.

Florida Reef Tract. A 3- to 5-mile-wide, discontinuous band of living coral reefs growing along the southern edge of the Florida Platform.

foraminifera. A very large group of marine protozoans that are closely related to the familiar freshwater amoeba. Most forams are housed in a small calcium carbonate shell called a *test*, and they are exceptionally common in the fossil record.

foredune. The most seaward sand dune on a beach.

fossil mold. See **mold**.

fringing reef. A coral reef that grows very close to the shoreline, so that essentially there is no lagoon between the reef and shore.

fulgurite. A sandy and glassy cylindrical tube that forms when lightning strikes the sand.

fuller's earth. Clay minerals that have especially good absorbent and decoloring qualities.

gastropod. The most diverse and globally widespread class of mollusks, including snails, conchs, whelks, and slugs. They generally have a single shell, though some are shell-less. They are mostly marine but also live in freshwater and on land.

geomorphic district. An area with distinctive topography that can be readily distinguished on topographic maps and is usually recognizable in the field. There are ten districts in Florida.

geomorphology. A discipline of geology concerned with describing the origin of Earth's surface features, or landforms. Processes that shape landforms include tectonic movements, weathering, erosion, and sediment deposition.

glacial stage. A period in which glaciers grow (advance) during an ice age, causing sea level to drop.

glyptodont. An extinct group of mammals that had large carapaces made of interlocking, ornamented, bony plates. Armadillos, sloths, and pampatheres are related to them.

gomphothere. An extinct group of proboscideans (elephants and their extinct relatives), some of which are called "shovel-tusks" because of their protruding lower tusks.

Gondwana. The southern portion of the supercontinent Pangaea. It consisted of Africa, South America, Antarctica, Australia, India, and the Middle East. The basement rock of Florida is a fragment of Gondwanan (African) crust.

gravel. A mixture of sediment particles that are greater than 2 millimeters in diameter, including pebbles, cobbles, and boulders.

Great American Biotic Interchange. The north-south exchange of mammals across the Central American Isthmus, which is especially evident in the vertebrate fossil record of Pliocene and Pleistocene age in Florida. Roughly 10 percent of modern North American mammals are of South American origin (for example, the armadillo, porcupine, and opossum), and over 50 percent of modern South American mammals are descendants of North American immigrants (for example, the jaguar and llama). Many of the South American immigrants, such as the ground sloth and glyptodont, died out in North America at the end of Pleistocene time, about 10,000 years ago.

Gulf Trough. A name used for the Suwannee Channel of Middle Eocene to Middle Miocene time, when the channel narrowed considerably. The channel opened widely into the Gulf of Mexico near Apalachicola, in the Apalachicola Embayment.

Halimeda. The genus name of tropical green algae that produce distinctive calcareous plates made of the mineral aragonite. These plates are extremely abundant and compose a substantial volume of the carbonate sediment produced along the Florida Reef Tract.

hammock. A forest composed of temperate to tropical hardwood trees, depending on location. In Florida, hammocks occur in narrow bands or patches on sandy slopes or uplands; in southern Florida, they grow on rocky surfaces.

hardbottom. A sea bottom composed of rock rather than loose sand or sediment and usually encrusted with various forms of marine life, including sponges, soft corals, barnacles, and oysters. Although hardbottom communities are popularly called "reefs," the marine life living on the hardbottom did not construct most of the rock on which it grows.

heavy minerals. Minerals with a higher specific gravity than quartz. In Florida, heavy minerals are scattered in most quartz sands, but they may also be concentrated as dark sand layers within quartz sand deposits. Heavy minerals of Florida include ilmenite, rutile, zircon, staurolite, garnet, tourmaline, corundum, spinel, and topaz.

highland. A geomorphic term referring to an area of high elevation compared to surrounding areas. Florida's highlands trend east-west across the northern panhandle and peninsula, and north-south down the central axis of the peninsula. Florida's highlands are modest by any standard, typically ranging from 50 to 300 feet above sea level.

Holocene Transgression. A dramatic global rise in sea level at the end of Pleistocene time and the beginning of Holocene time, from about 18,000 to 6,000 years ago. Sea level rose 390 feet or more to its present level because of global warming and the melting of most of the great continental glaciers in the northern hemisphere.

humate sand. A brown or black quartz sand loosely cemented together by organic material derived from the decay of plant debris. The organic material is transported by groundwater that percolates through soil and sediment.

Ice Age. A term that usually refers to Pleistocene time, about 1.8 million to 10,000 years ago, when continental glaciers were extensive in the northern hemisphere. During this time, sea level fluctuated as glaciers alternately grew and melted. There were several major "ice ages" in the geological past, including this one, as well as other lesser events.

index fossil. A fossil that is restricted in its stratigraphic occurrence and can therefore be used as an indicator for a particular period of geological time.

interglacial stage. An interval of time during an ice age when glaciers temporarily retreat (melt back), resulting in sea level rise. Ice ages are characterized by a regular advance and retreat of glacial ice, which is thought to be controlled by astronomical factors that affect Earth's climate.

intertidal zone. The coastal area between the average high and low tide marks. Also called the *foreshore* or *littoral zone.*

intraformational conglomerate. See **rip-up clasts**.

Isthmus of Panama. See **Central American Isthmus**.

joint. A fracture in rock along which there has been no movement.

karst. A terrain characterized by many sinkholes, depressions, cavities, caves, and related features that are the result of a weathering process called *dissolution*, in which acidic groundwater dissolves limestone. Karst terrains have limestone at or near the surface. Much of the surface of Florida is karstic.

karst window. A collapsed portion of a limestone cave system that reveals flowing water.

key. A small island.

lag deposit. A concentrated gravelly deposit of pebbles and/or fossils that remains after winnowing currents have removed finer sediment. These relatively larger particles are not easily moved by currents. The pebbles or fossils may be eroded from an older deposit or they may slowly accumulate from other sources. Lag deposits in Florida are often rich in shark teeth because sharks constantly shed teeth, which slowly fall to the sea bottom. They can accumulate in relatively large amounts before finally being buried.

lagoon. A body of water between the mainland and a barrier island or coral reef. A lagoon diluted by freshwater from a river is considered an estuary.

Lake Wales Ridge. A prominent ridge composed of the Pliocene-age Cypresshead Formation and Quaternary-age sands that rises 300 feet or more above sea level. It extends from central Florida, near Orlando, south to the vicinity of Lake Placid.

Laurasia. The northern portion of the supercontinent Pangaea. It consisted of North America, Greenland, and Eurasia.

living fossil. A modern organism that has an exceptionally long geological history and has persisted with little organic change over time.

longshore bar. A sandy shoal located offshore of a beach and commonly called a *sandbar.* Longshore bars are very dynamic and may grow, shrink, or migrate shoreward and become welded, or accreted, to the beach, forming a berm.

longshore current. A current that parallels the shore. It is generated as incoming waves strike the shore at a slight angle, usually less than 10 degrees, due to wave refraction, resulting in the deflection of some of the wave energy as a current along the shore. Longshore currents also form when water accumulates near the shore and cannot easily flow back to sea because incoming waves impound it. A longshore bar may also help hold the water near the shore. As a result, the mound of water flows laterally, parallel to the shore, sometimes in opposing directions. Longshore currents sweep sand along the shore in the shallow surf zone.

longshore drift. The shore-parallel movement of sand along the beach face and the shallow sea bottom near the shore. Both beach drift and longshore currents create longshore drift. Longshore drift may erode or build beaches depending on the amount of available sand and other factors, but it is almost constantly in progress. Also called *littoral drift.*

longshore trough. A channel between a beach and a longshore bar through which a longshore current flows.

Loop Current. A major clockwise-flowing current in the middle of the Gulf of Mexico. It flows around the end of the peninsula, becoming the Florida Current, and up Florida's east coast, contributing to the Gulf Stream. Such a clockwise-flowing current appears to have been in place as early as Cretaceous time and was probably the source for water that flowed through the Suwannee Channel.

mangrove. A tree uniquely adapted for growing in marine water along tropical and subtropical shores. They grow from the central peninsula to the Florida Keys.

micrite. A calcareous mud. It is forming today in Florida Bay in great abundance. See also **mud mounds**.

microkarst. Small-scale karst features, such as shallow dissolution depressions or small sinkholes. It is common in the southern peninsula where surface limestone is thin and geologically young (of Pleistocene age).

midden. An archaeological deposit of shell, bone, or other debris.

Mississippian Period. An archaeological time period from about AD 1,000 to 1,600 that is characterized by extensive ceremonial mound complexes, agricultural developments, and expanded populations. Mississippian Period cultures were in Florida during the time of European contact.

Mississippi Embayment. A large sedimentary basin that extends deep into the interior of the continental United States from the Gulf of Mexico. It contains a continuous sedimentary record from Jurassic time to the present.

mold. A fossil that is the hardened sediment that filled a cavity of an organism, or an impression of a fossil. Molds may be internal, as when sediment fills in the shell of a clam or a snail; or external, as when a shell is pressed into mud and removed, leaving an imprint. Footprints are commonly external molds. See also **cast**.

mollusk. A phylum of animals that includes bivalves (clams), gastropods (snails), cephalopods (squids, octopuses, nautiloids), and several lesser-known and/or extinct groups.

mud. A mixture of silt and clay.

mud mound. A shoal of calcium carbonate mud. They are common in Florida Bay. The mud is produced by living organisms, primarily by calcareous green algae of the genus *Penicillus*.

Northern Highlands. High ground of the northernmost peninsula and eastern panhandle that consists of erosionally dissected quartz- and clay-rich rocks and sediments and some limestone of Miocene and Pliocene age. It includes the Tallahassee Hills and Apalachicola Bluffs and Ravines.

nummulitic limestone. A densely fossiliferous limestone or coquina primarily composed of larger foraminifera of the genus *Nummulites*.

offshore zone. An area on the continental shelf that extends seaward of a longshore bar.

ooids. Tiny (0.3 to 1 millimeter in diameter) polished spheres of calcium carbonate that precipitate from shallow, warm, turbulent, tropical seawater. Ooids have a tiny nucleus, consisting of a sand grain, shell fragment, or other particle, that is coated by thin, concentric layers of the mineral aragonite.

Ophiomorpha nodosa. A trace fossil—not the actual animal—of a nearshore shrimp of the genus *Callianassa*. The fossil is a bumpy tube, up to 2 inches in diameter, that is constructed of mud balls of clay and sand. The tubes may connect in elaborate networks. Their presence indicates nearshore marine depositional conditions. They are especially common in formations of Miocene, Pliocene, and Pleistocene age in Florida.

oreodont. A group of short-legged artiodactyls (even-toed ungulates, or hoofed mammals) that lived in North America from Late Eocene to Miocene time. Their fossils are common in the Badlands of South Dakota and Nebraska but are also found in Florida.

Osceola Plain. A Pleistocene-age beach ridge plain composed of quartz sand and located in the eastern portion of central Florida.

outlier. An isolated erosional remnant of a formerly more-extensive rock formation. An example is Falling Waters Hill in Washington County.

oxbow. An abandoned river meander that forms when a cutoff isolates a river bend.

paleodunes. Relict (ancient) sand dunes.

Paleoindian Period. An archaeological time period from about 13,000 to 8,000 BC that is characterized by large, lanceolate projectile points and other artifacts. Includes distinctive points referred to as Clovis and Suwannee points. Paleoindians were the first human occupants of Florida and North America.

Pamlico Terrace. A relict ocean floor of the last interglacial interval of Pleistocene time. In the shallow Pamlico Sea, sea level was at least 25 feet higher than at present.

Panamanian Land Bridge. See **Central American Isthmus**.

patch reef. A relatively small, usually circular or oblong reef that typically grows in a lagoon or other low-energy waters.

peat. A deposit of partially decomposed plant debris in marshes, swamps, river floodplains, and related sedimentary environments. Peat formation is the first stage in the creation of coal.

pebble. A sediment particle that ranges between 4 and 64 millimeters (2.5 inches) in diameter.

peccary. A piglike animal that survives today primarily in the Central and South American tropics. Its fossils are found in Oligocene- to Pleistocene-age strata in Florida.

perched barrier. A barrier island that forms when sand accumulates on a rocky ridge. They are common along the east coast of Florida.

petrification. A fossilization process in which the original organic material (bone, shell, wood) is completely replaced by mineral material such as quartz.

phosphatic. A descriptive term referring to sediment or rock that is rich in calcium phosphate minerals.

photic zone. The part of the ocean that is well illuminated by sunlight. The depth of the photic zone varies depending on water clarity, latitude, and other factors.

plankton. Marine organisms that live in the water column, not on the sea bottom, and are primarily floaters and drifters rather than active swimmers, including single-celled algae such as diatoms and dinoflagellates (called *phytoplankton*), or protozoans and jellyfish (called *zooplankton*).

plate tectonics. The central geological theory that describes the creation, movement, and destruction of Earth's rigid surface layer, or *lithosphere*. The theory states that the lithosphere is fragmented into irregularly shaped pieces called *plates*. Plates move in response to forces generated by Earth's internal heat, and they are continually created (*tecton* means "to build") and destroyed at the boundaries where the plates are in contact.

platform. A large mass of rock, the top of which is usually flat.

platform margin reef. A reef that grows near the edge of a carbonate platform, usually near the edge of the continental shelf. Also called a *bank margin reef*.

point bar. A sediment deposit along the inside of a river meander, or bend, that usually consists of quartz sand in Florida.

reef. An organically built, rigid structure that rises above the seafloor. Reefs may be constructed by bacteria (stromatolites), coralline algae, sponges, corals, bryozoans, worm tubes, or oysters and other mollusks.

refugia. Limited microhabitats that are suitable for species that normally live in different regions or climates. The steephead ravines along the Apalachicola River are classic examples, as is the Lake Wales Ridge in the central peninsula.

rhodolith. A rounded nodule of limestone that is composed of fine concentric layers of coralline red algae. Rhodoliths are abundant in the Bridgeboro Limestone and in the deep fore-reef zone of the Florida Reef Tract.

rimmed shelf. A carbonate platform with a reef or shoal of skeletal debris at its edge, forming a rim, and a lagoon between the rim and the shore. The continental shelf offshore of the Keys is a good example. Rimmed shelves are almost exclusively associated with tropical continental shelves where coral reefs thrive.

rip current. A narrow, powerful current that flows away from the shore, perpendicular to it. They are usually generated by converging longshore currents. Rip currents typically occur as evenly spaced sets and only rarely occur singly. Many people refer to rip currents as *rip tides* or *undertow*, although some oceanographers prefer to apply these terms to other types of coastal currents.

rip-up clasts. Also called *intraclasts* or *inclusions*. Fragments of rock from a lower stratum that have been eroded and incorporated into an overlying stratum. The layer containing the rip-up clasts is typically a conglomerate, which is called an *intraformational conglomerate*.

rudist. An unusual oysterlike clam that lived primarily during Cretaceous time and commonly formed reefs or biostromes in shallow tropical waters. Rudists are found in the deep subsurface of Florida.

runnel. A very shallow channel of flowing water on a beach.

salt dome. A vertical subsurface pillar or mushroom-shaped structure composed of evaporite minerals. The dome forms as pressure from overlying strata forces the salt to flow and intrude upward into the strata. Salt domes commonly create traps for oil and natural gas.

salt marsh. A nontropical, muddy, vegetated coastal wetland that is usually covered by salt-tolerant grasses, such as those of the genera *Spartina* and *Juncus*.

sand. A sediment particle between 0.062 and 2 millimeters in diameter. Although it technically refers only to sediment size, the term *sand* is commonly used to describe quartz sand, the most common type of sand-sized sediment.

sandbar. A shallow-water, linear deposit of sand. Also called a *bar*, *sand shoal*, or *longshore bar*.

sand hill karst. Karst terrain characterized by a layer of unconsolidated sand over karstic limestone, resulting in many sandy depressions and lakes that form as sand falls into underlying sinkholes. Sand hill karst is well developed on the central peninsula. Also called *mantled karst* or *deep-cover karst*.

scarp. See **escarpment**.

sediments. Fragments of minerals, rocks, or organic particles that range in size from microscopic clay particles to giant boulders.

seep. An area where groundwater leaks or percolates onto the surface—essentially, a nonartesian spring. Often the term is used to describe the source of small streams, such as steephead ravines.

seepage lake. A lake that derives most of its water from groundwater seepage. A majority of Florida lakes are seepage lakes, which are very characteristic of *karst* terrains.

seepage stream. A stream that is largely supplied by groundwater that seeps through the riverbank.

silicification. A process of remineralization in which one mineral, rock, or other material (such as wood) recrystallizes as, or is replaced by, quartz minerals. Agatized coral forms when aragonite recrystallizes as chalcedony. Chert forms when limestone is replaced by chalcedony. Petrified wood forms when chalcedony replaces wood. See also **petrification**.

silt. A sediment particle between 0.004 and 0.062 millimeter in diameter.

sinkhole. A bowl- or cone-shaped depression that forms when limestone dissolves. It is a primary characteristic of karst. See also **solution sinkhole**, **cover collapse sinkhole**, and **cover subsidence sinkhole**.

solution sinkhole. A shallow sinkhole that forms very slowly (not by collapsing) as surface limestone dissolves. Also called a *doline.* Groupings of solution sinkholes create a gently rolling landscape.

sorting. A descriptive term referring to the average distribution of sediments of different size. A sediment deposit or sedimentary rock with particles that are nearly all the same size is said to be *well sorted.* A sample with a wide range of sediment sizes, fine sand to pebbles for example, is *poorly sorted.*

speleothem. Cave formations, such as stalactites and stalagmites, that form as water dissolves and reprecipitates calcium carbonate.

spit. An elongated sand ridge (not a sandbar) that forms as longshore drift deposits sand.

spring-fed stream. A stream primarily supplied by springs. Water is said to "rise" at a spring, or springhead, and may flow a short distance, called a *spring run*, to a larger stream channel.

spring magnitude. A spring classification system based on the volume of water discharged over a given time, usually measured as cubic feet per second. A first magnitude spring discharges 100 cubic feet per second (64.6 million gallons per day) or more.

springshed. A geographic region that discharges groundwater into a particular spring or group of springs.

spring tide. A tide that occurs twice a month and has the highest tidal ranges. It occurs around the time of the new and full moons, when the moon, sun, and Earth are aligned and their gravitational forces are combined.

spur-and-groove structure. A series of coral ridges (spurs) and erosional channels between the ridges (grooves). These structures are common in the fore-reef zone of the Florida Reef Tract.

steephead stream. A special type of seepage stream that erodes steep ravines as groundwater seeps through coarse, sandy soils.

stony coral. A common name for a group of corals of the order Scleractinia (also called Madreporaria) that secrete a hard skeleton that encases the polyps. Stony corals may be solitary or colonial, and colonial species are common reef formers. See also **coral**.

storm surge. An exceptional rise in sea level as a hurricane makes landfall. It is caused by low atmospheric pressure and the piling up of water as it is blown toward the coast.

strandline. The location of a former shoreline, usually detectable by an abrupt rise in elevation or by beach ridges. See also **escarpment**.

strand plain. A former coastal or shallow marine surface that is now above sea level. Many strand plains consist of numerous beach ridges, resulting in a washboardlike topography, of which St. Vincent Island is an example.

stratigraphy. The discipline of geology concerned with the description, correlation, and interpretation of Earth's strata, primarily the layered stacks of sedimentary rock of the upper crust. It is the study of Earth's geologic and paleontologic history.

stromatolite. A thinly layered sedimentary rock that was constructed by cyanobacteria, which grow as stringy, filamentous mats that entrap sediments in their fibers. The result is a laminated structure of bacteria-sediment-bacteria-sediment, and so on. When the bacteria decompose, the layered stromatolite rock remains.

surf zone. The area that extends from the high tide line to the last, most seaward longshore bar, and where waves encounter the bottom and breakers begin to rise. Also called the *nearshore zone* and *breaker zone.*

Suwannee Channel. A current-swept seaway that extended from the northern Gulf of Mexico to the Atlantic. Also called the Suwannee Strait, the channel was present from Late Cretaceous to Middle Miocene time. It is filled with rock and sediment now.

swale. A low area between two ridges, such as beach ridges.

swallet. Also called a *swallow hole*, it is a sinkhole or cave through which a stream disappears underground. The underground stream may later resurface downstream to form a spring, which is also called a *rise*, *river rise*, or *springhead.*

Tallahassee Hills. High ground of the eastern Florida Panhandle that consists of erosionally dissected carbonate and quartz-rich sediments of the Hawthorn Group and Miccosukee Formation, both of which were deposited in a shallow, submarine setting. The hills are part of the Tifton Uplands District.

tapir. An unusual perissodactyl (odd-toed ungulate, or hoofed mammal) with a proboscis similar to a very short elephant trunk. Their fossils are common in Florida in strata of Miocene to Pleistocene age. There are four modern species, none of which live in Florida.

terrace. In Florida, a terrace is a relatively flat surface that represents a former shallow sea bottom, the Pamlico Terrace for example. The landward extent of a terrace is marked by an escarpment, which represents the former shoreline.

terrain. A distinctive expression of Earth's crust at the surface, for example, sand hill terrain, karst terrain, and so on. A geological landscape.

terrane. A distinctive, fault-bounded body of rock or strata (not merely its surface expression) with a tectonic history that differs from surrounding rock.

Tethyan. A term that refers to fossils that have affinities with faunas of the ancient Mediterranean region. Many Eocene-age mollusks of Florida, for example, are very similar to Tethyan species.

tidal creek. A stream that flows landward from the ocean, transporting water during high tide. Salt marshes typically have many tidal creeks.

tidal range. The average difference between high and low tides on a given coastline. Microtidal is less than 2 meters (6.5 feet), mesotidal is between 2 and 4 meters (6.5 to 13.1 feet), and macrotidal is over 4 meters (13.1 feet). Most of the Florida shore is microtidal, with some mesotidal conditions along the northeastern coast.

trace fossil. A burrow, trail, trackway, or other sedimentary structure (such as a stromatolite) that indicates the former presence of an organism without actually preserving the remains of the animal.

Trail Ridge. A linear sand body (a barrier island and coastal wetland deposit of Late Pliocene or early Pleistocene age) that extends 130 miles from the Altamaha River in Georgia to around Gold Head Branch State Park in Clay County, Florida. It is rich in quartz and heavy-mineral sands and peat. Elevations are generally between 150 and 160 feet but can reach 240 feet.

tributary. A stream that discharges into a larger stream or river.

trough cross-stratification. A type of cross-stratification in which the inclined layers have a curved, half-circle appearance. This sedimentary structure commonly forms in longshore troughs or river channels.

type section. An exposure of a rock formation that geologists deem typical and exemplary of that formation. A type section is the standard by which other outcrops are compared and correlated.

washover fan. A fan-shaped layer of sand the ocean deposits during storms, such as a hurricane, as the water breaks through the dune field of a barrier island.

watershed. An area of land that drains into a particular river, including the river and its tributaries. Also called a *catchment* or *drainage basin.*

weathering. The breakdown or decomposition of minerals and rocks at or near the Earth's surface. Weathering may be mechanical (cracking, fracturing, breaking), chemical (dissolution, oxidation), or biological (burrowing, boring, growth of tree roots in rock fractures). All three processes weaken and destroy rock.

Western Highlands. High ground of the western panhandle that consists of erosionally dissected quartz-rich rocks and sediments of the Alum Bluff Group and the Citronelle Formation of Miocene and Pliocene age. These hills are part of the Southern Pine Hills District.

West Florida Shelf. The continental shelf offshore of the west coast of Florida and the submerged western portion of the Florida Platform. It is a carbonate platform with a ramp profile.

wetland. An environment characterized by standing water. Wetlands may be freshwater, such as marshes and swamps, or coastal. Coastal wetlands can be unvegetated mudflats, salt marshes, or mangrove swamps.

Wisconsinan glacial stage. The last glacial stage of Pleistocene time, between 100,000 and 10,000 years ago. During this stage, continental glaciers in North America reached their maximum size, and global sea level was as much as 400 feet below the present level.

Woodland Period. An archaeological time period from about 1,000 BC to AD 1,000 that is characterized by complex ceramics, highly social agricultural cultures living in villages with ceremonial mound complexes, and extensive regional trade networks with other tribes throughout the southeastern United States.

Woodville Karst Plain. A karst surface south of Tallahassee that developed over Oligocene- and Miocene-age limestone. It is the western end of the Ocala Karst District.

zooxanthellae. A microscopic, single-celled dinoflagellate algae that lives symbiotically with many stony corals. The algae live in the gastrodermal cells of the coral, which lie just below a surface layer of cells called *ectodermal cells*, and provide nutriment to it through photosynthesis. The coral provides protection for the algae.

Appendices

Sedimentary Rocks of Florida

Sedimentary rocks are the most common rocks exposed at Earth's surface and are the only type of rock found at the surface in Florida. Igneous rocks, which form by the crystallization of magma, and metamorphic rocks, which form by the mineralogical transformation of other rocks under conditions of high heat and pressure, are found only deep in the subsurface of Florida. Some well-known sedimentary rocks, such as coal, are not found in the state, but some more unusual varieties, such as phosphorite, are common. Many sedimentary rocks of Florida are mixtures of some of these categories, for example, sandy limestones or phosphatic sands.

CARBONACEOUS ROCKS. Rocks rich in carbon, usually made from plant debris.

lignite. A soft, brown coal. Found in the Trail Ridge sands.

peat. Decayed, compressed plant debris; appears woody. Common in Holocene-age deposits, such as under barrier islands, in river deposits, and in the Everglades.

CARBONATE ROCKS. Composed of calcium carbonate; most are biogenic in origin, the sediment particles being composed of the fragments of the shells of marine organisms such as foraminifera, calcareous algae, corals, mollusks, and bryozoans. Some formations in Florida, such as those in the Hawthorn Group and the Tamiami and Caloosahatchee formations, are mixtures of carbonate and quartz-rich sediments.

boundstone. Made of overlapping and intergrowing coral or other calcareous skeletal material, such as calcareous algae; common in reef deposits. The Key Largo Limestone is a coral boundstone. Also found in the Ocala, Suwannee, and Bridgeboro limestones.

brecciated limestone. Also called *limestone breccia.* A breccia made of limestone fragments. Common in sinkhole deposits. Sometimes called *intraformational conglomerate,* especially if the clasts are broken mud-crack fragments. These usually form in the intertidal zone, where mud layers are exposed, dry up, and fragment into elongated, flat mud pebbles. Found in the Avon Park Formation, Ocala and Suwannee limestones, and Tampa Member of the Arcadia Formation.

coquina. Nearly 100 percent fossil shell material. The Anastasia Formation is a coquina, and coquinas are found in parts of the Ocala and Suwannee limestones.

dolostone. Primarily composed of the mineral dolomite (calcium magnesium carbonate). Many formations in Florida contain dolostone. The Avon Park Formation is a typical dolostone, and dolostones are found in the Ocala, Marianna, and Suwannee limestones, in the Hawthorn Group, and in subsurface formations. A dolostone can also be classified as a chemical rock.

fossiliferous limestone. Any limestone with abundant fossils. May also be classified as a wackestone, packstone, or grainstone. Common in most limestone formations.

grainstone. Coarser-grained limestone with no mud. The lower part of the Ocala Limestone and much of the Suwannee Limestone is grainstone.

lime mudstones. Composed of very fine lime sediment. Two types: micrite, or micritic limestone, and marl. Micrite is a common name for lime mudstones. It is common in parts of most limestone formations and actively forming in Florida Bay. Marl generally forms in lake sediments and may contain abundant clay. Found in some sediments of Pleistocene and Holocene age.

limestone conglomerate. A limestone composed of rounded pebbles of limestone. Found in the Tampa Member of the Arcadia Formation.

oolite. Also called *oolitic limestone.* A limestone rich in ooids. Common in the Miami Limestone (formerly called the Miami Oolite).

packstone. A limestone with nearly equal mixtures of lime mud and larger sediment particles. The most common rock type in the Ocala Limestone.

wackestone. A muddy limestone with some larger sediment particles. Found in several limestone formations but common in the Marianna Limestone.

CHEMICAL ROCKS. Usually form by the precipitation of minerals out of seawater or groundwater. Some chemical rocks form through diagenesis, a process in which sedimentary rocks are chemically altered.

chalcedony. Microcrystalline variety of quartz that is smooth to glossy in appearance, is very hard (will easily scratch glass), and occurs in several colors: brown, gray, or white (chert); black (flint); and red or orange (jasper). Found in many limestone formations and in parts of the Peace River Formation (in this formation it is mainly in a form called *opal,* which contains water in its crystal structure).

dolostone. A carbonate rock that has been chemically altered, from calcite or aragonite to the carbonate mineral dolomite. A dolostone can also be classified as a carbonate rock.

evaporites. Form by evaporation of seawater. Rock salt (rich in the mineral halite), rock gypsum (rich in the mineral gypsum), and rock anhydrite (rich in the mineral anhydrite) are found in some subsurface formations in Florida.

geode. A generally spherical concretion, usually partly hollow, with an interior filled with banded chalcedony (called *agate*) and pointed quartz crystals projecting into the interior (called *drusy quartz*). Forms in cavities of some limestone deposits and fossil mollusks. Much agatized coral of the Suwannee Limestone and Tampa Member of the Arcadia Formation forms geodes.

ironstone. A rock rich in iron oxide minerals such as limonite, hematite, or goethite; sometimes occurs in odd-shaped concretions that form as sediments are cemented by the percolation of iron-rich groundwater. Common in the Cypresshead, Miccosukee, and Citronelle formations.

phosphorite. A rock rich in calcium phosphate; usually gray, blue, or black and contains sand, silt, and clay. The Bone Valley Member of the Peace River Formation is a typical example.

travertine. Banded layers of calcite or aragonite crystals. Found in many caves.

DIATOMACEOUS EARTH. A very-fine-grained sediment composed of the tiny, silicon-rich shells (called *tests*) of unicellular diatom algae; typically a freshwater deposit, it is associated with peat in the Everglades.

QUARTZ-RICH (SILICICLASTIC) ROCKS. Also called *clastic* or *detrital* rocks. Composed of sediment particles consisting of quartz or other silicon-rich minerals.

breccia. Contains large, angular particles, or clasts, more than 2 millimeters in diameter. Occasionally found in the Suwannee Limestone and Hawthorn Group and commonly associated with sinkhole-fill deposits (see **brecciated limestone**).

conglomerate. Contains large rounded clasts more than 2 millimeters in diameter. Found at the top of the Suwannee Limestone and in the Cypresshead, Miccosukee, and Citronelle formations. Conglomeratic gravels are especially common in some riverbeds of the western panhandle.

quartz sandstones. Many types; classified by average size of the sand grains and the amount of mud matrix surrounding the sand grains: coarse grained (0.5 to 2 millimeters diameter grains), medium grained (0.25 to 0.5 millimeter diameter grains), fine grained (0.0625 to 0.25 millimeters diameter grains). Sandstones are found in many formations, especially the Cypresshead, Miccosukee, and Citronelle. Quartz sands are also common in soil, lake, river, dune, beach, and nearshore sediments. Many fossil-rich formations in Florida are fossiliferous sand deposits.

Sedimentary Environments, Sediment Types, and Representative Surface Formations in Florida

Sedimentary Environment	Sediment Types	Examples in Florida
Continental Environments		
glacial	coarse, unsorted sediments	not found in Florida
volcanic	ash, clays	some Oligocene- and Miocene-age clays may contain small amounts of altered volcanic ash
windblown (dunes)	fine-grained quartz sand	Quaternary-age sands
soils	organic-rich sand/sandstone	primarily in Quaternary-age sands
lakes	organic-rich mud/mudstone fine to medium sands/sandstone	Quaternary-age sediments Hawthorn Group
rivers	coarse sand, gravel (channel, levee, bars) fine sand and mud (floodplain)	Citronelle and Miccosukee formations Holocene-age river deposits
freshwater wetlands, swamps, marshes	muds, sands, peat, diatomaceous earth, marl	Quaternary-age sediments
karst (fissures, caves, sinks, spring runs)	clays, sands, breccias	Hawthorn Group (fills many fissures and sinks), Santa Fe Marl
Transitional Environments		
deltas	clay to sand	Hawthorn Group, Citronelle and Miccosukee formations
estuaries/lagoons	muds, sands, calcareous sands	Alum Bluff Group, Hawthorn Group
tidal flats/coastal wetlands	lime mud	Avon Park Formation, Marianna and Suwannee limestones (in part)
beaches/barriers islands	quartz sands, shell debris	Cypresshead and Anastasia formations, Trail Ridge sands, Quaternary-age sands
Marine Environments		
inner- to mid-continental shelf	sands, various limestones	Avon Park Formation, Lower Ocala Limestone, Marianna Limestone, Suwannee Limestone, Chattahoochee Formation, St. Marks Formation, Hawthorn Group, Alum Bluff Group, Jackson Bluff Formation, Intracoastal Formation, Citronelle Formation, Miccosukee Formation, Cypresshead Formation, Tamiami Formation, Caloosahatchee Formation, Miami Limestone
mid- to outer-continental shelf	sands, limestones, phosphatic sediments	Ocala Limestone, Hawthorn Group
reefs	coralline limestone, coarse-grained limestone (grainstone), algal limestone	Bridgeboro Limestone, Suwannee Limestone, Tampa Member of the Arcadia Formation, Chipola Formation, Tamiami Formation, Key Largo Limestone, modern reef tract
oceanic (deep marine)	various muds, gravels, ash, etc.	Not exposed in Florida

Museums

Bailey-Matthews Shell Museum
3075 Sanibel-Captiva Road
Sanibel, FL 33957

Florida Association of Museums
459 Cedar Hill Road
PO Box 10951
Tallahassee, FL 32302-2951

Florida Museum of Natural History
University of Florida Cultural Plaza
SW 34th Street and Hull Road
PO Box 112710
Gainesville, FL 32611-2710

Gillespie Museum of Minerals
234 East Michigan Avenue
Deland, FL 32723

Mulberry Phosphate Museum
101 SE First Street (Highway 37)
Mulberry, FL 33860

Museum of Florida History
R. A. Gray Building
500 South Bronough Street
Tallahassee, FL 32399

Geological, Paleontological, Oceanographic, and Archaeological Organizations

Florida has a great many professional and/or amateur organizations devoted to geology, paleontology, oceanography, archeology, and other areas of natural history. Many of these organizations are quite active, others less so. State and national institutions provide many educational resources and programs. With local organizations permanent addresses change regularly, so it would be nearly impossible to keep a current listing. Fortunately, an Internet search will normally lead you to an organization of interest. We apologize if we overlooked an organization.

Geological

Florida Geological Survey
Gunter Building MS# 270
903 W Tennessee Street
Tallahassee, FL 32304-7700

Northwest Florida Water
Management District
81 Water Management Drive
Havana, FL 32333-4712

South Florida Water
Management District
PO Box 24680
West Palm Beach, FL 33416-4680

Southwest Florida Water
Management District
2379 Broad Street
Brooksville, FL 34604-6899

St. Johns River Water
Management District
4049 Reid Street
Palatka, FL 32177

Suwannee River Water
Management District
9225 County Road 49
Live Oak, FL 32060

Auburndale	Imperial Bone Valley Gem, Mineral & Fossil Society, Inc.
Cocoa	Central Brevard Rock & Gem Club, Inc.
Davie	Florida Gold Coast Gem & Mineral Society
Deland	Gem & Mineral Club of Deland
Fort Myers	Everglades Geological Society Southwest Florida Gem, Mineral & Fossil Club
Fort Pierce	Saint Lucie County Rock & Gem Club
Fort Walton Beach	Playground Gem & Mineral Society
Gainesville	Florida Speleological Society Gainesville Gem & Mineral Society
Hernando Beach	Withlacoochee Rockhound Club
Jacksonville	Jacksonville Gem & Mineral Society
Lake Butler	Lake City Gem & Mineral Society
Melbourne	Canaveral Mineral & Gem Society, Inc.
Miami	Miami Mineral & Gem Society Miami Mineralogical & Lapidary Guild, Inc. Miami Geological Society Tropical Mineral & Gem Society
Ocala	Mid-Florida Gem & Mineral Society
Orlando	Central Florida Mineral & Gem Society
Ormond Beach	Tomoka Gem & Mineral Society
Panama City	Gulf Coast Gem & Mineral Society, Inc. Panama City Gem & Mineral Society, Inc.
Pensacola	Florida Panhandle Gem & Mineral Society
Sarasota	Manasota Rock & Gem Club, Inc.
Sebring	Highlands Gem & Mineral Club
St. Petersburg	Pinellas Geological Society Suncoast Gem & Mineral Society, Inc.
Summerland Key	Keys Gem & Mineral Society
Tallahassee	Florida State University Geological Society Southeastern Geological Society
Tampa Bay	Tampa Bay Mineral & Science Club
Venice	Gulf Coast Mineral, Fossil & Gem Club, Inc.
Vero Beach	Treasure Coast Rock & Gem Society
West Palm Beach	Gem & Mineral Society of the Palm Beaches, Inc.
Wimauma	Rockbusters Fossil, Rock, Mineral Club

Paleontological

Florida Paleontological Society, Inc.
Florida Museum of Natural History
University of Florida
Gainesville, FL 32611

Cape Coral	Paleontological Society of Lee County
Fort Myers	Fossil Club of Lee County Southwest Florida Conchologist Society
Miami	Fossil Club of Miami
Orlando	Florida Fossil Hunters
Punta Gorda	Southwest Florida Fossil Club
Sanibel Island	Sanibel-Captiva Shell Club
Sarasota	Manasota Fossil Club
St. Petersburg	Sun Coast Archeological & Paleontological Society
Tampa	Tampa Bay Fossil Club
West Palm Beach	Palm Beach County Shell Club, Inc.

Oceanographic

Apalachicola National Estuarine
Research Reserve
Florida Department of
Environmental Protection
350 Carroll Street
Eastpoint, FL 32328

Florida Oceanographic Society
890 NE Ocean Boulevard
Stuart, FL 34996

Guana Tolomato Matanzas National
Estuarine Research Reserve
Marineland Office
9741 Ocean Shore Boulevard
St. Augustine, FL 32080

Gulf Specimen Marine Laboratory
PO Box 237
222 Clark Drive
Panacea, FL 32346-0237

Harbor Branch Oceanographic Institution
J. Seward Johnson Marine Education
and Conference Center
5600 US 1 North
Ft. Pierce, FL 34946

Keys Marine Laboratory
PO Box 968
68486 US 1
Layton/Long Key, FL 33001

Mote Marine Laboratory
1600 Ken Thompson Parkway
Sarasota, FL 34236

Newfound Harbor Marine Institute
1300 Big Pine Avenue
Big Pine Key, FL 33043

Rookery Bay National Estuarine
Research Reserve
Florida Department of
Environmental Protection
300 Tower Road
Naples, FL 34113

Archaeological

Florida Bureau of Archaeological Research
R.A. Gray Building
500 South Bronough Street
Tallahassee, FL 32399-0250

Dania	Broward County Archaeological Society Florida Anthropological Society Florida Archeological Council
Freeport	Emerald Coast Archeological Society
Jacksonville	Northeast Florida Anthropological Society
Merritt Island	Indian River Anthropological Society
Miami	Archaeological Society of Southern Florida
Naples	Southwest Florida Archaeological Society
North Port	Warm Mineral Springs Archaeological Society
Orlando	Central Florida Anthropological Society
Ormond Beach	Volusia Anthropological Society
Pensacola	Pensacola Archaeological Society
Pineland	Southwest Regional Center, Florida Public Archaeology Network
Sarasota	Time Sifters Archaeological Society
Sebring	Kissimmee Valley Archaeological & Historical Conservancy
St. Augustine	St. Augustine Archaeological Association Sun Coast Archaeological & Paleontological Society
Stuart	Southeast Florida Archaeological Society
Tallahassee	Panhandle Archaeological Society at Tallahassee
Tampa	Central Gulf Coast Archaeological Society

References and Suggested Reading

Archaeology

Brown, R. C. 1994. *Florida's First People: 12,000 Years of Human History*. Sarasota, Fla.: Pineapple Press.

Bullen, R. P. 1975. *A Guide to the Identification of Florida Projectile Points*. Rev. ed. Gainesville, Fla.: Kendall Books.

Doran, G. H., ed. 2002. *Windover: Multidisciplinary Investigations of an Early Archaic Florida Cemetery*. Florida Museum of Natural History: Ripley P. Bullen Series. Gainesville, Fla.: University Press of Florida.

Griffin, J. W. (ed.), J. T. Milanich (au.), and J. J. Miller (au.). 2002. *Archaeology of the Everglades*. Florida Museum of Natural History: Ripley P. Bullen Series. Gainesville, Fla.: University Press of Florida.

Milanich, J. T. 1994. *Archaeology of Precolumbian Florida*. Gainesville, Fla.: University Press of Florida.

Perry, I. M. 1998. *Indian Mounds You Can Visit: 165 Aboriginal Sites on Florida's West Coast*. 2nd ed. St. Petersburg, Fla.: Great Outdoors Publishing Co.

Purdy, B. A. 1981. *Florida's Prehistoric Stone Technology*. Gainesville, Fla.: University Press of Florida.

———. 1996. *How to Do Archaeology the Right Way*. Gainesville, Fla.: University Press of Florida.

Webb, S. D., ed. 2006. *First Floridians and Last Mastodons: The Page-Ladson Site in the Aucilla River*. Topics in Geobiology, vol. 26. Dordrecht, The Netherlands: Springer.

Willey, G. R. 1998. *Archaeology of the Florida Gulf Coast*. Southeastern Classics in Archaeology, Anthropology, and History. Gainesville, Fla.: University Press of Florida. (Orig. pub. 1949.)

Ecology, Natural History

Davis, J. E., and R. Arsenault, eds. 2005 *Paradise Lost?: The Environmental History of Florida*. The Florida History and Culture Series. Gainesville, Fla.: University Press of Florida.

Fernald, E. A., and E. D. Purdum, eds. 1998. *Water Resources Atlas of Florida*. Tallahassee, Fla.: Institute of Science and Public Affairs, Florida State University.

Fernald, E. A., and E. D. Purdum, eds., and Anderson, J. R., Jr., and P. A. Krafft, cartographers. 1992. *Atlas of Florida*. Rev. ed. Gainesville, Fla.: University Press of Florida.

McCally, D. 1999. *The Everglades: An Environmental History*. The Florida History and Culture Series. Gainesville, Fla.: University Press of Florida.

Miller, J. J. 1998. *An Environmental History of Northeast Florida*. Florida Museum of Natural History: Ripley P. Bullen Series. Gainesville, Fla.: University Press of Florida.

Myers, R. L., and J. J. Ewel, eds. 1990. *Ecosystems of Florida*. Orlando, Fla.: University of Central Florida Press.

Whitney, E. N., D. B. Means, and A. Rudloe. 2004. *Priceless Florida: Natural Ecosystems and Native Species.* Sarasota, Fla.: Pineapple Press.

Winsberg, M. D. 2003. *Florida Weather.* 2nd ed. Gainesville, Fla.: University Press of Florida.

Geology and Paleontology (Field Guides and Technical Literature)

Abbott, R. T. 1974. *American Seashells.* 2nd ed. New York: Van Nostrand.

Agassiz, A. 1883. *The Tortugas and Florida Reefs.* Memoirs of American Academy of Arts and Sciences, vol. 11, pp. 107–34. Cambridge, Mass.

———. 1896. The Elevated Reef of Florida. *Bulletin of the Museum of Comparative Zoology at Harvard College* 28:1–62.

Agassiz, L. 1880. *Report on the Florida Reefs.* Memoirs of the Museum of Comparative Zoology at Harvard College, vol. 7, no. 1. Cambridge, Mass.

Aronson, R. B., ed. 2007. *Geological Approaches to Coral Reef Ecology.* Ecological Studies, vol. 192. New York; London: Springer.

Balsillie, J. H. , and J. F. Donohue. 2004. High Resolution Sea-Level History for the Gulf of Mexico Since the Last Glacial Maximum. *Florida Geological Survey Report of Investigations* 103.

Berry, E. W. 1929. A Palm Nut of *Attalea* from the Upper Eocene of Florida. *Journal of the Washington Academy of Sciences* 19(12). Reprinted in 1931 in *Twenty-First–Twenty-Second Annual Report of the Florida State Geological Survey*, pp. 120–25.

Bush, D. M., N. J. Longo, W. J. Neal, L. S. Esteves, O. H. Pilkey, D. F. Pilkey, and C. A. Webb. 2001. *Living on the Edge of the Gulf: The West Florida and Alabama Coast.* Durham, N. C.: Duke University Press.

Campbell, K. M. 1986. The Industrial Minerals of Florida. *Florida Geological Survey Information Circular* 102.

Cheetham, A. H. 1963. *Late Eocene Zoogeography of the Eastern Gulf Coast Region.* Geological Society of America Memoir 91. New York: Geological Society of America.

Cooke, C. W. 1939. Scenery of Florida Interpreted by a Geologist. *Florida Geological Survey Bulletin* 17.

———. 1945. Geology of Florida. *Florida State Geological Survey Bulletin* 29.

Cooke, C. W., and S. Mossom. 1929. Geology of Florida. *Twentieth Annual Report of the Florida State Geological Survey.*

Dall, W. H. 1887. Notes on the Geology of Florida. *American Journal of Science* (3) 134:161–70.

———. 1890–1903. *Contributions to the Tertiary Fauna of Florida: With Special Reference to the Miocene Silex-Beds of Tampa and the Pliocene Beds of the Caloosahatchie River.* Transactions of the Wagner Free Institute of Science, vol. 3, pts. 1–5. Philadelphia.

———. 1915. A Monograph of the Molluscan Fauna of the *Orthaulax pugnax* Zone of the Oligocene of Tampa, Florida. *Bulletin of the U.S. National Museum* 90.

Dall, W. H., and G. D. Harris. 1892. Correlation Papers: Neocene. *United States Geological Survey Bulletin* 84.

Davis, J. H, Jr. 1946. The Peat Deposits of Florida: Their Occurrence, Development, and Uses. *Florida Geological Survey Bulletin* 30.

Doyle, L. J., D. C. Sharma, A. C. Hine, O. H. Pilkey Jr., W. J. Neal, O. H. Pilkey Sr., D. Martin, and D. F. Belknap. 1984. *Living with the West Florida Shore.* Durham, N. C.: Duke University Press.

Enos, P., and R. D. Perkins. 1977. *Quaternary Sedimentation in South Florida.* Geological Society of America Memoir 147. Boulder, Colo. Geological Society of America.

Gardner, J. A. 1947. The Molluscan Fauna of the Alum Bluff Group of Florida. *United States Geological Survey Professional Paper* 142.

Gillette, D. D., and C. E. Ray. 1981. *Glyptodonts of North America.* Smithsonian Contributions to Paleobiology 40. Washington: Smithsonian Institution Press.

Healy, H. G. 1975. Terraces and Shorelines of Florida. *Florida Geological Survey Map Series* 71.

Heilprin, A. 1887. *Explorations on the West Coast of Florida and in the Okeechobee Wilderness.* Transactions of the Wagner Free Institute of Science of Philadelphia, vol. 1. Philadelphia.

Huddlestun, P. F. 1993. A Revision of the Lithostratigraphic Units of the Coastal Plain of Georgia: The Oligocene. *Georgia Geological Survey Bulletin* 104.

Hurlbert, R. C., Jr., ed. 2001. *The Fossil Vertebrates of Florida.* Gainesville, Fla.: University Press of Florida.

Hurlbert, R. C., Jr., G. S. Morgan, and S. D. Webb, eds. 1995. Paleontology and Geology of the Leisey Shell Pits: Early Pleistocene of Florida. *Bulletin of the Florida Museum of Natural History* 37, pts. 1 and 2.

Johnson, R. A. 1989. Geologic Descriptions of Selected Exposures in Florida. *Florida Geological Survey Special Publication* 30.

Kurz, H. 1942. Florida Dunes and Scrub, Vegetation, and Geology. *Florida State Geological Survey Bulletin* 23.

Lane, E. 1986. Karst in Florida. *Florida Geological Survey Special Publication* 29.

Moore, R. C., ed. 1955. *Treatise on Invertebrate Paleontology, Part P: Arthropoda 2.* Kansas: Geological Society of America and University of Kansas Press.

Perry, L. M., and J. S. Schwengel. 1955. *Marine Shells of the Western Coast of Florida.* Ithaca, N. Y.: Paleontological Research Institution.

Pilkey, O. H., Jr., D. C. Sharma, H. R. Wanless, L. J. Doyle, O. H. Pilkey Sr., W. J. Neal, and B. L. Gruver. 1984. *Living with the East Florida Shore.* Durham, N. C.: Duke University Press.

Pratt, A. E. 1990. Taphonomy of the Large Vertebrate Fauna from the Thomas Farm Locality (Miocene, Hemingfordian), Gilchrist County, Florida. *Bulletin of the Florida Museum of Natural History* 35.

Pumpelly, R. 1893. An Apparent Time-Break Between the Eocene and Chattahoochee Miocene in Southwestern Georgia. *American Journal of Science*, 3rd ser., 46:445–47.

Puri, H. S., and R. O. Vernon. 1964. Summary of the Geology of Florida and a Guidebook to the Classic Exposures. *Florida Geological Survey Special Publication* 5.

Randazzo, A. F., and D. S. Jones, eds. 1997 *The Geology of Florida.* Gainesville, Fla.: University Press of Florida.

Scott, T. M. 1992. A Geologic Overview of Florida. *Florida Geological Survey Open File Report* 50.

———. 2001. Text to Accompany the Geologic Map of Florida (MS 146). *Florida Geological Survey Open File Report* 80.

Scott, T. M., and W. D. Allmon, eds. 1993. Plio-Pleistocene Stratigraphy and Paleontology of Southern Florida. *Florida Geological Survey Special Publication* 36.

Scott, T. M., K. M. Campbell, F. R. Rupert, J. D. Arthur, T. M. Missimer, J. M. Lloyd, J. W. Yon, and J. G. Duncan. 2001. Geologic Map of Florida. *Florida Geological Survey Map Series* 146.

Scott, T. M., G. H. Means, R. P. Meegan, R. C. Means, S. B. Upchurch, R. E. Copeland, J. Jones, T. Roberts, and A. Willet. 2004. Springs of Florida. *Florida Geological Survey Bulletin* 66.

Smith, F. G. W. 1948. *Atlantic Reef Corals: A Handbook of the Common Reef and Shallow-Water Corals of Bermuda, Florida, the West Indies, and Brazil.* University of Miami, Marine Laboratory, Special Publications. University of Miami Press.

Vaughan, T. W. 1910. A Contribution to the Geologic History of the Floridian Plateau. In *Papers from the Marine Biological Laboratory at Tortugas, Carnegie Institution of Washington Publication* 4(133):99–185.

Vernon, R. O. 1951. Geology of Citrus and Levy Counties, Florida. *Florida Geological Survey Bulletin* 33.

Vilas, C. N., and N. R. Vilas. 1970. *Florida Marine Shells: A Guide for Collectors of Shells of the Southeastern Atlantic Coast and Gulf Coast.* 2nd. ed. Rutland, Vt.: Charles E. Tuttle Company.

Voss, G. L. 1976. *Seashore Life of Florida and the Caribbean: A Guide to the Common Marine Invertebrates of the Atlantic from Bermuda to the West Indies and of the Gulf of Mexico.* Miami, Fla.: E. A. Seemann Pub.

White, W. A. 1958. Some Geomorphic Features of Central Peninsular Florida. *Florida Geological Survey Bulletin* 41.

———. 1970. The Geomorphology of the Florida Peninsula. *Florida Geological Survey Bulletin* 51.

Zullo, V. A., W. B. Harris, T. M. Scott, and R. W. Portell, eds. 1993. The Neogene of Florida and Adjacent Regions: Proceedings of the Third Bald Head Island Conference on Coastal Plains Geology. *Florida Geological Survey Special Publication* 37.

Geology and Paleontology (Popular)

Andrews, J. 1994. *A Field Guide to Shells of the Florida Coast.* Houston, Tex.: Gulf Publishing Company.

Brayfield, L., and W. Brayfield. 1986. *A Guide for Identifying Florida Fossil Shells and Other Invertebrates.* 3rd ed. Gainesville, Fla.: Florida Paleontological Society.

Brown, R. C. 1996. *Florida's Fossils: Guide to Location, Identification, and Enjoyment.* 2nd ed. Sarasota, Fla: Pineapple Press.

Campbell, K. M. 1984. A Geologic Guide to the State Parks of the Florida Panhandle Coast: St. George Island, St. Joseph Peninsula, St. Andrews and Grayton Beach Parks and Recreation Areas. *Florida Geological Survey Leaflet* 13.

Campbell, K. M., and R. W. Hoenstine. 1982. The Geology of Torreya State Park. *Florida Geological Survey Leaflet* 11.

Comfort, I. T. 1998. *Florida's Geological Treasures.* Baldwin Park, Calif.: Gem Guides Book Co.

Hoenstine, R. W., and S. Weissinger. 1982. A Geologic Guide to Suwannee River, Ichetucknee Springs, O'Leno, and Manatee Springs State Parks. *Florida Geological Survey Leaflet* 12.

Hoffmeister, J. E. 1974. *Land from the Sea: The Geologic Story of South Florida.* Coral Gables, Fla.: University of Miami Press.

Lane, E. 1987. Guide to Rocks and Minerals of Florida. *Florida Geological Survey Special Publication* 8 (revised).

———. 1994. Florida Geological History and Geological Resources. *Florida Geological Survey Special Publication* 35.

———. 1998. The Florida Geological Survey: An Illustrated Chronicle and Brief History. *Florida Geological Survey Special Publication* 42.

Murray, M. 1975. *Florida Fossils: A Guide to Finding and Collecting Fossils in Florida.* Tampa, Fla.: Trend House.

Neal, W. J., O. H. Pilkey, and J. T. Kelley. 2007. *Atlantic Coast Beaches: A Guide to Ripples, Dunes, and Other Natural Features of the Seashore.* Missoula, Mont.: Mountain Press Publishing Company.

Olsen, S. J. 1959. Fossil Mammals of Florida. *Florida Geological Survey Special Publication* 6.

———. 1965. Vertebrate Fossil Localities in Florida. *Florida Geological Survey Special Publication* 12.

Renz, M. 1999. *Fossiling in Florida: A Guide for Diggers and Divers.* Gainesville, Fla.: University Press of Florida.

Romashko, S. D. 1975. *Coral Book: A Guide to Collecting and Identifying the Corals of the World.* Fla.: Windward Publishing, Inc.

Rupert, F. R. 1989. A Guide Map to Geologic and Paleontologic Sites in Florida. *Florida Geological Survey Map Series* 125.

———. 1994. A Fossil Hunter's Guide to the Geology of Panhandle Florida. *Florida Geological Survey Open File Report* 63.

———. 1994. A Fossil Hunter's Guide to the Geology of the Northern Florida Peninsula. *Florida Geological Survey Open File Report* 65.

Rupert, F. R., and E. Lane. 1992. The Geology of Falling Waters State Recreation Area. *Florida Geological Survey Leaflet* 16.

Schmidt, W. 1982. Florida Caverns State Park, a Nature-Made Geologic Wonderland. *Florida Geological Survey Leaflet* 10.

Scott, T. M., and F. R. Rupert. 1994. A Fossil Hunter's Guide to the Geology of Southern Florida. *Florida Geological Survey Open File Report* 66.

Simpson, G. G. 1932. Miocene Land Mammals from Florida. *Florida State Geological Survey Bulletin* 10.

Thomas, M. C. 1968. *Fossil Vertebrates: Beach and Bank Collecting for Amateurs.* M. C. Thomas.

Voss, G. L. 1988. *Coral Reefs of Florida.* Sarasota, Fla.: Pineapple Press.

Waterbury, J. P. 1993. *Coquina (ko-kee-na).* St. Augustine, Fla.: St. Augustine Historical Society.

Yon, J. W., Jr. 1965. Adventures in Geology at Jackson Bluff. *Florida Geological Survey Special Publication* 14.

Historic and Literary Interest

Andrews, E. W., and C. Andrews, eds. 1985. *Jonathan Dickinson's Journal, or, God's Protecting Providence.* Port Salerno, Fla.: Florida Classics Library.

Douglas, M. S. 2007. *The Everglades: River of Grass,* 60th anniversary edition. Sarasota, Fla.: Pineapple Press.

Gannon, M., ed. 1996. *The New History of Florida.* Gainesville, Fla.: University Press of Florida.

Gannon, M. 2003. *Florida: A Short History.* Rev. ed. Gainesville, Fla.: University Press of Florida.

Jahoda, G. 1978. *The Other Florida.* Port Salerno, Fla.: Florida Classics Library.

McPhee, J. 1967. *Oranges.* New York: Farrar, Straus and Giroux.

Rawlings, M. K. 1938. *The Yearling.* Charles Scribner's Sons.

Rowe, A. E. 1992. *The Idea of Florida in the American Literary Imagination.* Florida Sand Dollar Books. Gainesville, Fla.: University Press of Florida.

Van Doren, M., ed. 1955. *Travels of William Bartram.* Unabridged ed. New York: Dover Publications.

Waitley, D. 2005. *Florida History from the Highways.* Sarasota, Fla.: Pineapple Press.

Travel: Canoeing, Hiking, Diving, and Nature Guides

Carter, E. F., and J. L. Pearce. 1985. *A Canoeing and Kayaking Guide to the Streams of Florida,* vol. 1, *North Central Peninsula and Panhandle.* Birmingham, Ala.: Menasha Ridge Press.

DeLoach, N. 2004. *Diving Guide to Underwater Florida.* 11th ed. Jacksonville, Fla.: New World Publications.

Delorme, ed. 1997. *Florida Atlas and Gazetteer.* 4th ed. Yarmouth, Maine: DeLorme.

Friend, S. 2004. *The Florida Trail: The Official Hiking Guide.* Englewood, Colo.: Westcliffe Publishers.

Glaros, L., and D. Sphar. 1987. *A Canoeing and Kayaking Guide to the Streams of Florida*, vol. 2, *Central and South Peninsula*. Birmingham, Ala.: Menasha Ridge Press.

Godown, J. A. 1998. *Scenic Driving in Florida*. Helena, Mont.: Falcon Publishing.

Logan, W. A. 1998. *Canoeing and Camping the 213 Miles of The Beautiful Suwannee River and History of the Okefenokee Swamp*. 2nd ed. Cocoa, Fla.: William A. Logan.

McCarthy, K. M., and W. L. Trotter. 1990. *Florida Lighthouses*. Gainesville, Fla.: University Press of Florida.

Molloy, J. 2001. *The Hiking Trails of Florida's National Forests, Parks, and Preserves*. Gainesville, Fla.: University Press of Florida.

Ohr, T., P. Carmichael, and W. Williams. 2004. *Florida's Fabulous Canoe and Kayak Trail Guide*. Hawaiian Gardens, Calif.: World Publications.

Strutin, M. 2000. *Florida State Parks: A Complete Recreation Guide*. Seattle, Wash.: The Mountaineers Books.

Suwannee River Water Management District (Fla.). 2000. *Lower Suwannee River Boat Ramps and Canoe Launches/Upper Suwannee River Boat Ramps and Canoe Launches* (double-sided sheet). Live Oak, Fla.: Suwannee River Water Management District.

Index

Note: Page numbers in *italic type* indicate photographs, illustrations, or information in captions.